Two Hundred Years of Geology in America

Cecil J. Schneer, Editor

Two Hundred Years of Geology in America

Proceedings of the New Hampshire Bicentennial Conference on the History of Geology

Published for the University of New Hampshire by the University Press of New England Hanover, New Hampshire 1979

Library of Congress Catalogue Card Number 78-63149
International Standard Book Number 0-87451-160-7
Printed in United States of America
Library of Congress Cataloging in Publication data
will be found on the last printed page of this book.

**The University Press
of New England**

Brandeis University
Clark University
Dartmouth College
University of New Hampshire
University of Rhode Island
University of Vermont

The New Hampshire Bicentennial Conference on the History of Geology was held at the University of New Hampshire, Durham, New Hampshire, October 15–19, 1975.

Acknowledgment

The New Hampshire Interdisciplinary Conferences on the History of Geology in 1967 and 1976 were supported by grants from the National Science Foundation and the University of New Hampshire at Durham. Of the many colleagues and students whose long and effective labors made these volumes possible, I am particularly indebted to Aurele LaRocque, Ronnie Harrington, Helen K. Langone, and my wife, Mary Schneer.

At Sea Point
(18 October 1976)

Gratuitously escaped from self-wrought
Tumults of fragile erudition
(The litter of older, dying leaves)
We now confront obstreperous remains
Of what once were molten—basalts,
Intrusive like ourselves, but
Annoyingly more durable.
When they remain,
And we are gone, may there be still
Free men, New England, and the United States
To celebrate in further centuries.
May these three, and truth—
The integrity of fact to which
We dedicate ourselves—likewise endure
Time's battering calamity of indifferent abuse.

Dennis R. Dean

Contents

Affiliations

Michele L. Aldrich
American Association for the Advancement of Science
1776 Massachusetts Ave. N.W.
Washington, D.C. 20038

Nancy Alexander
505 Cypress Station No. 2007
Houston, Texas 77090

Eugene A. Alexandrov
Department of Earth & Environmental Sciences
Queens College
Flushing, New York 11367

Harold L. Burstyn
Office of the Historian
U.S. Geological Survey
Reston, Virginia 22092

André Cailleux
Laboratoire de Geomorphologie
9, Avenue de la Tremouille
94100 St-Marie-Des-Fosses
France

Dennis R. Dean
Humanistic Studies Division
The University of Wisconsin–Parkside
Kenosha, Wisconsin 53140

R. H. Dott, Jr.
Department of Geology and Geophysics
University of Wisconsin–Madison
Madison, Wisconsin 53706

Arthur Donovan
Department of History
West Virginia University
Morgantown, West Virginia 26506

Val Dusek
Department of Philosophy
Hamilton Smith Hall
University of New Hampshire
Durham, New Hampshire 03820

Henry Frankel
Department of Philosophy
University of Missouri
Kansas City, Missouri 64110

Fritiof M. Fryxell
Augustana College
1331 42nd Avenue
Rock Island, Illinois 61201

Patsy A. Gerstner
Howard Dittrick Museum of Historical Medicine
11000 Euclid Avenue
Cleveland, Ohio 44106

Stephen Jay Gould
Department of Geological Sciences
Harvard University
Cambridge, Massachusetts 02138

Joseph T. Gregory
Museum of Paleontology
University of California
Berkeley, California 94720

Margaret H. Hazen
4410 Clearbrook Lane
Kensington, Maryland 20795

Robert M. Hazen
Carnegie Institution of Washington
Geophysical Laboratory
2801 Upton Street, NW
Washington, D.C. 20008

Marie-Louise Hemphill
Museum of Natural History
1 Place de Wagram
Paris, France 75017

William M. Jordan
Department of Earth Sciences
Millersville State College
Millersville, Pennsylvania 17551

Stephen F. Lintner
U.S. Geological Survey
MS 760
Reston, Virginia 22092

Thomas G. Manning
Department of History
Texas Tech University
Box 4529
Lubbock, Texas 79409

H. William Menard
Office of the Director
U.S. Geological Survey
Reston, Virginia 22092

Arthur Mirsky
Department of Geology
Indiana University–Purdue University at Indianapolis
Indianapolis, Indiana 46202

Clifford M. Nelson
Office of the Historian
U.S. Geological Survey
Reston, Virginia 22092

Ellwood C. Parry III
School of Art and Art History
University of Iowa
Iowa City, Iowa 52242

Thomas E. Pickett
Delaware Geological Survey
University of Delaware
Newark, Delaware 19711

Stephen J. Pyne
U.S. National Park Service
North Rim, Arizona 85015

Susan B. Schlee
Marine Biological Laboratory
Woods Hole, Massachusetts 02543

Cecil J. Schneer
Department of Earth Sciences
James Hall
University of New Hampshire
Durham, New Hampshire 03824

Hubert C. Skinner
Department of Geology
Tulane University of Louisiana
New Orleans, Louisiana 70118

Darwin H. Stapleton
Case Western Reserve University
Cleveland, Ohio 44106

Kenneth L. Taylor
Department of History of Science
University of Oklahoma
Norman, Oklahoma 73019

C. Gordon Winder
Department of Geology
University of Western Ontario
London, Ontario
Canada

Two Hundred Years of Geology in America

Introduction

Cecil J. Schneer

The history of geology is a subject currently pursued with diligence by geologists interested in the origins of the science and by historians of science who see in geological thought a significant influence in Western society. At first glance, geology may appear deceptively simple, requiring little or no background in mathematics or the exact sciences. To the beginner, it is a branch of natural history with principles no more difficult to master than those of bird-watching, say, or collecting wild flowers. The principle that in a sequence of layered rocks deposited as sediments settling out of an ancient sea, the stratum which is above is the younger and that which is below is older appears to be almost as much as one must grasp to become a geologist. This is comparable to Poincaré's dismissal of formalism—shall we allow that one and one makes two and this is all there is to mathematics? The intricacies of a natural history make for a thicket of complexity, and although in the exact sciences, progress consists of relating more and more phenomena to fewer and more powerful general principles, the opposite may be the case for a historical science like geology, where the individual event and the unique record are commonly the intrinsic concern of the investigator. It is not always appreciated by students of other disciplines that the geologist as much as the good physician is practicing an art in which the intuition born of long experience and patient observation is a prerequisite of skill.

The professional scientists on the other hand, those who have mastered the subject sufficiently to contribute to its progress, are by the very nature of their training all too frequently unprepared to submit to the historian's canons of scholarship and to recognize in books and papers in the library the same disciplinary autonomy as that presented by rocks and minerals in the field. Obviously we need both the insight born out of direct experience with the science and that arising from the practice of historical scholarship to produce cooperatively a viable history.

It was my aim at the New Hampshire Interdisciplinary Conference in 1967 and again in 1976 to bring together scholars with a common interest in the evolution of geology but with backgrounds and training as diverse as possible, so that the subject seen in manifold perspective would take on depth and illumination and reach the maturity of a considered discipline. The 1967 Conference and the Proceedings volume that resulted were intended to serve as a record of the state of the art, to provide standards for further developments and to identify and focus on what Gerald Holton has termed *themata* in the history of the earth sciences. The history of geology must go beyond its empirical or phenomenological content and the analysis of that content, to consider the fundamental presuppositions, perhaps religious or philosophic, social or political; the notions and linguistics; the methodology and judgments that have accompanied the evolution of geology through time. It was my hope that out of the diversity of our backgrounds and the stimulus of our collective debates, our Bicentennial Conference would raise the history of geology in America to a new level both quantitative and qualitative.

The papers published here were not delivered at New Hampshire in October 1976. They were distributed in preliminary form to the conferees in advance of the Conference. The themes of these papers were discussed in five formal sessions and a great number of informal sessions over a four-day period of almost total immersion. Out of this concentration on the history of geology two kinds of themata, in Holton's sense of the term, emerged. The first kind had to do with the methodology of the history of geologic science, the other with the objects of that history, geology itself. A tension clearly developed between, on the one hand, conferees primarily concerned with the origin of ideas, their interconnections, their sources and their further evolution—the internalists, viewing science as an autonomous intellectual discipline (not always the geologists among us)—and, on the other hand, the externalists, who see their function as placing the science in its living context in the society and world in which it was nourished. This antithesis—the dependence or independence in the United States of science and general culture, of science and technology, even of science, technology and the natural world which it purports to represent—emerged clearly in the discussions. The debate is continued in the final versions of the papers that are included here and, it is to be hoped, will influence the level of future discussion of the history of geology.

The second kind of themata that engaged the attention of the conferees were intrinsic to the science itself. Just as observations of marine remains on the tops of mountains had inspired early protogeological ideas, the detailed field work of the first state surveys in the early nineteenth century led to the appearance in the United States of theoretical generalizing concepts of originality and great power. The geosynclinal hypothesis and its relationship to isostasy and to the new plate tectonics is the most obvious of these major subjects. This is not to say that the forty-odd scholars who came to the New Hampshire Conference would agree on the identification and classification of the most significant themes which engaged our attention last October. The papers published below reflect the diversity of their views and ideas as they emerged in discussion and as they were considered in the preparation of these final versions.

Some explanation of the circumstances leading to the calling of the Conference is in order. The history of geology in recent time dates from the inauguration at Yerevan, Armenian S.S.R., of the International Commission for the History of Geology (INHIGEO) in the midst of the Arab-Israeli War of 1967, and the call for the formation of national committees. Dispersed by the invasion of Prague from its next meeting in 1968, INHIGEO held its first formal meeting in 1970 at Freiberg D.D.R., meeting again in 1971 at Moscow, in 1972 at Montreal, in 1974 at Madrid, in 1975 at London, in 1976 at Sydney, and in 1978 at Muenster. A United States National Committee for the History of Geology was established. Meeting in Miami in 1974 with the Geological Society of America, the U.S. Committee called for a Bicentennial Conference to celebrate the national birthday and to provide a commemorative of our profession. Our conference, convened by the University of New Hampshire in 1976 with the sponsorship of the Geological Society of America and the History of Science Society and with support from the National Science Foundation, was the result.

Of all the sciences, geology seems to be the one most closely associated with the United States—a land of mountains and rivers, of glaciers and prairies, atolls, hot springs, badlands, mines, and gushers. This is the country of the Mammoth Cave and the Teapot Dome, the

San Andreas Fault and LaBrea tarpits, the Delaware Water Gap and the Great Okefenokee, Niagara Falls, the Grand Canyon of the Colorado, the Homestake Mine and Texas Gulf Sulphur, the Catskill scarp, the Mississippi delta, the Palisades, the Great Salt Lake and Mount Rainier. "I have fallen in love with American names," wrote Stephen Vincent Benét. Between the United States and the science of the earth there is a particularly close attachment born out of America's love affair with the wide open spaces; born too out of its history of successive waves of westing that carried the frontier from the Atlantic Coast to the Pacific; born out of eighteenth-century utilitarianism and the attitudes toward science, education, and industry of such men as Benjamin Franklin and Thomas Jefferson. The geologic resources of the thirteen colonies at the time of the Revolution were described at the 1976 New Hampshire Conference by Arthur Mirsky. The intellectual resources available to deal with the new world were discussed by Kenneth L. Taylor in his account of geology in 1776 as an incipient science. The point is made by comparing our present maps with Guettard's map of 1752, discussed by André Cailleux.

The geologic resources available to the original thirteen United States were not necessarily the same as those available to the aboriginal Indian inhabitants or to the present citizens occupying the same ground. Resources are a function of culture—of technology and society—determined also by such things as mercantilist trade restrictions imposed by Britain and yet influencing in turn the further development of science and society. The conference debated this problem of the relationship between nature and the science of nature and the parallel problem of the relationships between science and technology. William Jordan used the early to mid-nineteenth-century industrial and transportation revolution in Pennsylvania, to illuminate the contemporary development of geology as a science. Jordan's paper compliments Taylor's account of the state of the art in 1776, an historian's essay that provoked heated argument over the charge of Whiggery or "presentism"—the historian's characterization of the scientist's tendency to appraise the ideas of the past in the light of the understanding of the present instead of recreating the life of the past in full dimensions and all of its color. Yet in New Hampshire in 1967 and at London in 1975, historians rather than geologists criticized as whiggish the historical chapters of Lyell's *Principles*. Lyell in 1830 failed to write the history of geology as we would see it today! No matter—it is from such exchanges that a discipline derives the standards and its practitioners the tools with which to proceed.

The American love affair with geology was not unfruitful. Early geological ideas of the European center were applied and confirmed and extended in America and achieved here dramatic and unparalleled successes. The principle of faunal succession which William Smith worked out in the Secondary rocks of Southern England and Cuvier and Brongniart derived for the Paris Basin was established as the basis for a worldwide geological time scale when the New York State Board of Geologists commenced in 1836 the detailed description of the Transition Series overlying the primary crystalline rocks of the Adirondacks—stretching in unbroken layers succeeded by ever younger formations, like the pages of some vast book, to the Mississippi and beyond, to the Quaternary outwash of the Rockies on the High Plains of Nebraska Territory. Pallas' Wernerian law of mountains was replaced by a science of orogeny and dynamical geology in 1842, when the Rogers brothers described the folded Appalachians as a subsurface architectural construct of harmonic folds governing the topography from Vermont to Alabama.

The Rogers' efforts to establish what James Hall had once called for —an American nomenclature for American rocks—are the subject of Patsy Gerstner's essay. It was a system of nomenclature based on lithological characteristics, with time-sequential terms: for the periods of the day, dawn, morning, sunrise . . . twilight, dark. How the students of today would have blessed them if they had succeeded in stopping cold the system of type locality which Lardner Vanuxem proposed for the New York Board of Geologists! Michele Aldrich discussed the era of the state surveys—a provincial assumption of responsibility that corresponded to the prevalent sectionalism but which was transcended by the field geologists who worked and thought together to correlate and standardize a national geology. In cooperation with men like William Logan, Elkanah Billings, and T. Sterry Hunt of the Canadian Survey, or Europeans like Charles Lyell, they helped to build an international geology.

Until the establishment of the U.S. Geological Survey Bureau in 1879, the major burden of geological research was born by the state surveys. There were exceptions, of course. The tradition of federal surveys of the territories which Jefferson inaugurated with the expedition of Lewis and Clark, 1804–1806, included the 1835 geological reconnaissance of G. W. Featherstonhaugh and W. W. Mather (the latter was to play a major role in the New York Survey) and David Dale Owen's massive survey of the territory of Wisconsin, Minnesota, Iowa, and part of Nebraska. This last, Owen wrote in 1852, was the most extensive geological survey ever conducted in this country, covering four times the territory of New York State and two and one half times the territory of Great Britain. Nevertheless the detailed attention which the geologists of the state surveys were able to bring to bear on relatively limited problems provided Owen with the personnel and much of the intellectual paraphernalia essential to his purposes. The relationships of the federal and state surveys, their mutual interdependence, and the influences of organizational and political details on the development of geological science are subjects for further research.

Local investigations at the state level or even narrower, entered in part into nearly every discussion. Detailed studies of the work of Rogers and J. P. Lesley of the Pennsylvania Survey, in William M. Jordan's paper; of Raymond Thomassy (who fought a duel over the size of the Mississippi River) and his geology of Louisiana, in Hubert C. Skinner's paper; of James C. Booth and the first Delaware Geological Survey, in Thomas E. Pickett's paper; as well as the work of E. T. Dumble of the Texas Survey on lignite deposits, in the paper of Nancy Alexander, provide constraints in the form of materials and evidences to confine further historical generalization. A view also of the transformation of American geology with the establishment of the United States Geological Survey is furnished in Stephen Pyne's study of the mind of G. K. Gilbert and in Thomas G. Manning's portrait of the scientist turned Washington bureaucrat. Manning's subject, George Otis Smith, was the impeccably colorless successor, as director of the U.S.G.S., to J. D. Walcott and his four freight-car loads of soft-bodied fossils from the Burgess shale, to the daring Major Powell of the Colorado, and to Clarence King, who led a secret life in Brooklyn's Black community.

The close relationships of Canadian with American geological science were discussed at the Conference by C. Gordon Winder, whose own career illustrates some of the interaction he described. At the Conference, Winder was elected Vice-President of the newly formed Division for the History of Geology of the Geological Society of America (whose President at the time was R. R. Folinsbee, also a Canadian).

In the antebellum period, men such as James Hall, David Dale Owen, William Logan, and their successors traced individual formations which they correlated across all political boundaries. These formations started with the great supracrustal masses of the mountains and extended halfway across the continent to thin, scarcely disturbed leaves of rock in the Ohio and Mississippi valleys. James Hall conceived of a crustal lineament marginal to a vast internal sea, subsiding under the weight of accumulating shallow water deposits and then crumbled and uplifted to form mountains. This was the geosynclinal concept (named by J. D. Dana). It was one of the great unifying ideas;

an organizing and explanatory principle with ramifications in every branch of earth science. Geosynclinal theory in one way or another dominated American and world geology until very recently, when it received its first serious modifications. The geosynclinal concept combined geological and physical ideas and provoked studies of crustal mobility and measurements of crustal shortening (first on the Rogers' folded Appalachians). Dana drew from it the contraction hypothesis. The origin of the geosynclinal concept, its uniquely American stamp, and its position in the geological science of the past century was discussed at the Conference by Robert Dott.

Successive expeditions preceding and accompanying the restless westward expansion of the nation uncovered the materials for major developments in vertebrate paleontology—the science which in Cuvier's dramatization had captured the imagination of early nineteenth-century Europe and provided much of the basis for the debate over the origin of species. In post Civil War America paleontology reached new levels of discovery and popularization, providing abundant materials and paving the way for the eventual acceptance of Darwinism. This was federal science, which came to center on the Smithsonian Institution and later on the U.S. Geological Survey. Joseph T. Gregory terms it the heroic period in North American vertebrate paleontology and summarizes the paleontological discoveries of the Missouri River basin and the Rocky Mountains. The development of this federal science out of the State Survey movement and the scope of these geological explorations is illustrated in the paper by C. M. Nelson and Fritiof M. Fryxell on the antebellum collaboration of F. B. Meek and F. V. Hayden, pupils and successors of James Hall. Amos Eaton had remarked in 1824 that no Cretaceous rocks had yet been found in America. Meek could describe them in 1856 as "so grandly developed over an area of country more extensive than all that of Great Britain." The geology of William Smith was on a scale fitting the close-grained English countryside of lanes and villages; the geology of North America was scaled to fit a continent.

From its first precarious beginnings, clinging to the coast in the shadow of the Appalachian highlands, colonial America was uniquely dependent on the sea—its only link of refuge to the parent culture and for a long time its primary internal line of communication. The North Atlantic became the world's busiest sea lane early in the nineteenth century, and American whalers and merchant vessels reached every part of the globe. The sciences of oceanography and geodesy (beginning initially as coastal surveying) were very early a serious concern of government. The always ingenious Dr. Franklin applied folk understanding of the North Atlantic currents early in the development of this science. It was to receive a further American character with the publication of Matthew Maury's *Physical Geography of the Sea*. Jefferson founded the United States Coast Survey in 1807. Under Ferdinand Hassler (and his successor, Franklin's great-grandson Alexander Bache) the Coast Survey provided the impetus for the triumphs of American geodesy in the latter half of the nineteenth century. The paper by Harold Burstyn and Susan B. Schlee discussed the rivalry between Maury, whose concern with practical results made him contemptuous of theoretical science, and Bache, the latter representing an emerging maturity and sense of professionalism in American science. Bache and his successor, Benjamin Peirce, encouraged William Ferrel, who first explained the relationship between the earth's rotation and the circulation of the world's seas and atmosphere. Ferrell provided the theoretical basis for physical oceanography and for the domination of this field by the United States in the mid-nineteenth century.

It was in the Coast Survey that Charles Sanders Peirce, founder of the uniquely American philosophy of pragmatism, spent most of his productive years and it was for such practical geodetic purposes as instrumentation that he embarked upon the mathematical studies that gained him his reputation—a relationship between geodesy and pure thought with a parallel in the life of C. F. Gauss, who played a similar role in the geodetic survey of his native Bavaria. When Gauss wrote that space—geometry—unlike number, has a reality outside of the mind, was this the product of his experience as land surveyor or was it an imposition of his fundamental presuppositions on the triangulation which he supervised? The influence of experience with the problems of geodesy and the earth on the founder of pragmatism were discussed by Val Dusek. Peirce's emphasis on experimentation, his views on the nature of scientific truth and the logic of scientific discovery, his emphasis on the cumulative character of science and the collective nature of the scientific enterprise may very well have derived from the unique characteristics of the earth sciences and Peirce's lifetime association with the Coast Survey.

The development of instrumental, mathematical, and theoretical geodetics at the highest levels of sophistication in the second half of the nineteenth century in the United States marked its entry on the international scientific scene and a gravitational shift to the Atlantic of a hitherto dominantly European intellectual activity—the pursuit of basic science. The appearance at International Geodetic congresses of men like Peirce, W. F. Bowie, and J. F. Hayford, with results of major significance, gained a respect for American intellectual attainment which had been conceded only grudgingly to Joseph Henry earlier in the century. When the naturalists of the new post Civil War Geological Survey combined the geophysics of the Coast Survey personnel with the geosynclinal concepts of Hall and Dana, isostasy acquired both its name and its decidedly American caste. Geologists such as G. K. Gilbert, J. D. Barrell, and T. C. Chamberlin provided the field observations of structures and the disposition of rocks and of mountains, sedimentation, erosion, faulting, and deltas that gave rise to their ideas about the strength and mobility of the crust. Gilbert's application of formal mechanics and thermodynamics to the Henry Mountains was the subject of Stephen Pyne's contribution to the Conference. With Captain Clarence Dutton, who coined the term "isostasy," and the one-armed Major Powell of the Colorado, Pyne claims Gilbert formed a collaboration comparable in the history of geologic thought to that of Hutton and Playfair.

While an Atlantic science was emerging to replace the European science of the eighteenth century, the post Civil War period saw the rise to a dominant position of a Washington-centered federal science out of the scientific clusters of Philadelphia, Albany, Boston, and the universities. Geologists returned for the winter from the Missouri basin and the Rockies to study their specimens at the Smithsonian Institution or the laboratories of the newly established U.S. Geological Survey, to meet and interact with the geodesists of the Coast Survey or the astronomers of the Nautical Almanac and the other federal naturalists and scientists. The theme of the localization of geological activities and the identification of geology with specific scientific communities was treated vertically by Carol Faul in her account of the institutions and scientists starting with Benjamin Franklin who made Philadelphia a nexus of geological activity. Many of the individuals whose relationships to geology form the subject of other papers of this volume were in the Philadelphia milieu. America's first commune, New Harmony on the Wabash, was an artificial center of natural science for a time. It was founded by Robert Owen, the Fourier socialist, as an answer to the social blight of the new industrial Britain. Owen's partner in the New Harmony project was William Maclure, the father of American geology and an educational reformer. Maclure had launched a famous "boatload of knowledge" down the Ohio to New Harmony in 1825—a flat-boat of books and instruments and for passengers the naturalists Troost, Say, Rafinesque, Lesueur, and Owen's son, David Dale Owen. Mary R. Fleming, a resident of New Harmony, brought illustrations and accounts of the current restoration of the community.

M. L. Hemphill, whose father, Adrien Loir, headed the Museum d'Histoire Naturelle du Havre, prepared a study of Charles-Alexandre Lesueur, the first head of the Havre museum. Brought by Maclure to America in 1815, Lesueur was among those who initiated Americans into the science of paleontology. He accompanied Maclure as artist and naturalist on a tour of the United States from New England to Maryland, and settled at New Harmony for many years.

The glacial hypothesis which Louis Agassiz had appropriated from Jean de Charpentier became with Agassiz' translation to New England and his transformation to a proper Bostonian, a talisman resolving the genesis of the American landscape from the beaches of Cape Cod or Niagara Falls which had so fascinated Charles Lyell, to the drumlins of Boston Bay and even the Great Salt Lake. Primarily a zoologist, Agassiz appeared like a meteor on the Boston-Cambridge scene, lending to science and paleontology a social prestige and an influence not reached again until our own times. There was a negative side of this influence. It led to the Brahminization of transcendentalists like Emerson and the rise of a baronial professoriate to complement the political and economic establishment of the Gilded Age. The scientific rivalries of American geologists over the Taconic System or vertebrate paleontology or the appointments of Clarence King and later J. W. Powell to Survey positions were followed in the latter nineteenth century by emerging railroad and land interests and their involvement with scientists as mining and petroleum experts—events presaging our modern conflicts over technological warfare and nuclear power and the biochemical degradation of the earth and sea.

In a perceptive essay on Louis Agassiz' marginal notes on his copy of Haeckel's evolutionary manifesto, Stephen Jay Gould illuminates the record of this famous scientist, bypassed at the height of his career by the triumph of Darwinism. The popular history that portrays Agassiz as frustrated and embittered—retreating into dogmatic adherence to demonstrably false doctrine and bolstering his lost position with appeals to religious prejudice—is a caricature. In his analysis of Agassiz' private reactions to Ernst Haeckel and the rational basis for his rejection of evolution, Gould raises a theme which has become central to modern historiography of science: what is the basis for the acceptance or rejection of a major change in our scientific *Weltanschauung*? Have we in our past histories of science blinded ourselves by concentration on roots, to the neglect of soil and climate?

Within the last decade the geological sciences have undergone a transformation in outlook as dramatic and as radical as the transformation undergone by the life sciences in the acceptance of Darwinism.

The new *plate tectonics* has demonstrated its power as an organizational and explanatory theory in every branch of earth science, overwhelming doubters and outright opponents, authorities and the public, and capturing completely the students who are the next generation and the future. To live through a scientific revolution is for a philosopher of science an unparalleled opportunity. One of the participants in the making of the geological revolution, H. W. Menard opened the meeting with his recollections of what has now in retrospect taken on the attributes of a golden age of science. He spoke of the static world picture with which our generation of geologists were indoctrinated in the graduate schools. It was a view of most of the world—the ocean basins—as featureless, uniform, and geologically dead. He spoke of what it was like to explore this world and to begin to realize that every observation, however tentative, was in contradiction to this view and in fact to much of geology's theoretical basis. The exception was the heresy of continental drift. In the fable of the blind men attempting to study the elephant, Menard found a point of departure for his consideration of what is the real miracle of science—the creation of a new idea.

Professional philosophical analysis of this process has been largely confined to examples from Renaissance astronomy or theoretical physics. Henry Frankel is a philosopher of science who has applied himself to this scientific revolution of our time. Using the formalist analysis of Imre Lakatos, Frankel undertakes to demonstrate that confirmation of Harry Hess's concept of sea floor spreading was the novel fact that established the drift heresy. How does the scientific community distinguish between the surd and the rational? When does an idea make the transformation from the incredible to absolute truth? How does the process called science maintain growth—which requires receptivity—and maintain standards? In short how does science remain alive without degenerating? Understanding the birth pains of plate tectonics, this most recent example of mass conversion, is essential in answering this question.

Philosophers and historians have usually concentrated on the exact sciences. The successes of physics in particular have influenced our view of the nature of science and our view of history. It seems probable, however, that in America more than anywhere else except perhaps Australia, the earth, and sea, our landscape, and our mineral resources, including our perceptions of them (geological science) have influenced our culture and our history and our perceptions of ourselves and world society. Dennis Dean writes that this orientation goes back to the gold lust of the Conquistadores. He argues that geology entered into the theological consciousness of the Puritans with the New England earthquakes of 1727 and 1755. Utilitarian and cosmological influences on such American authors as John Winthrop and Phillip Freneau foreshadowed a transcendentalist compromise between religion and science which Oliver Wendell Holmes saw ended with the publication of Darwin's *Origin*. The geology in Mark Twain's *Tom Sawyer*, *Life on the Mississippi*, *Innocents Abroad* (and in fact almost everything that he wrote) is well known. Charles Wilson Peale's excavation of two mastodon skeletons and his painting "Exhuming the Mastodon" are relatively less familiar. His son Rembrandt Peale's "Account of the Skeleton of the Mammoth" (London, 1802) and a photograph of the C. W. Peale painting, which hangs in the Philadelphia Museum, were on display at the Conference. Nor is it generally known that B. H. Latrobe, the architect of the Capitol and the White House and a major figure in the Greek Revival movement in American architecture, was an associate of William Maclure and carried on extensive geological studies of his own. The paper by Stephen F. Lintner and Darwin H. Stapleton describes some of this involvement. Latrobe was the engineer of canals and water supplies as well as architect. His journals illustrate the extent to which geology, both in theory and application, permeated his thought.

The close involvement with geological science of Thomas Cole, the founder of the Hudson River School of landscape painting in America, was brought out at the Conference. Ellwood C. Parry III traced connections between the painters of the period and geologists like Benjamin Silliman. The problem of the dependence of the ideas of science on nature is the same problem as that of the dependence of the picture on reality—the image and the object. Parry argues that the transformations of Cole's painting throughout his career reflect his ideological shift from catastrophism to uniformitarianism, and in fact his work is "probably unthinkable except against an international cultural background in which the ideas of [geology] were being fervently debated."

The Conference ended without a consensus on the principal themes of the history of geology in America, even without agreement on what these themes might be or their relative significance. We could not agree on the distinctions between Parry's acts of God and acts of man—between science and nature; still less could we agree on a methodology for the history of geology. We could not declare ourselves Whigs or Tories, nor in fact agree on canons of history or science, or even the standards for publication in this volume or in the journals. It seems that history itself, as much as science, is an act of man—in Einstein's term, a free construction of the human mind.

A consensus did appear—unspoken, unstated, but deeply felt. When we cease to be receptive, when we fix finally our methodology and our canons of historiography, it will be because we have degenerated. Our history will be written finally and therefore it will be over, it will be dead. The history of geology in America is another aspect of the science of geology in America. It lives to the extent that it grows and develops and therefore to the extent that it is receptive. It must be both inquisitive and acquisitive not only within the bounds of the earth sciences but also outside. We shall never agree to confine it or to limit it. That would be to confine the mind itself.

In this volume are to be found studies of some of the complex relationships between the development of a particular branch of science—the study of the earth and the development of the nation. Materials for that consensus which by design escaped the forty conferees are to be found below. The volume does not require the conventional disclaimer—the management is not responsible for the opinions found below. One way or another, they emerged from four days at New Hampshire, thanks to the University officials and the professional societies which sponsored the conference and most of all to the geologists, theologians, artists and poets, the settlers and mountain men, politicians and jailbirds (Amos Eaton for one) of the past two hundred years. The reader should know at whom to aim his brickbats or his praise. It is our country and our history, and in a way we are all responsible.

I

Themes for the History of Geology in America

2

Very Like a Spear
H. William Menard

When Professor Schneer invited me to address the Bicentennial Conference on the History of Geology, it was with the understanding that I would not document a manuscript for publication. On this basis I prepared an outline and spoke at the Conference. Afterward he asked me if I could reconstruct for publication what I had said there. I agreed to do so provided it was understood to be the equivalent of a tape recording. What follows should not be regarded as a documented history. It is a transcript of an oral history of what I remember as important discoveries and of what I remember as influential ideas during the 1950's.

My subject this morning is the history, as I personally recall it, of the conceptual revolution that occurred in marine geology, and indeed in all geology, in the 1950's and 1960's. My focus is on the timing of ideas, why people had some ideas when they did, and why they did not have others. During and immediately after a scientific revolution, those who are involved, and in this revolution we all were, tend to be impressed by the great victory of the tentative and timorous new ideas over the solidly entrenched facts and theories that had been accepted for decades. After a while, however, the world picture that we accepted is exposed as a flimsy tissue of misinterpreted observations and ill-founded speculations whereas the new world picture is miraculously capable of quantitative predictions. In retrospect, the great question is not how and why a scientific revolution occurred but why it took so long. I shall address myself to this question.

My title is from the well-known fable of the blind men who were taken to study an elephant. One feels the elephant's leg and says it is like a tree. Another feels the side and says it is like a wall. A third feels its tusk and says it is very like a spear. A scientific revolution is not unlike a group of blind men examining an elephant, except that no one tells the scientists that they are all examining the same creature, or, indeed, that there *is* such a creature as an elephant.

Let me propose a somewhat more complicated fable to describe the history of a scientific revolution. Photographs of the Imperial City in Peking show kneeling elephants and camels. Let us imagine a time when these animals were unknown in China. After careful study of imported manuscripts, the scholars at the emperor's court inform him that the camel exists but that the elephant is a myth. The emperor immediately orders a frontier garrison to send out an expedition to procure a camel; but despite an exhaustive search none can be found. Something must be sent to the court, or the emperor will be very angry, so the garrison leader finds a large animal, covers it with a tent and states that the camel is too delicate a creature to be exposed to the pernicious airs of Peking.

The court scientists begin to study this tent-shrouded marvel and one of them finds that the leg is much larger than the scholarly manuscripts say it would be. Being a scientist, he naturally dashes off a note to *Nature* describing the true dimensions of the legs of camels. However, one of his colleagues conceives the "sick camel" hypothesis, and he also writes to *Nature* saying that the emperor has been given a defective camel with an abnormally swollen leg.

The examination, and the controversy, continue. The back of the camel is examined, and the "hard field data versus scholarly speculation" school says that camels, or at least some camels, do not have humps. The "sick camel" school says that just as expected this camel is so sick that it has absorbed its hump. They examine the shrouded mouth. One school says that the description of camels has been sadly defective, and the other school says that one could hardly ask for a better confirmation of its theories than the actual discovery that the poor camel has such malformed teeth that it cannot eat. No wonder it has absorbed its hump.

Finally a scientist cannot stand the uncertainty and one night he creeps in and looks under the corner of the tent. The next morning he says what he has done, lifts the corner and proclaims it is not a camel but an elephant. That is not quite the end, because at first many people have not had a chance to look under the tent and they continue to proclaim that even if it is not a camel, it can't be an elephant because they don't exist. Finally almost everyone looks, and the scientific revolution is nearly over. There remain only a few people who would rather not look at the elephant but prefer instead to look at the literature on sick camels and elephants.

Let us turn from fable to the myth-agreed-upon of history. What was the shrouded camel of the 1950's which the geologists perceived? Looking backward I would say it was an old and relatively static world. No beginning, no end, slow erosion and sedimentation, changes gradual, and so on. It wasn't entirely static; a few geologists proposed scores of kilometers of overthrusting in the Alps and Scotland. A few radicals suggested tens of kilometers of offset on the San Andreas Fault but the more generally accepted offset was only 1-2 kilometers.

There was a different perception of the world, one in which continents drifted about, which had advocates in the southern hemisphere —where the supporting evidence existed—and in Europe, but which was rejected in the United States by a generation of geologists. In retrospect, considering that the structural and paleontological evidence for drift looks so convincing, we can ask ourselves why this perception of the world was not accepted. It is my speculation that Wegener probably set the acceptance of continental drift back by several decades rather than advancing it as is generally supposed. He took an idea which was supported by or compatible with the geological evidence available at the time, tied it into a driving mechanism, and proposed a critical test in the form of remeasurement of the longitude in Greenland. The critical test was a wretched failure, and the driving mechanism was rejected by geophysicists. I suspect that these facts colored the interpretation of geological and paleontological data. Something did.

In any event, when I was a student and began to attend scientific meetings I witnessed the public humiliation of one of my young professors by one of the most famous geologists of the time, because he had the temerity to ask how the great man's paper tied into continental drift.

The other people who, in 1950, were about to explore the deep ocean basins from institutions in the United States had similar backgrounds to mine with regard to continental drift. Consequently we expected to find that the basins were as old and even more static than the continents.

At that time the most imaginative and profound ideas about the deep ocean basins were being published by Philip Kuenen. He calculated from the observed rates of erosion on land and the known age of the continents that some 5 to 10 kilometers of sediment covered the basement rocks of the sea floor. It followed that any original features of the basement were concealed. Moreover there were few if any earthquakes within the basins, so folding and faulting were not deforming the thick cover of sediment. In short, except for a little active vulcanism, it appeared that the great ocean basins were pretty unlikely prospects on which to build a professional career. To the student these days it may seem impossible that anyone ever thought of the ocean basins as aseismic. This feat was generally achieved by definition, that is, if there were earthquakes the crust was considered to be continental instead of oceanic. The andesite line, for example, was generally taken as the western boundary of the Pacific crust even though it took several days by ship to reach it from China. This may seem a strange definition but in fact it is not very different from the modern definition of plates, namely an aseismic region bounded by a line of earthquakes. What no one realized was that one of the important lines was on mid-ocean ridges.

In addition to a paradigm of a quiescent, stable earth, there were many ideas current in the 1950's about the specific nature of marine sedimentation. First, geologists knew that waves were only capable of stirring sediment, and thus eroding, down to a limiting depth called "wave base." An extensive oceanographic literature existed on the depths at which waves had been observed to stir sand. Likewise geologists commonly interpreted the environment of deposition of sedimentary rocks in terms of wave base. If there were ripple marks or current scours, or sorted sands, then it followed that a sediment was deposited above wave base—perhaps 100 meters.

Likewise, in physical oceanography, it was assumed that a level of no motion occurred simply because one was necessary when making dynamic calculations of ocean currents from observations of water density. As I recall, the depth was 1 to 2 kilometers.

In short, sediment that fell to the deep sea floor remained eternally at rest until it was buried by younger material. There was also a concept of a "veil of lutite," meaning essentially that clay settled so slowly that it was homogeneously mixed in the ocean. Thus the red clay in one part of the ocean was like all other parts.

I am now at the point when the first modern expeditions began to put to sea from Denmark, Sweden, and the United States, but I would like to make an aside about the nature of oceanography in this country in the 1950's. Because of the foresight and organizational genius of a few people, of whom Roger Revelle is most familiar to me, we suddenly had available a flood of money, ships, facilities, equipment, and students. Nature provided the only other thing necessary for happy science, namely an abundance of unknown problems which we could solve with the new instruments and techniques.

The scientists who were blessed with this opportunity were not oceanographers because, for all practical purposes, there were no such people. Instead they were physicists, geologists, chemists, biologists, and applied mathematicians—as unlikely a combination to be herded onto a small ship for a month as can be conceived. Still, that is what happened. The ships were enormously expensive according to the standards of most science. In marine geology we spent the same amount of money in a day at sea as customarily in a summer field season on land. Thus we had to carry out many different programs at the same time on a ship in order to justify the expense. A wonderful thing occurred. All the various specialists were driven to communicate to keep the ship going, and they had eighteen hours a day to do so. That is when oceanographers were born. We learned to view marine sediment in terms of its nutrient supply for benthonic organisms and its physical properties that affected acoustic transmission as well as in terms of sedimentation as understood by geologists.

As examples of the golden opportunities that were ready for gathering in the oceans about 1950, let me cite two personal experiences. In 1949 I went on my first oceanographic cruise, out of San Diego on the only ship Scripps owned, with Bob Dietz and his former mentor Fran Shepard. It was a great moment for a young man who had just received his doctorate a month before. On that cruise of a single day, Dietz and Shepard each had familiar problems to study but for me everything was new. Imagine my amazement when we discovered the first known deep-sea channel at a depth of about 1200 meters. My thesis had been on flume studies of sediment transportation and though I don't remember for sure, it seems to me that I should have immediately realized that most of what I had learned about deep sea sedimentation was wrong. Fortunately Dietz and Shepard were always generous colleagues, and I was the only one who knew any hydraulics, so I had the priceless opportunity to describe the channel and its implications.

In 1950 Revelle organized and led the Mid-Pac Expedition, with Dietz in charge of the Navy Electronics Laboratory contingent, of which I was a part. Such now well-known oceanographers as K. O. Emery, Russ Raitt, and Ed Hamilton participated. Within a few months the expedition came up with the wholly unexpected results that heat flow is the same in ocean basins as on continents, that sediment is only a few hundred meters thick at most, that Tertiary sediment outcrops in some places, that guyots are drowned ancient islands, that Cretaceous reefs existed in the central Pacific, and that fault scarps are abundant in the interior of the basin.

For myself, I obtained a few days of ship time at the end of the expedition to deflect one ship somewhat north of a direct track home and make a few zigzags across a region where questionable soundings suggested an anomaly of some sort. What I had the good luck to discover with my first ship time was the Mendocino fracture zone, the first of what have become known as transform faults in plate tectonics terminology, and the longest fault trace on earth. Those were grand times.

In the early 1950's marine geology had a choice of revolutions. It could either revolutionize modern sedimentation and the geological interpretation of ancient rocks or it could revolutionize tectonics. Maurice Ewing, who founded Lamont Geological Observatory, reached, I believe, a conscious decision to concentrate on sediment as offering the greater opportunities for certain and important results. I don't recall any similar decisions at Scripps. Indeed, we continued to investigate tectonic problems along with sedimentation. This was in part because we had trenches to intrigue Bob Fisher, and the great fracture zones could hardly be missed once their existence was conceived. Moreover, Fisher and I had some time to make those crucial zigzags. Bruce Heezen, working in the Atlantic, did not. Topographically, the Atlantic was a mess, and with the instrumentation available in the early 1950's it probably would have been fruitless to try some of the things that were easy in the Pacific. Moreover, Ewing had set the priorities. He controlled the only thing that really counts in an oceanographic institution—access to ship time—and he used it to collect a matchless suite of long piston cores.

The first few deep sea expeditions equipped with cameras discovered ripple marks and current scours at all depths in the ocean. Their gravity and piston corers contained sorted sands where the bottom was flat, regardless of depth. Clearly the concepts of wave base and level of no motion were wrong. Photos and cores proved that animals were eating and stirring abyssal sediment, and micropaleontologists could not find continuous sedimentation anywhere. Moreover the geochemists found significant regional variations in red clay. The veil of lutite concept was wrong, and there was no haven of eternal rest.

The discoveries I have been recounting mainly apply to abyssal sediment, which we could not, at that time, identify in many places on land. The greater revolution in geological thinking resulted from the proof that turbidity currents are the principal agents distributing sediment around continental margins, and that they produce, at any depth, phenomena that had been identified as characteristic of shallow water. Only a few oceanographers had access to the ships and samples, but they all found about the same things at about the same time. From the mouths of submarine canyons, leveed channels cross deep-sea fans to gently sloping abyssal plains and ultimately to flat ponds of sediment. The sediment of the plains contains graded beds and shallow-water benthonic microfossils. The hills projecting from the plains are covered with normal pelagic red clay. Sequential cable breaks occasionally occur on the fans and plains and always downslope.

The evidence accumulated within a few years, a controversy raged, and then the first revolution in marine geology was over. As a sociological measure of what happened, we can consider the fraction of papers in marine geology devoted to turbidity currents. This rose in a few years to about 5 percent, as I recall, and then declined when acceptance was general. Thereafter the number of papers increased but the fraction relative to all papers in marine geology was constant and very small. In Kuhn's terms the paradigm was being elaborated.

That brings us to the second half of the 1950's and the second revolution in marine geology. At this point you may well want to say, "But you already had the big revolution in 1950. The heat flow, the thin sediment, and the fracture zones showed conclusively that the static model of the earth was wrong." Well, in retrospect, what was discovered in 1950 might have told us we were in need of a revolution in our thinking. However, that was not the way the blind men operated. A separate solution was proposed for each problem when it was discovered. No one realized that we were all looking at different parts of the same camel. Certainly no one told us that there were elephants.

By the mid-1950's we knew that heat flow was high throughout the ocean basins, and we knew a lot about trenches, and both were attributed to convection in the mantle. We knew that the whole northeastern Pacific was cut by a subparallel system of giant fracture zones, and they too were attributed to mantle convection. No such fracture zones had been found in the Atlantic, although Heezen and I always assumed that ultimately the same features would be found in all ocean basins. He found one at last, without, as far as I know, ever having time to zigzag, by detecting a narrow tongue of abyssal plain along a transverse rift in the surrounding mass of mountains.

In the northeastern Pacific Ron Mason and Art Raff discovered the next important feature of the ocean basins that led to plate tectonics, namely the magnetic stripes. These were surveyed quite incidentally in the course of the first detailed topographic survey of the deep sea, using precise electronic navigation and precision sounders. The soundings, collected for a long-range SONAR project, were classified and therefore unpublishable for a while but the Navy didn't care about the magnetics.

Looking backward, you may wonder why no one at Scripps, where the data were long unpublished, noticed what is now obvious, namely that the magnetics on the Gorda and Juan de Fuca ridges are symmetrical. Why was the great opportunity left waiting the vision of Vine and Wilson? Well, maybe we just lacked the vision. However, the history of the discovery of the magnetic stripes may suggest why no one was really looking.

The stripes were discovered off central and southern California in the first phase of the survey. The fact that the earth's magnetic field reverses had not been established at that time and no one had a clue about the real origin of the mysterious stripes. In that region the stripes are not symmetrical, because the ridge crest was subducted (in modern terms). What was evident was that there was an 80-kilometer offset on the Murray fracture zone. This may not seem big news now, but at the time it was a phenomenal offset for a fault. Marland Billings, for example, devoted part of his presidential address to the Geological Society of America to questioning whether the apparent offset was a fault displacement. He argued that the offset changed from place to place and that this was not the nature of faulting. Of course he was correct; it was only much later that Wilson began to explain the nature of transform faults.

So the controversy and the excitement were related to the offsets in the pattern of stripes, especially after Vacquier made additional traverses and discovered the then incredible 1170-kilometer offset on the Mendocino fracture zone. Meanwhile the detailed survey near shore continued northward and the slow construction of magnetic anomaly maps proceeded. North of the Mendocino fracture zone it was clear that the character of the stripes was different, but no one knew why, except perhaps it was partially because the shallower sea floor was nearer the detectors in the ship.

After several years of bafflement about the anomalies, it is my memory that even Mason and Raff had grown tired of them. Certainly it was difficult to get them to publish their results. Perhaps if the survey had started in the north and worked south, there would have been enough interest and someone would have noticed the symmetry. You might expect that the topography would have provided the necessary clues. However, at that time no one knew that the whole northeastern Pacific is the flank of a midocean ridge which once had a symmetrical flank. Just looking at the topography of the Gorda and Juan de Fuca ridges would not have provided many useful clues. The former ridge is grossly distorted by the confining structures of the continental margin and the Mendocino fracture zone; the latter ridge is not symmetrical because the eastern flank is buried by turbidites. Moreover, at that time, the existence of the Blanco fracture zone connecting the two ridges was unknown. That area had been surveyed by a Navy ship whose captain ran his engines at the most efficient speed regardless of whether the ship noise prevented the acquisition of the soundings that the ship was trying to collect. It was only sometime later that the soundings were unscrambled.

The next discovery which turned out to be crucial was made by Tharp, Heezen and Ewing, namely that a line of earthquakes and a rift followed the crest of the midocean ridge system throughout the world. This was announced in a very exciting paper which arrived in 1960, just in time to shape the planning of our Downwind expedition to the southeastern Pacific, and it influenced many other programs of the ongoing International Geophysical Year.

The authors later disagreed on the cause of what they identified as a splitting earth. Ewing, if I remember correctly, had a limited or conservative view that the splitting was caused by elevation, but Heezen opted for an expanding earth. They agreed, however, that the cracking was permanent rather than transient.

Many of us who thought more about the Pacific than the Atlantic tended to believe that the midocean ridges were transient features. Hess, among others, reasoned that they were transient from the existence of guyots and from the interpretation from high heat flow that convection currents existed. From Griggs's analysis of such currents it was reasoned that the overturn time for one convection would be 100-200 million years.

Thinking on the same lines, I looked for direct evidence of an ancient midocean ridge. I attempted to show that the region with guyots was in the center of the Pacific and thus the likely site for such a ridge, but much later this turned out to be a false clue. I also drew a paleobathymetric map of the guyot region and found that a giant bulge, which I called the "Darwin Rise," had existed about 100 million years ago, but now had subsided. This seemed to be exactly the evidence we were seeking. The place, the dimensions, and the timing were just right. People sometimes ask me "What ever happened to the Darwin Rise?" as though the idea was wrong. Well, it is quite true that this ancient rise was not just what I visualized, and it has stubbornly resisted any attempt to relate it to a fossil-spreading center. However, I like to think that the original concept of the subsided rise was just ahead of its time. By now we all accept the evidence that the whole sea floor was once a shallow midocean ridge and has now subsided, and that includes the area of the Darwin Rise.

We had another false alarm in trying to work out the pattern of heat flow in the ocean basins. I had occasion to talk to Francis Birch at a meeting sometime in the mid-fifties, and he remarked that if convection really existed, there would be a systematic pattern of regions with high and low heat flow. Our Downwind expedition was accordingly designed to test this hypothesis as well as look for the opening crack in the crest of the East Pacific Rise and many other things.

It was a very successful expedition despite high winds and giant waves that made operations marginal most of the time. I seem to recall, for example, that one ship ran aground at Rapa and slowly flooded its echo-sounder transducer, thereby losing the ability to do anything until repairs were made. However the main problems were related to the big winch with heavy tapered wire which we had to use to take piston cores, dredge, and measure heat flow. Every time we stopped the winch, it would throw a kink in the wire. We had no way of repairing it in those seas and we could not pass the kink through a sheave. I gave top priority to the heat flow measurements from then on, and this required that the whole expedition be reorganized so that each lowering for some months ahead would be shallower than all that had gone before. Well, to end this tale, after harrowing months, von Herzen got his heat flow data, and they were in a pattern with very high values on the crest of the East Pacific Rise and low values on each flank. I showed a map of our results at the next opportunity to Francis Birch, and you can imagine my reaction when, after inspecting it quite carefully, he said "Wrong pattern." As we now know, the pattern results from cooling of a leaky lithosphere.

Let me now focus on the most vexing characteristic of the various tectonic features of the ocean basins in the late 1950's. This characteristic was that the best data on any given feature tended to be in a region where the other features were little-known or absent. Thus we had fracture zones and magnetic stripes in the northeastern Pacific but no obvious midocean ridge. We had a midocean ridge, a pattern of heat flow, and some conjectural fracture zones in the southeastern Pacific but no magnetic anomalies or central rift. We had a midocean ridge with a rift in the Atlantic but no heat flow, fracture zones, or magnetic stripes. We had dated guyots and atolls, and outcrops of Tertiary sediment in the western Pacific but they were unrelated to the undated tectonic features. The list could be continued indefinitely. Nature was putting us to the test.

I attempted to relate some of these apparently unrelated features in a paper on the East Pacific Rise in 1960. By that time we had enough data to map the ridge crest and the associated seismicity through the south Pacific to the mouth of the Gulf of California, into western North America, and out to sea again off Oregon and Washington. We could show that the crest so defined was elevated and hot, and that the flanks were deeper and cool. We knew that the mantle formed a bulge parallel to the sea floor under the oceanic part of the rise. A genetic relationship between the rise and fracture zones was indicated by the fact that they were in the same place, on a similar scale, and they interacted. The sloping flank of the rise was offset the same distance as the magnetic anomalies.

There the tectonics of ocean basins stood, about 1960. It was time to attempt to integrate many features of the sea floor into a single hypothesis instead of having a different explanation for each feature. But which features were important, and which ones were related? What of the thin sediment? Was it related to the tectonic features? What of the age of the basins? One of the most mysterious aspects of marine discoveries in the fifties was that we could find no evidence that the ocean basins were very old. At first this was dismissed as an expected result because our paradigm said that all old rocks would be deeply buried. When it turned out that the sediment was very thin, it was argued that we were merely involved with a sampling problem. Moreover, some ancient rocks had been dredged from the Mid-Atlantic Ridge, and one had to dismiss them as ice-rafted in order to say that the bedrock was all young. Slowly but steadily we explored until hundreds and then thousands of samples were collected. None of them was more than 200 million years old, and the probability that this was a statistical fluke was diminishingly small.

Doubtless many hypotheses then existed which related the newly discovered features of the sea floor, but as a marine geologist I recall chiefly one by Bruce Heezen and one of my own. Mine, to dispose of it first, was based on the data from the East Pacific Rise and ignored continental drift because there was no evidence that seemed to call for it in the Pacific. I also dismissed the thin sediment on the grounds that sedimentation was much slower than had once been thought. That left me with the hypothesis that giant convection cells rose under midocean ridges, thinned the crest by block faulting, thereby producing the fracture-zone offsets, and compressed or downwarped the crust at the outer edges in the trenches bounding the eastern Pacific. This was a quantitative and complete tectonic system but regional rather than global.

Heezen also had a quantitative and complete tectonic system. I suppose that after the global viewpoint provided by the IGY, everyone realized that whatever was done on a grand scale in one region had to be balanced by some reverse phenomenon on a similar scale elsewhere. What he proposed was a model related to the region where he had been working for a decade, and emphasizing, as I had done, those features that he had himself helped discover. For some time his hypothesis was offered more in talks than in print but, as I recall, he proposed that the Atlantic was somehow splitting along the central rift, thereby allowing the continents to move apart without drifting through the mantle. He closed his tectonic system not by disposing of an equal area, or volume, of crust but by expanding the earth. Thus his hypothesis had the capability of explaining the youth of the ocean basins and the thin sediment.

There the matter stood at the end of the glorious decade of the fifties, locked in confusion and great argument, when Harry Hess walked over to the tent, lifted a corner, looked at the creature inside, and said, "It's an elephant."

3

Peripatetic Perambulation on Publication

C. Gordon Winder

Pierre Teilhard de Chardin once wrote, "The greater and more revolutionary an idea, the more does it encounter resistence at its inception." The problem, of course, is that once an orthodoxy has been established, any challenge, revision, or deviation from the "truth" will be resisted because the authorities who subscribe to the dogma cannot evaluate the new concept without prejudice. The methods of suppression can include ridicule, obfuscation, rejection on publication, and warnings to the author that his personal reputation as a scientist will be placed in jeopardy.

The present foremost concept in the earth sciences is plate tectonics, an outgrowth of continental drift, which itself was subjected to suppression or ignored. The breakthrough on plate tectonics came with the interpretation of geomagnetic data: (1) the magnetic striping on the ocean floor, first thought to be variation in intensity and then recognized as reversals, and (2) the apparent wandering of the magnetic poles as revealed by field determination from oriented specimens, considered in a stratigraphic setting. Two scientists, L. W. Morley and A. Larochelle were the first to make the initial contribution to the second premise. Their attempts to present results to the scientific community are revealed in the following letter:

Geological Survey of Canada
601 Booth Street
Ottawa Canada
K1A OE8

October 8, 1976

Mr. C. G. Winder
The University of Western Ontario
London, Ontario

Dear Mr. Winder,

Thank you for your letter of October 4, where you were asking me for details relating to my early modest contribution to plate tectonics. The article in the Royal Society of Canada volume on Geochronology (1964) to which you refer is indeed the only record published by L. W. Morley and myself on the subject. It is a fact that L. W. Morley had in vain previously tried to get the concept of sea-floor spreading in the literature: Once in *Nature*, when he was told that there was no space available in a scientific journal such as *Nature* to print fantasies of this type; and later in the *Journal of Geophysical Research*, where the referee suggested that perhaps the subject would be appropriate for a cocktail party discussion but could hardly meet J.G.R. standards. I am aware that Prof. J. Tuzo Wilson has since that time made reference to our contribution in a number of talks and papers of which I have no reprints. Someone brought to my attention recently a science review article in *Le Monde* of Paris (July 14, 1976), where full credit was given to us as well as to Vine and Matthews.

I appreciate your concern about restating the facts for the record, and I am thankful to you for this. On the other hand, I don't think that Vine and Matthews can be stripped of any merits in the development of plate tectonics, nor that the matter as to who had first the idea of sea floor reversals is highly relevant.

Sincerely yours,
A. Larochelle

The title of their article is "Palaeomagnetism as a Means of Dating Geologic Events," in the Royal Society of Canada volume *Geochronology in Canada* (1964), edited by F. F. Osborn.

4

Neglected Geological Literature: An Introduction to a Bibliography of American-Published Geology, 1669 to 1850 (Abstract)

Robert M. Hazen and Margaret H. Hazen

The literature of geology is extensive and diverse. The great range of endeavors in the earth sciences constitutes a major problem for the historian of geology, who must familiarize himself with numerous publications dealing with such varied subjects as fossils, soils, mines, and topography. Many of the early American publications relating to these topics have been long neglected and are now difficult to locate.

A bibliography containing 14,000 references has been compiled for earth science literature printed through 1850 in the United States. Articles, notes, and reviews on geology from 700 journals, as well as textbooks, geological maps, government documents, mining company reports, and other, non-newspaper, published sources have been included. The bibliography contains thousands of references not cited in previous compilations of early American geology, and consequently encompasses a large body of neglected geological literature. The first edition of the *Bibliography of American-Published Geology: 1669 to 1850* is now available on microcards, and an index to this edition is available from the authors on request.

The number of publications on geological phenomena increased dramatically from the mid-eighteenth to the mid-nineteenth century. Most of the 70 pre-1760 references are descriptive accounts or religious interpretations of natural phenomena, especially the earthquakes of New England. By 1800 approximately 400 more works had been published; earth-science writing had diversified, and articles on local geology, mineral and fossil localities, and mining techniques had appeared. From 1800 to 1850 the publication rate for geology approximately doubled every ten years; in 1850 nearly 1,000 references on the earth sciences appeared in American books and periodicals.

A variety of subjects were treated in early writings on geology. Scientific geology, including both theoretical and field studies, is the best documented and most studied aspect of American geological history. Several hundred earth science book reviews are important sources that not only call attention to geological publications of the day, but also provide useful summaries of the contents of these publications. United States mining became big business in the early nineteenth century, and corporate publications include a wealth of details on mine geology, mining techniques, and production statistics. Agricultural geology was another subject of earth science writing; by 1850 hundreds of texts and periodical articles were published to help the farmer evaluate and improve the soil with which he worked. Theologians of the eighteenth century proposed religious explanations for natural phenomena, and religion continued to figure prominently in nineteenth-century discussions of the Bible and interpretations of geological processes. Medical doctors contributed to the literature of geology through treatises on mineral waters and medical springs. "Medical topography," or the influence of local soils and rock formations on the prevalence of disease, also received consideration from several American scientists and physicians.

Geology was of importance to many Americans: farmers, miners, physicians, and theologians all had reasons to study aspects of the earth sciences in their professional lives. Yet geology appealed to an even wider audience as a result of the compelling curiosity with which man views his world. As a consequence, hundreds of popular American periodicals contained short articles and notes on subjects of geological interest. Discoveries of new mines or unusual fossils were widely publicized, as were the occurrences of volcanoes and earthquakes. The basic principles of geology were also described in the popular media, where theories of Lyell, Cuvier, and others received widespread circulation. By the mid-nineteenth century, published information about geological phenomena was widely disseminated.

A rich literary heritage has been left to the historian of American geology. From the first account in 1669 to the hundreds of 1850 publications, the changing intellectual and social climate, westward expansion, and growing prosperity of the United States are mirrored in the extensive earth science literature. The lives of most Americans were affected by geological developments, whether in scientific, industrial, agricultural, medical, religious, or popular contexts; and this influence, too, is reflected in the diversity of publications on geology. It is hoped that consideration of the complete scope of American earth science publications will lead to further understanding of the founding and development of geology in the United States.

Reference

Robert M. Hazen, assisted by Margaret H. Hazen, *Bibliography of American-Published Geology: 1669 to 1850*, Microcard Publications, No. 4 (Boulder, Colorado, Geological Society of America, 1976), 984 pp.

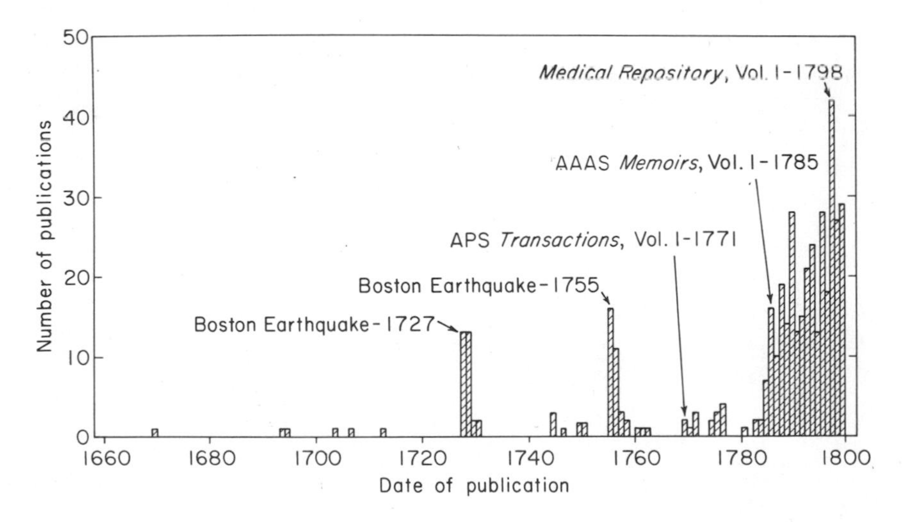

1a. The growth of earth science publications in the United States, 1660–1800.

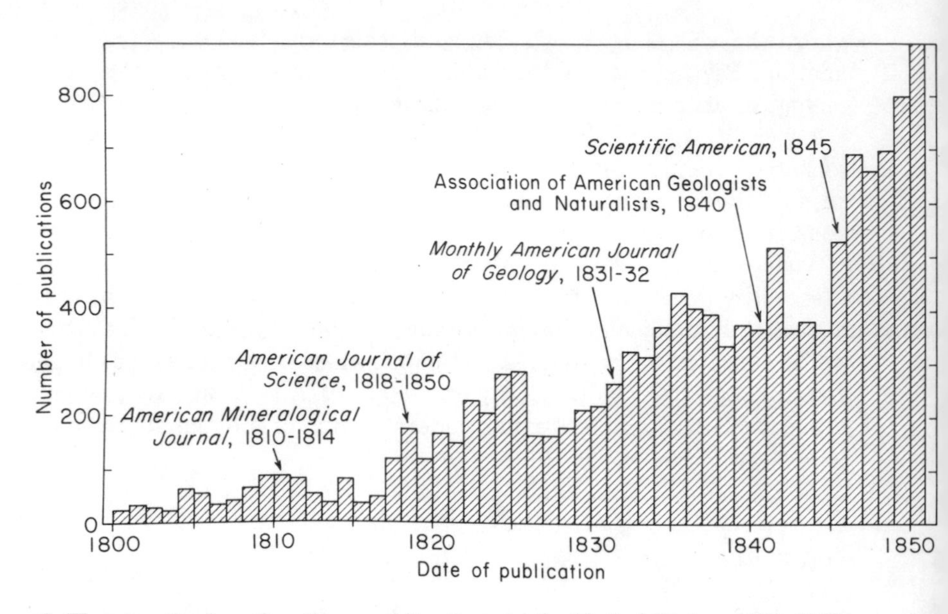

1b. The growth of earth science publications in the United States, 1800–1850. Note that the vertical scale of 1b is fifteen times greater than of 1a.

II

Acts of God, Acts of Man

5

Geologic Resources of the Original Thirteen United States (Summary)

Arthur Mirsky

The growth and development of any new nation depends to a great extent either on the natural resources contained within its boundaries or on the availability of such resources elsewhere through trade or military conquest. Geologists have known this for a long time, but other citizens have only recently begun to appreciate it, primarily in connection with present-day oil shortages.

During the colonial period in North America, all the English colonies used geologic resources in varying degrees. The earliest natural advantages stemming from local geology to be utilized by the colonists were the excellent harbors and deep, navigable rivers along the Atlantic coast, features that were formed when otherwise normal stream valleys were "drowned" with the rise in sea level as the meltwaters of waning continental glaciers emptied into the ocean. Mineral resources in use at this early period included soil for raising crops, clay for pottery and bricks, hematite for red pigment, salt for seasoning, and thermal springs for health.

By the eve of the American Revolution, the colonists were using clay, coal, cobalt, copper, feldspar, garnet, glass, gold, granite, gypsum, iron, kaolinite, lead, limestone, marble, mica, nickel, peat, salt, saltpeter, sandstone, silver, slate, soapstone (talc), thermal springs, tin, whetstone, and zinc (Fig. 1).

Mining activity was a chief factor in establishing some colonial settlements, but it is clear that, except for iron and building materials, at the beginning of the American Revolution the original thirteen states were not using their mineral resources to any significant extent. Building materials, of the types available locally, were used everywhere, but with little if any trade in these items. Among activities based on mineral resources, the iron industry was the best developed, but it never reached its potential for two reasons: one was political and the other was geological. The political reason, which is generally acknowledged, stems from the repressive parliamentary acts whereby England prohibited the colonies from developing any mineral industries that might be competitive with those in the mother country, and that also prohibited the colonies from exporting such products as iron to any place but England. Certainly England did encourage the colonies to develop iron deposits and the manufacture of pig iron and bar iron, but after 1750 it forbade the erection of any new mills for slitting and rolling iron, preferring that the colonists purchase such finished products from English manufacturers because of the enormous profit.

But the geological reason why the colonies were not developing important mineral industries, as the new country would do in the next century, is generally overlooked. This reason is related to the location of the colonies along the Atlantic coast, and the nature of the geology of this area, which is not especially favorable for significant mineral deposits. Thus the minerals known in the original colonies occurred in relatively small deposits compared to what would be discovered elsewhere beyond the Appalachians during the nineteenth and twentieth centuries. Even if England had not passed the adverse laws, development of the colonial iron industry would have been restricted, as indeed it was for some time after the Revolution freed the colonies from English control. Although charcoal was used successfully in iron-making, to develop the iron industry fully required that the colonists locate and use coking coal in iron-making, which did not occur until half a century after the Revolution. So the coke and iron deposits that would be the mainstay of industrialization, and the other mineral deposits that would support major industries, had to await the exploration and the exploitation of new parts of the growing nation beyond the Appalachian Mountains, which had been not only a physical barrier to colonization but a geological barrier to minerals discovery and development as well.

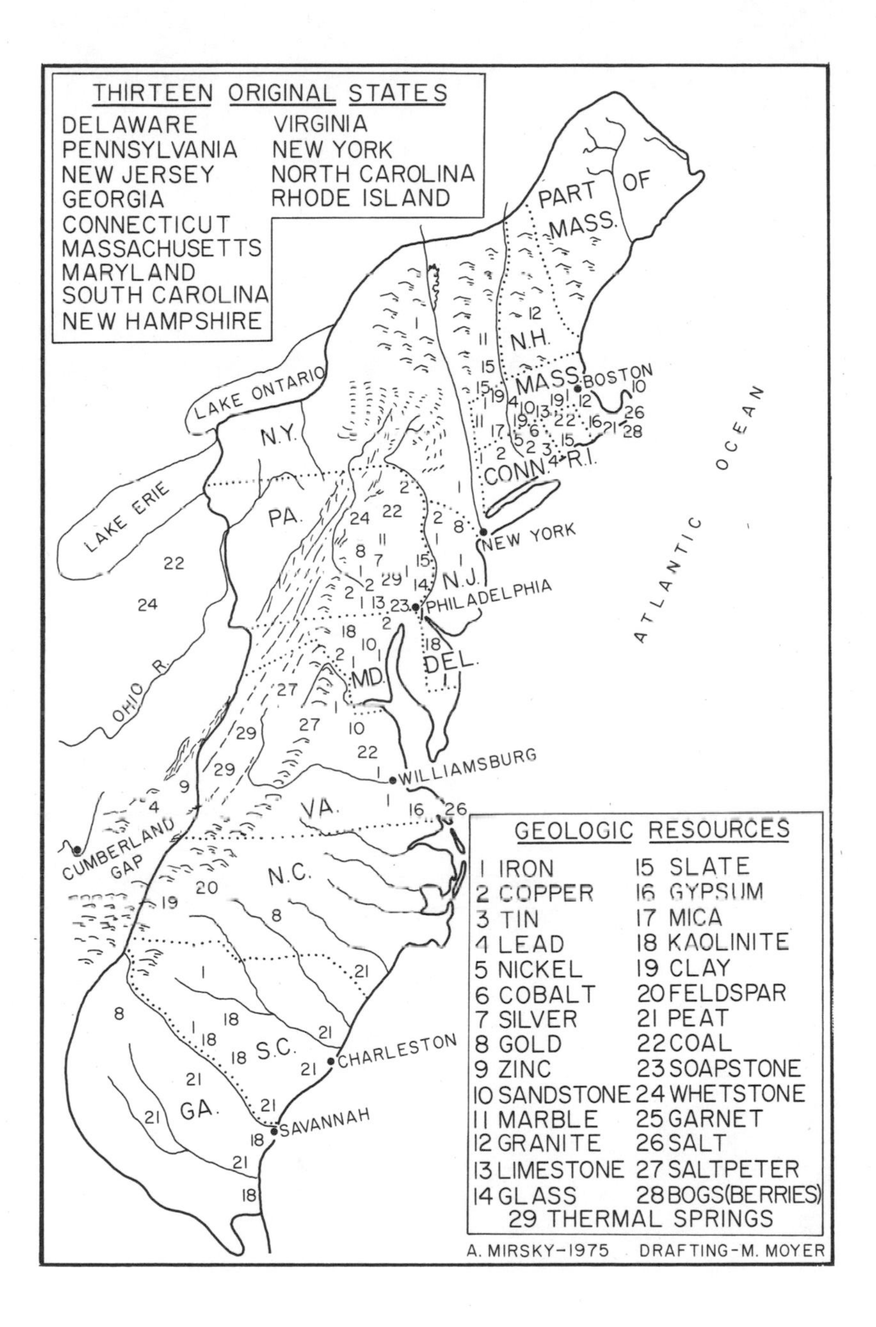

1. Geologic resources of the thirteen original states.

6

The Geological Map of North America (1752) of J.-E. Guettard

André Cailleux

The first geological map of North America is probably the map published by Jean-Etienne Guettard, in the *Mémoires de l'Académie Royale des Sciences* (Paris, 1752), plate VII.[1]

This map was drawn by Philippe Buache. It is 31 cm. broad and 27 cm. high. The projection is de Delisle, i.e. conical with two standard parallels, first used by Mercator for his map of Europe (1554) and much later by Delisle for his map of Russia (1745). Since that time it has been widely used in, for instance, Germany, Great Britain, and Italy. The meridians are straight lines, the parallels are concentric and equidistant circles. On the meridians, the scale is about 1:19,300,000. Neither on the map nor in the text are the two standard parallels (i.e. with the same scale) defined. Measurements made on the map show that these parallels most probably are the 60° and 30° N. For comparison, let us remember that for his map of Russia, Delisle used 62° 30′ and 47° 30′. For America, Buache wisely chose the more southern 30° parallel. His map extends from 25° to 74° N, i.e. from the Gulf Coast to Baffin Bay.

As in most French and German maps of this time, the prime meridian is fixed by the western Hierro Island (Canary Archipelago; meridian numbered 360°). Guettard's map stretches to 250° E, or 110° W of Hierro. Since Hierro is 18° 07′ West of Greenwich, the map covers from approximately 20° to 130° W, from Iceland through Greenland to the Pacific Coast (Mer de l'Ouest), where the so-called "Fleuve de l'Ouest," located near 45° N, probably is the Columbia River.

Taking into account the scale, we may neglect the 7′. Then the corresponding longitudes read as shown in Table 1.

Table 1.

Guettard-Buache	250	260	270	280	290	300	310	320	330	340	350	360
W of Greenwich	128	118	108	98	88	78	68	58	48	38	28	18

The geological symbols extend from the Gulf Coast to Greenland and westward from the Mississippi Valley and Hudson Bay. In the lower right corner an inset shows the St. Lawrence region at the scale of about 1:3,000,000.

In the legends are shown 36 symbols (Table 2), 34 of which are quite clear. From the accompanying text it appears that "Fer dissous" (dissolved iron) means ferruginous spring or water. Only "Terre tremblante" (quaking earth), remains somewhat equivocal and no mention of it can be found in the text.

Besides these 36 symbols, the word "marneuse" (marly) also appears on the map. All in all, the 37 designations are distributed as follows (percentages): Fossils 3, Petroleum (in Ohio) 3, Waters 12, Metals 13, Minerals 23, Rocks 46.

Table 2

Explication des caracteres	*Legend*
Ardoise	Slate
Argent	Silver
Caillou roulé	Rolled pebbles
Charbon de terre	Coal
Coquilles ou corps marins fossiles	Shells or other marine fossils
Cristal	Crystal
Cuivre	Copper
Emeric [*sic*]	Emery
Fer	Iron
Fne Minérale chaude	Warm mineral spring

Fne Minérale froide	Cold mineral spring
Fne Salée	Salt spring
Fne Sulphureuse	Sulphurous spring
Glaise	Clay
Gyps [*sic*] ou Pierre a Plâtre	Gypsum or Plaster-stone
Marbre	Marble
Marcassite ou Pyrite	Marcasite or Pyrite
Marcassite ferrugineuse	Ferruginous (oxydized) Marcasite
Mine ou substance Metallique	Mine or Metalliferous substance
Nitre ou Salpêtre	Nitre or Saltpeter
Ocre ou Fer dissous [*sic*]	Ochre or dissolved iron
Or roulé	Gold placer
Pétrole	Petroleum
Pierre Blanche	White Stone
Pierre à chaux	Limestone
Pierre à éguiser [*sic*]	Oilstone
Pierre Olaire	Potstone
Pierre Talqueuse	Talcose stone
Plomb	Lead
Sable	Sand
Schiste	Schist/or shale
Sel gemme	Rock salt, halite
Spath	Spar
Talc	Talc
Terre Tremblante	Quaking earth (?)
Fer en Sable	Ferruginous sand

Though minerals form less than one quarter, the title of the map reads:

Mineralogical Map
Where the nature of terraines
in Canada and Louisiana can be seen.
Drawn by Philippe Buache of the Ac. of Sciences
upon the researches and for a Memoire
by Mr Guettard of the same Acad.
1752

(This translation and the following ones are by the author of this paper.)

As is well known, using the word "Mineralogical" instead of "Geological" is normal in the eighteenth century. The word "Geology," though coined in its Italian form as early as 1603 by Aldrovandi, was printed in French for the first time in the *Encyclopédie* of Diderot in 1751, and it was ignored by the *Dictionnaire de l'Académie* as late as the sixth edition, 1835.

Four texts concerning North America were published in the same 1752 volume of the *Histoire de l'Académie Royale des Sciences*, all four by Guettard.[2] The first, on pages 12–16, is a comparison of the minerals of Canada with those of Switzerland. The next, pages 189–220 and plate VII, is a memoir in which the minerals of Canada are compared with those of Switzerland. Pages 323–360 and plates VIII–12 (plates IX–12 are numbered I-IV in the legend on pp. 359–361) bear the same title and comprise the second and last part. It is mainly about Switzerland, but plate IV and some paragraphs are important for North America as well. The text on pages 524–538 is an addition to the preceding two texts. Guettard writes (p. 524): "It will be more practical to find the map and all the observed facts upon which it was drawn in the same volume."

This geological map is one of the oldest in the world (Table 3).[3]

Table 3. Some of the Earliest Geological Maps

Year	*Author*	*Thema*
1739	Piccoli	Italy. Fossiliferous localities
1743	Packe	England. Geology of Canterbury and neighborhood
1746	Guettard	France and England. Geology
1752	Guettard	North America. Geology
1752	Guettard	Switzerland. Geology

The accompanying text and plates are also interesting. In the second part of the memoir (plate III, fig. I, and plate IV, fig. I) a molar tooth (probably a Mastodon) from Canada is figured. On plate III, fig. 2, a beautiful spirifer is interpreted by Guettard as an imprint of a butterfly! On plate IV, fig. 2, we easily recognize a brachiopod. In the legend it is called "poulette striée," striated pullet; in that time and even later on in the country, such brachiopods were commonly named poule or poulette. This fossil also comes from Canada.

Ferruginous mineral waters are considered by Guettard (p. 189) as a possible sign of iron ore. Guettard is right to emphasize (p. 15), "the great amount of asbestos of excellent quality and with very long fibers" and when concluding (p. 189) that Canada and—if we judge from the map—what was to become the United States are "a land similar to the lands of Europe which contain mines, even the most precious ones."

Guettard's maps have for us another and perhaps deeper interest. In 1746, in his "Memoir and mineralogical map on the nature and situation of the terrains that traverse France and England"[4] he mapped what we now call the Anglo-Parisian Basin and neighboring older eroded chains as composed of three more or less concentric zones: 1. Schistose (Pre-Mesozoic), 2. Marly (mostly Mesozoic), and 3. Sandy (mostly Cenozoic). Thirteen years before Ferber and fourteen before Michell, Guettard was the first man to state and to demonstrate on a map that different kinds of rocks are not randomly scattered but show a global organization in space. The Anglo-Parisian Basin is obviously a small region, a good but limited example. North America offers the benefit of a continental scale. Here again Guettard distinguishes three zones (1752, p. 14):

1. Schistose metalliferous zone "with different metals, bitumens, sulphur, slates, marbles, crystal, mines, cold and warm mineral waters."
2. Marly zone: "No more animal [*sic*], except iron. Only limestone, calcinable stones, chalk, marl, fossil shells, plasterstone."
3. Sandy zone.

Of course, as in many preliminary works, this one is full of minor and major inaccuracies. The distinction of the three zones is very rough, as also is the map. The schistose zone stretches from Texas to western Ontario, Ungava Peninsula, Greenland, and even Iceland. The marly zone (grey tone) extends from the Gulf Coast through the Carolinas to New England and Quebec. The sandy zone, cautiously qualified "suspected," is mostly located under the sea.

In the Anglo-Parisian Basin, Guettard (1746) rightly mapped the three zones as being roughly concentric, the sand lying in the middle. But in the western half of North America—a continental border—he recognizes that the zones are elongated. That is probably one of the reasons why he took, for his comparison as a European counterpart, not France and England but Switzerland and Germany, where he finds again, going from the interior of the continent toward its border, three parallel zones, the last and sandy one lying in northern Germany, as everyone knows.[5]

On pp. 355–356 Guettard stresses the geological similarities between East Greenland and both sides of Hudson Bay—one of the first hints of the concept of a North-American shield (his schistose zone). Faithfully he reports observations by Ellis (1748 and 1750), with evidence favoring his own views.[6]

Guettard himself never went to America. His map and memoir are mainly based on the information and samples—minerals, rocks, and fossils—sent to him from North America, mostly by four Frenchmen: R. M. de la Galissonière, (1693–1756) governor of Canada, J. F. Gaulthier, physician, M. de Lotbinière, and F. Picquet, clergyman.[7] In his memoir, Guettard also quotes the *History of New France*, by Charlevoix.[8] Though using information sent or written by others, Guettard was very cautious and critical. Although Charlevoix believes that near Quebec there are true diamonds, Guettard (pp. 196–197), studying samples received from Canada, rightly concludes that the so-called diamonds are only quartz.

Philippe Buache (1700–1773), who drew the map, was first an architect. At the age of 19 he won the first prize of the Academy of Architecture. Then he devoted himself to geography. He worked as a collaborator of Delisle, and he was his successor at the Academy. This explains the fact that he used the same kind of projection. He is described as "a rather small man, not handsome," but "sweet, gentle, humane and always ready to help others whenever the opportunity occurred." His panegyric is dated 1772, though he did not die until 1773. In fact he died in January; probably the *History of the Academy* for 1772 was not yet in print, and a zealous secretary joined the panegyric to that volume.[9]

Jean-Etienne Guettard (1715–1786) was born in Etampes, 50 kilometers south of Paris, where he had the opportunity of collecting fossil shells in the highly fossiliferous Upper Oligocene Fontainebleau sands. He was encouraged to study natural sciences by his grandfather, a physician. Beside his geological maps of 1746 and 1752, he is renowned for having recognized extinct volcanoes in the mountains of the French Massif Central (1752), which was a strong argument against Werner's exaggerated neptunism and in favour of Hutton's vulcanism. Guettard worked hard; he wrote thirteen books and more than ninety memoirs, not only on geology, but also on botany, veterinary art, and zoology. He was a friend and great admirer of Linnaeus. Needing a collaborator for geological mapping north of Paris, he chose a yet unknown young man: Lavoisier.[10]

He felt the need of helping others and encouraging them to work. In his 1752 memoir (p. 190) he advocates observations to increase our knowledge of North America. In the Archives of the Académie des Sciences an unpublished booklet, possibly a proof, gives pieces of advice for such purpose, under the title "Short Reflexions on the Means of Multiplying Mineralogical Observations."[11]

What might be in the future the fruit of such observations and mapping is clearly felt and expressed by Guettard with regard to North America (p. 190): Observations would help to bind together our knowledge, "to set them into a kind of systematic order which might even throw some light on the structure of our globe." Thus Guettard predicted what is now happening more than two centuries later. He was the Keeper of the duc d'Orléans' natural history collection. Despite this relatively high position, he was modest. In his panegyric Condorcet says that Guettard "never conceived that the place a man occupies in the society might add new needs to the needs imposed by Nature to every man."[12] His frankness often verged on roughness. To a man for whom he had given his voice in a vote, he said, "If I did not think it right to give you my voice, you would not have it, for I don't like you." Once when Condorcet was to pronounce a panegyric, Guettard said to him, "You will lie well!" He added, "When my turn comes, I want only the truth."

Guettard was a strong opponent of the death sentence. His friend Condorcet tells us that, being a bachelor, he adopted his servant's children, reared them, and took care of even the smallest details of their education. He could not see a poor man without helping him and even weeping with him. He extended these feelings to animals, forbidding that any be killed for him or in his house. Pity he considered useful and almost necessary to keep with all its purity—this strongest feeling and perhaps the only effective barrier opposed by Nature to selfishness and to anger.[13]

Such was the man to whom we owe the first essay at the geological mapping of the North American continent.

Acknowledgments

The author is grateful to Dr. Robert Courrier and Paul Germain, both Secrétaires Perpétuels de l'Académie des Sciences, and to Mrs. Louis Hautecoeur, Conservateur de la Bibliothèque de l'Institut de France, for having authorized him to consult their archives and manuscripts, as well as J. G. Affholder, for his help in determining the standard parallels of the map, and Martine de Cayeux, for revising the English text.

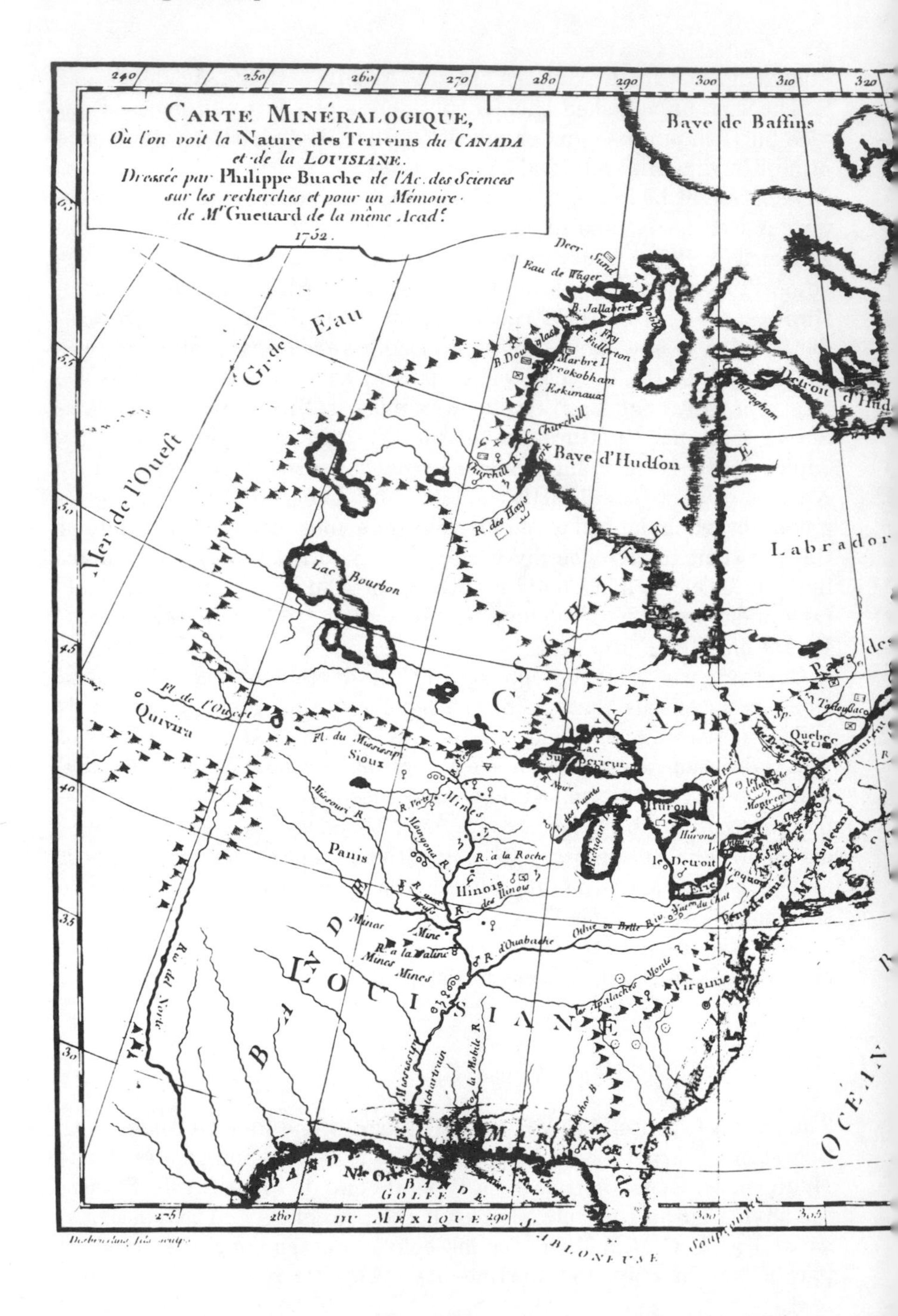
Carte Minéralogique,
Où l'on voit la Nature des Terreins du CANADA
et de la LOUISIANE.
Dressée par Philippe Buache de l'Ac. des Sciences
sur les recherches et pour un Mémoire
de Mr Guettard de la même Acad.
1752.
Baye de Baffins
Mer de l'Ouest
Gr. de Eau
Baye d'Hudson
Labrador
Lac Bourbon
Quivira
Sioux
Panis
Ilinois
le Detroit
Quebec
Virginie
Pensilvanie
N. York
N. Angleterre
Golfe du Mexique
Ocean

Mem. de l'Ac. des Sc. 1752. p. 220. Pl. III.
340
350
360
Hekla
Islande
Groenland
Détroit de Davis
Forbisher
Explication des Caracteres.
Ardoise
Argent.
Caillou roule
Charbon de Terre.
Coquilles ou Corps marins fossilles.
Cristal.
Cuivre.
Emeric.
Fer.
F.ne Minérale chaude.
F.ne Minérale froide.
F.ne Salée.
F.ne Sulphureuse.
Glaise.
Gyps ou Pierre à Platre.
Granit.
Marbre.
Marcassite ou Pyrite.
Marcassite Ferrugineuse.
Marne
Marne Pierreuse
S.M. Mine ou Subst.ce Metallique.
Nitre ou Salpètre.
Ocre ou Fer dissous.
Or roule.
Petrole.
Pierre Blanche.
Pierre à Chaux.
Pierre à eguiser.
Pierre Olaire.
Pierre Talqueuse.
Plomb.
Sable.
Schiste.
Sel Gemme.
Sp. Spath
Talc.
Terre Tremblante.
Fer en Sable.
60
55
50
45
40
B. Phelipeau
Terre Neuve
C. de Ray
Grand Banc
Montagnes du Nord
les Trois Rivieres
Fleuve St Laurent
d'Orleans
Lac St Pierre
Montreal
Detail particulier des Rivagas du Fl. St Laurent depuis la Malbaye jusqu'a Montreal.
1 Degre
5 10 15 20 25 Lieues.

Notes

1. Jean-Etienne Guettard, "Mémoire dans lequel on compare le Canada à la Suisse, par rapport à ses minéraux," *Hist. Acad. R. des Sci.* (1752), *B*: 189–220, pl. VII; (1752), *C*: 323–361, pls. VIII–XII.
2. Ibid. Also, "Sur la comparaison du Canada avec la Suisse, par rapport à ses minéraux," ibid. (1752), *A*: 12–16. "Addition au Mémoire . . . , ibid. (1752), *D*: 524–538.
3. André Cailleux, "Histoire de la géologie," *Que sais-je?, n° 962* (Paris, Presses Univ. de France, 1968), p. 73.
4. J.-E. Guettard, "Mémoire et carte minéralogique sur la nature et la situation des terrains qui traversent la France et l'Angleterre," *Hist. Acad. R. des Sci. (1746)*, *B*: 363–392, pls. 31, 32.
5. Guettard, "Géographie," *Hist. Acad. R. des Sci.* (1746), *A*: 105–107.
6. Henri Ellis, *A voyage to Hudson-Bay, The Dobbs galley and California, in 1746–47, for discovering a north-west passage* (London, 1748), and *Considerations on the north-western passage and a clear account of the most practical method of attempting that discovery* (London, 1750).
7. Roland Lamontagne, "La participation canadienne à l'oeuvre minéralogique de Guettard," *Rev. d'His. des Sci.* (1965), *18*, 4: 385–388.
8. P. F. X. Charlevoix, *Histoire de la Nouvelle-France* (Paris, 1744).
9. J. P. Grandjean de Fouchy, "Eloge de M. Buache," *Hist. Acad. R. des Sci.* (1772), 135–150. Cf. anonymous, "Buache," in Ferdinand Hoefer, *Nouv. Biogr. Gén.* (1853), *7*: 676–678.
10. Guettard and Monnet, *Atlas et Description minéralogique de la France, Ière partie* (Paris, 1778–80).
11. Guettard (no date), "Réflexions abrégées sur les moyens de multiplier les observations minéralogiques," *Arch. Acad. des Sci., Paris*, Guettard, n° 63 S 1618, 12 pp.
12. J. A. N. Condorcet, "Eloge de M. Guettard," *Hist. Acad. R. des Sci.* (1786), 47–62. Cf. A. E. C. Dareste, "Guettard," in Hoefer, *Nouv. Biogr. Gen.* (1858), *22*: 472–477; and J. B. Soreau, "Notice historique sur la vie littéraire du Docteur Guettard," *Mag. Encycl.* (1803), *9*, 2: 471–485, Paris.
13. Condorcet, "Eloge."

7

Acts of God, Acts of Man: Geological Ideas and the Imaginary Landscapes of Thomas Cole

Ellwood C. Parry III

Although a few landscape painters were active in the United States between the 1790's and the 1820's, they tended to work in the older, eighteenth-century mode, delighting mainly in pleasant pastoral prospects and topographical views of human settlements, especially of gentlemen's country seats. In contrast to this valley-dwelling tradition, the first major artist-spokesman for the wilder, dangerous, untamed features of American mountain scenery was Thomas Cole (1801–1848). Often described as a founder of the Hudson River School, meaning the first mature phase of Romantic landscape painting in America, Cole was actually born in Bolton-le-Moors, Lancashire, not far from the English Lake District, and did not emigrate to the United States with his parents and sisters until after the Napoleonic Wars.[1]

It is known that Cole's formal schooling in Lancashire was fairly limited, but it is still safe to assume that he arrived in Philadelphia in July of 1818 with a mind already steeped in the conventional English theories of the Beautiful, the Sublime, and the Picturesque; a mind eagerly predisposed to the more refined arts, to the act of reading as a means of sharing in vast segments of the world's experience, and to the process of playing music or writing poetry as the expression of a sensitive spirit. His was a decidedly Romantic mind given to the contemplation of "Nature" on frequent hikes or hill-climbs in the country where, with a "loving eye," he could gaze on the transcendent harmony of "the pure creations of the Almighty."[2]

On the other hand, the extent of Cole's scientific, rather than aesthetic, interests when he crossed the Atlantic for the first time is difficult to determine. During his apprenticeship as an engraver's assistant, he might have gained some elementary knowledge of certain metals or minerals principally through working with steel and copper plates, etching acids, and carbon blacks for ink, as well as a few earth pigments and possibly even lithographic limestone. But whether he ever started a collection of minerals on his own, as so many other youths were beginning to do, and brought it with him to America is simply not recorded. Nevertheless, at least some ability, no matter how rudimentary, to distinguish among different rock formations would have been indispensable for anyone planning to become a landscape artist in that era.

Admittedly, Cole's earliest surviving nature studies, done in Ohio and western Pennsylvania (1823), have nothing to do with geology in any form. They were factual drawings not of impressive local outcroppings or examples of fluvial erosion, but of small saplings, twisted stumps, and unusual tree trunks—picturesque details that might have been useful either for his father's wallpaper printing business or else, saved in a portfolio, for his own initial attempts at painting in oils. Significantly, however, Cole's fame as a portrayer of truly American scenes did not begin until after he had moved to New York City, at the time of the opening of the Erie Canal (spring 1825). And a summer sketching trip up the "North River," through the Hudson Highlands and on to the Catskill Mountains, resulted in a set of pictures that quickly captured critical and popular attention because they openly proclaimed the artist's discovery of the Sublime in the American wilderness, particularly at the tops of inaccessible peaks and precipitous waterfalls.

There are a number of Cole's mountain-top drawings, moreover, that can testify to his interest in discernible strata and his awareness of basic rocks and minerals. For example, in several drawings of this type—which are often no more than rapid outline sketches of large boulders dotting the foreground, played off against the waves of receding hills in the middle and deep background, with handwritten notations at the top or bottom of the sheet as a key to the colors and distances from point to point—one can still find a few phrases like "gray granite with black moss" in the foreground of a "View looking southeast from Chocorua Peak" and "limestone belts" in a later rendering of the "Dent de Jaman" in Switzerland.[3] These scant indications, although infrequent, may be enough to suggest that Cole, like John Ruskin (1819–1900), the English art critic who became an international influence in the 1840's and '50's, could look at a thrilling mountainscape with eyes that acted as both telescope and microscope, taking in the sweeping 360-degree panoramic view, while also noticing the interesting bedding patterns or intrusive veins in the rocks at his feet.

It goes without saying that an artist and a geologist on the same expedition would tend to work in radically different ways. A Romantic painter like Cole would be predominantly interested in the outward appearances of rugged mountain scenery and in the train of powerful associations engendered by thoughts of beauty and sublimity; whereas a natural scientist, like Benjamin Silliman (1779–1864) at Yale or Edward Hitchcock (1793–1864) at Amherst, would be more professionally concerned with taking accurate measurements or detailed notes and collecting samples (instead of sketches) as a way of studying the structure of a given mountain system. At the same time, there was at least one important area of common ground.

In retrospect, it seems true that the idea of the "Panorama"—either as a direct landscape experience or else as an enjoyable visit to a panorama theater inside of which the enchanted viewers, after climbing to an elevated platform in semidarkness, could suddenly look outward in all directions at a huge, cylindrical, uninterrupted landscape painting that seemed magically real—may have been one of the most fundamental visual concepts shared by American artists and geologists in the nineteenth century.[4] Benjamin Silliman was certainly aware of the panorama impulse when he devoted six pages at the start of his book, *Remarks Made, On a Short Tour, Between Hartford and Quebec in the Autumn of 1819* (1820), to a long description of Monte Video, the country estate (with a summer house in the new Gothic Revival style) atop Talcott Mountain at Avon, Connecticut, owned by his brother-in-law, Daniel Wadsworth (1771–1848). Silliman obviously admired the situation of the house, near the edge of a precipice overlooking the Farmington Valley, but the finest view was afforded by a tower (55 feet high) that Wadsworth had added at the summit of the northern ridge:

> From the top which is nine hundred and sixty feet above the level of the Connecticut River, you have at one view, all those objects which have been seen separately from the different stations be-

> low. The diameter of the view in two directions, is more than ninety miles, extending into the neighboring states of Massachusetts and New-York, and comprising the spires of more than thirty of the nearest towns and villages. The little spot of cultivation surrounding the house, and the lake at your feet, with its picturesque appendages of boat, winding paths, and Gothic buildings, shut in by rocks and forests, compose the foreground of this grand Panorama.[5]

The Silliman-Wadsworth connection is especially interesting because it seems to represent the most intimate possible link between art and science in that period. First of all, their wives—Faith Trumbull Wadsworth (1769–1846) and Harriet Trumbull Silliman (1783–1850)—were nieces of the important painter Colonel John Trumbull (1756–1843), whose collection of portraits of Revolutionary War heroes, battle scenes, and political events, such as the Declaration of Independence, was eventually acquired by Yale College through Silliman's intercession.[6] Beyond their first-hand acquaintance with Trumbull and other key figures in the development of art and architecture in the new nation, Wadsworth and Silliman were clearly the best of friends as well, often going on short and long trips together for the purposes of relaxation, health, and "the gratification of reasonable curiosity" about geological features along the way. In fact, since it was Wadsworth's habit, "when travelling, of taking hasty outlines of interesting portions of scenery, and of finishing them after his return,"[7] Silliman simply used his brother-in-law's sketches as illustrations, both in his *Short Tour* book (Fig. 1) and later in his *American Journal of Science*.[8]

In addition to being an amateur artist and geologist, Daniel Wadsworth was also an active art collector, thanks to the fortune he had inherited from his father's banking and manufacturing interests. And he, like other prominent men in other cities along the east coast in the later 1820's, quickly bought a number of paintings from the rising young landscape artist who was being discussed so glowingly in the New York press, Thomas Cole. More to the point, Wadsworth's tastes were catholic enough for him to appreciate Cole's imaginary scenes as well as his topographical views. In 1828, for example, when he commissioned Cole to paint a scene as specific as a view of his own estate, Monte Video, glimpsed from the north ridge, he had already purchased from the artist two landscape compositions, a *St. John in the Wilderness* (1827) and a *Scene from "The Last of the Mohicans," Cora Kneel-*

ing at the Feet of Tamenund (see Fig. 4), to go with several other distinctly American views he already owned (all now in the Wadsworth Atheneum, Hartford).[9]

Cole's friendship with Wadsworth—surviving letters between them date from the 1820's to the 1840's—holds further implications for the interaction of art and geology. For instance, even in the absence of evidence to prove that Cole and Benjamin Silliman were directly acquainted, one can at least assume that, through Wadsworth, they were aware of each other's work. This would go a long way toward explaining why, when the English writer, John Howard Hinton, published his two-volume study of *The History and Topography of the United States* (London, 1830–32), a study in which the aid of American geologists such as Amos Eaton (1776–1842) and Benjamin Silliman was openly acknowledged, so many of the topographical illustrations, including one of *Monte Video* (Fig. 2), were supplied by Thomas Cole, who was then living in London (1829–31).

Furthermore, given Cole's contacts with many men interested in geological issues, one can also assume that he must have been aware of the concepts and ideas being discussed in the pages of Silliman's *Journal* all through the 1820's up to the time of his departure for Europe in June 1829. Significantly, one of the more popular topics among Silliman's contributors was the puzzling question of New England's "rocking stones," which often served as local landmarks and objects of intense curiosity from the colonial era well into the nineteenth century. Although one incautious author cited these intriguing boulders (along with dolmen-like structures supposedly found by hunters deep in the New England woods) as "druidical remains," positive proofs that the race of men who built Stonehenge and Avebury Circle had also come to these shores, most of the articles Silliman published were short descriptive notices submitted by amateur geologists who, in addition to recording the basic facts and supplying a sketch or two (Fig. 3), frequently decried the fact that vandals armed with crowbars and levers were mindlessly destroying the geological evidence simply by overturning these delicately balanced rocks merely for the pleasure of watching them go hurtling down the mountain sides and hearing them crash through the trees below.[10]

Cole must have shared in this serious community of interest because rocking stones appear, perched on promontories and pillars of rock, in several of the pictures he completed in 1827–28, based on episodes from James Fenimore Cooper's latest novel, *The Last of the Mohicans* (1826). In the version that Daniel Wadsworth purchased, a rounded boulder on top of a tall pillar seems to mark a special and perhaps sacred site in the landscape, a magic circle where the Indians might gather for an important tribal conference in total security, far from the eyes of civilization. The fact that Wadsworth allowed the painting he owned to be engraved and published in a giftbook annual (Fig. 4) tends to show that Cole's vision of the untouched American wilderness, where these boulders or "erratics" took on a special presence of their own, was a popular image, one that appealed to many different levels of society at the same time. No doubt, it was the inexplicability of the rocking stones, more than a decade before the promulgation of a coherent theory of glacial transport, that made them such a popular subject for American artists and scientists, both amateur and professional.[11]

Comments in Cole's letters and diaries make it clear that he was an extremely ambitious artist for whom the act of painting topographical views and single landscape compositions was never enough of a challenge. One of the ways he found to escape the tyranny of having to imitate or imagine a single aspect of nature on a single canvas was by producing companion pictures, which offered a more satisfying range for his pictorial imagination. As early as 1828, for example, he completed and exhibited "two attempts at a higher style of landscape" which he described to a prospective buyer as follows:

> The subject of one picture is The Garden of Eden [now known only from an engraving]. In this I have endeavoured to conceive a happy spot wherein all the beautiful objects of Nature were concentrated. The subject of the other is The Expulsion from the Garden [Fig. 5]. Here I have introduced the more terrible objects of Nature, and have endeavoured to heighten the effect by giving a glimpse of The Garden of Eden in its tranquility.[12]

To Cole's mind, what elevated these compositions to a higher plane was that they embodied differing aesthetic principles—the Beautiful versus the Sublime—in terms of the landforms alone. It has long been known that Cole was directly influenced by John Martin's mezzotints for a new edition of Milton's *Paradise Lost* (1825–27), but the American artist went beyond these English illustrations by throwing in extra geological ideas that were not in the original. Thus in Cole's *Expulsion*, which a Boston critic praised as "a world made terrific by tokens of the anger of God,"[13] shafts of light can be seen coming from the flaming sword at the Gates of Paradise; but in addition to this conventional symbol of the Lord's wrath, a frightening rift seems to have opened suddenly at the viewer's feet (as if in response to the earthshaking events that have just taken place in human history), and in the far distance Cole inserted a volcano in violent eruption, even though he had never seen such a thing in person on his travels in England or North America.

The truth is that by the 1820's eye-witness accounts and accurate images of volcanic pyrotechnics were available everywhere and in countless different forms. Cole could have seen the latest book on the *History of Vesuvius* which contained John Martin's image of the eruption of 1822 as a frontispiece; as a child he might have owned and played with one of the many transparency prints of volcanos at night that were so popular after 1800 (Fig. 6); or, at some point in his life, he might have visited the sort of landscape display theater with cork models that Benjamin Silliman saw when he was in London in 1805:

> We were conducted behind a curtain where all was dark, and through a door or window, opened for the purpose, we perceived Mount Vesuvius throwing out fire, red hot stones, smoke and flame, attended with a roaring noise like thunder; the crater glowed with heat, and, near it, the lava had burst through the side of the mountain, and poured down in a . . . real flood of glowing and burning matter, which the ingenious artist contrives to manage in such a manner as not to set fire to his cork mountain. . . . He has not forgotten to appeal to the sense of smell as well as to those of sight and hearing, for, the spectator is assailed by the odour of burning sulphur, and such other effluvia as volcanoes usually emit.[14]

Since lava flows and eruptions had become part of the basic repertory of landscape ideas in the public mind long before Cole painted his version of the *Expulsion*, it may be unnecessary to try to link this particular picture with contemporary theories that emphasized the role of volcanic activity or extreme heat deep inside the earth in the formation of rocks and in most geologic processes. Yet it would be very hard to believe that Cole could have remained ignorant of the controversy between the Vulcanists and the Neptunists. Indeed, a circumstantial case could be made that where the *Expulsion* (1828) could be claimed by one side of the argument, a picture entitled *The Subsiding of the Waters of the Deluge*, which Cole was inspired to paint in 1829 after seeing and making drawings after a John Martin image of the rising waters of *The Deluge* (Fig. 7), could just as openly be claimed in evidence either by the fervent believers in the Biblical Flood or by the more rational defenders of Wernerian contentions that virtually all the earth's rocks were precipitated out of a once universal ocean. Since this canvas has not been located, a reviewer's description will have to suffice:

> . . . the latest effort by Mr. COLE; his genius has, in this instance, soared to an exalted flight; his subject is the "Subsiding of the Waters of the Deluge," and the manner in which it has been treated is worthy of his established reputation. The waste of the waters is seen to be diminished; the bare and rugged peaks of the loftiest mountains are left uncovered; in the distance, the ark of safety is dimly visible floating in lonely security; and, over all, the glow of the returning sun is spread like a promise of hope in the midst of so great desolation.[15]

Ultimately, of course, the essential thing from Cole's point of view was not the moot issue of exactly how rocks were formed at the beginning of the world, but how he might produce the most dramatic and effective compositions possible. Sublimity obviously took precedence over geology. Yet, more than John Martin (1789–1854), J. M. W. Turner (1775–1851), Francis Danby (1793–1861), or any other historical landscape painter of the period, Cole further reduced the relative size (hence the role) of his Biblical figures compared to the awe-inspiring convulsions of nature, the acts of punishment by a vengeful, Old Testament God, which seem to occur in perfect concert with the stories from Genesis that Cole was retelling for his American audience.

In the end, because of the active part played by the scenery, it may be that Cole's renderings of the *Expulsion from the Garden of Eden* and *The Subsiding of the Waters of the Deluge* have to be seen and interpreted against an international cultural backdrop in which the ideas of James Hutton (1726–1797), Abraham Gottlob Werner (1750–1817), and later theoreticians in geology were being ardently debated. These two Biblical landscapes strongly suggest that in 1828–29 Cole's geological views leaned toward catastrophism; although as an artist, never having to subscribe to any one particular theory of the history of the earth, he could freely adopt whatever details or concepts he needed in order to make his ambitious pictures as powerful and morally instructive as possible.

It is tempting to think that the arch-romantic, catastrophist phase of Cole's career ended abruptly when he went to England and then Italy for an extended sojourn, beginning in the spring of 1829 and lasting until late in 1832. The fact is that he was working and studying in London when the first volume of Sir Charles Lyell's *Principles of Geology* appeared (January 1830), and it is more than possible that

Lyell's forceful defense of uniformitarianism had a significant effect on the early planning stages for Cole's first major series of historical landscapes, *The Course of Empire*, now in the collection of the New-York Historical Society (Fig. 11).

Visual evidence that Cole may have read and absorbed part of Lyell's *Principles* several years in advance of the scientific community in the United States can be found by comparing the well-known frontispiece to Lyell's first volume, a view of the *Present State of the Temple of Serapis at Puzzuoli* (Fig. 8), with one of Cole's most interesting drawings, entitled *Ruins or the Effects of Time* (Fig. 9). This juxtaposition shows that Cole took a similar idea of columns in a row, the ruins of an ancient edifice vulnerable to the sea, and stretched them out to the horizon where a Vesuvian cone helps to mark the location as Mediterranean. It should be pointed out that just as Lyell had visited Puzzuoli, eleven miles west of Naples, in 1828, Cole, too, might have traveled to that popular site in 1832 when he was in Naples for a month and made several excursions to places like Pompeii and Paestum.[16] Furthermore, knowing that Lyell used the three surviving columns at Puzzuoli as visible "Proofs of the Elevation and Subsidence in the Bay of Baiae,"[17] it is not surprising to discover that Cole, in writing the following caption to his drawing, would stress the idea, not of lithodomic holes only at certain places on the shafts of the standing columns, but of the more easily representable change in the relationship of land and sea:

> Sun setting in the quiet ocean. Pyramids rising in the distance. Water & remains of stupendous edifices. The sea having encroached since they were erected. Broken mountains with huge rocks which seem to have descended from there and overturned columns in their downward course to the sea where a fragment is yet seen above the waves. A Bridge over the dry bed of a river. The river, having changed its course, flows a little distance. On the beach the wreck of a vessel. Broken aqueduct. In the foreground scattered trees. A fountain flowing into a broken cistern. The remains of a human skeleton seen in an uncovered sarcophagus. Broken swords, vases, &c.[18]

This passage makes it clear that Cole thought in additive ways—which was perfectly typical for the time. He made these lists for himself of all the powerful yet appropriate symbols that could be wedged, one beside the other, into the foregrounds of many of his imaginary scenes. Fortunately, however, he rarely attempted to compress everything on such a list into a single canvas; some degree of editing was always required. Nevertheless, what makes the caption to *Ruins of the Effects of Time* so fascinating is the remarkable olio of conventional artistic or archaeological emblems of death and decay (broken columns, fallen aqueducts, open sarcophagi), coupled with new geological concerns about discernible changes in the surface features of the land (subsidence and earthquake activity in the vicinity of volcanos).[19] Short of finding a written reference to the *Principles of Geology* in the artist's diaries or notebooks, the fact that he shows the sea encroaching on a row of columns in his drawing is the strongest proof one has that he had learned something from Lyell.

Finally, it is in the five paintings of *The Course of Empire* (Fig. 11), begun in the fall of 1833 and finished late in 1836, that Thomas Cole appears to have adopted something approaching uniformitarian principles. In this series, which tells the epic saga of the rise, culmination, and decline of a great seaport civilization, no natural cataclysms, no earthquakes, landslides, or volcanic convulsions are allowed to intrude. The disaster that occurs in the fourth picture, *Destruction*, is an act of man; human beings are here the cause of their own extinction.

A collocation drawing survives (Fig. 10) to show what Cole had planned for the fireplace wall in Luman Reed's Gallery, 13 Greenwich Street, New York City. The pictures naturally divide themselves into two pairs—both of the turbulent, sublime scenes (the *Savage State*, upper left, and *Destruction*, upper right) were meant to hang above the calmer, more beautiful ones (the *Pastoral State* and *Desolation*)—leaving the larger *Consummation* in the center over the mantelpiece. A closer look at the quick sketch that Cole inscribed "The State of Desolation" (Fig. 10, lower right) reveals a setting sun on the left and a few standing columns that seem to be an echo of the *Ruins or the Effects of Time*; but one cannot tell if Cole was still planning to show the sea encroaching.

Ultimately, it seems, the artist decided to eliminate any sense that landscapes can change as quickly as empires rise and fall. The actual viewpoint in *The Course of Empire* moves from picture to picture, as if the onlooker were constantly changing seats in a natural amphitheater, always hoping to get a better glimpse of the dramatic action, but Cole deliberately put in a fixed reference point in all five paintings. A huge boulder (if not necessarily a rocking stone) sits rather precariously atop a great cliff (whose steeply inclined sedimentary strata are clearly indicated) in the deep middleground of every picture (Fig. 12). Cole wrote that this unusual feature was intended to be a mariner's landmark; yet its very presence also serves to reinforce the idea that Cole, more than any earlier painter of before and after scenes, while using archaeological remains and appealing to the universal pleasure of ruins, was particularly mindful of the differences between the traditional framework of human history and the new geological time scale of worlds without end. Compared to the helpless transience of man and his works so quickly turned to ruins, Cole made certain that the landscape, serving as a temporary stage for the human melodrama, goes on unchanged—and what few architectural alterations man had the folly to erect are quickly eclipsed, save for the single column in the last scene, which has not yet been totally overtaken by the slow and yet inexorable processes of erosion in nature.

Acknowledgments

Part of the initial research for this paper was conducted under a National Endowment for the Humanities Fellowship for Independent Study and Research in 1975–1976. I gratefully acknowledge my debt to the N.E.H. for its generous support.

Notes

1. Basic biographical information on Cole is available in Louis Legrand Noble, *The Life and Works of Thomas Cole*, ed. Elliot S. Vesell (Cambridge, Mass., Harvard University Press, 1964).
2. See Cole's "Essay on American Scenery," reprinted from *The American Monthly Magazine*, n.s. 1 (1836), 1–12, in John W. McCoubrey, ed., *American Art, 1700–1960* (Englewood Cliffs, Prentice-Hall, 1965), pp. 98–110.
3. The drawings cited are in the collection of The Detroit Institute of Arts.
4. For a discussion of American artists' involvements with the panorama, the diorama, and the moving panorama, see my article, "Landscape Theater in America," *Art in America*, 59, no. 6 (1971), 52–61.
5. Silliman, *Remarks Made, On a Short Tour* (New Haven, 1820), p. 15.
6. See Theodore Sizer, "Benjamin Silliman and his Uncle-in-law Colonel Trumbull," *Art Quarterly*, 22 (1959), 335–344.
7. Silliman, *Remarks Made*, pp. 4, 9.
8. Lithographs after Wadsworth's drawings of the "Junction of Trap and Sandstone, at Rocky Hill near Hartford, Connecticut"

and of a "View Taken from the Upper Falls of the Genessee River" were published in the *American Journal of Science*, 17 (1830) and 18 (1830), respectively.

9. See *The Hudson River School: 19th Century American Landscapes in the Wadsworth Atheneum* (Hartford, Wadsworth Atheneum, 1976), pp. 16–26.
10. Descriptions of rocking stones appear in numerous places in the pages of Silliman's *Journal*, 2 (1820), 200–201; 6 (1823), 243–244; 7 (1824), 59–61, 201–203; 9 (1825), 27; and 10 (1826), 9–10. Also see John Finch, "On the Celtic Antiquities of America," *American Journal of Science*, 7 (1824), 149–161.
11. A list of Rocking Stones was published in Edward Hitchcock, *Final Report on the Geology of Massachusetts* (Northampton, 1841), II, 374–378, with credit given to various correspondents.
12. Letter from Cole to Robert Gilmor, Jr., of Baltimore, dated May 21, 1828, quoted in Noble, *Life and Works of Thomas Cole*, p. 64.
13. Editor, "The Garden of Eden," *Boston Recorder and Religious Telegram*, 13 (August 22, 1828), 135.
14. Silliman, *A Journal of Travels in England, Holland, and Scotland, and of Two Passages Over the Atlantic in the Years 1805 and 1806* (New Haven, 1820), I, 246–249.
15. *New York Mirror*, 6 (May 16, 1829), 354.
16. Noble, *Life and Works of Thomas Cole*, pp. 119–120.
17. Lyell, *Principles of Geology*, 9th ed. (New York, Appleton, 1856), pp. 507–519.
18. Thomas Cole, *Sketchbook No. 1*, p. 42. The Detroit Institute of Arts.
19. For a brief discussion of the idea of volcanic subsidence in relation to Cole's most unusual imaginary landscape, see my article, "Thomas Cole's 'The Titan's Goblet': A Reinterpretation," *Metropolitan Museum of Art Journal*, 4 (1971), 123–140.

1. Daniel Wadsworth, Monte Video, *1820. Engraving by S. S. Jocelyn. Published in Benjamin Silliman,* Remarks Made on a Short Tour, Between Hartford and Quebec in the Autumn of 1819 *(New Haven, 1820).*

2. Thomas Cole, Monte Video, The Residence of D. Wadsworth, Esqr., Near Hartford, Connecticut, *1831. Engraving by Fenner Sears & Co. Published in John Howard Hinton,* The History and Topography of the United States *(London, 1832), 2, no. 39.*

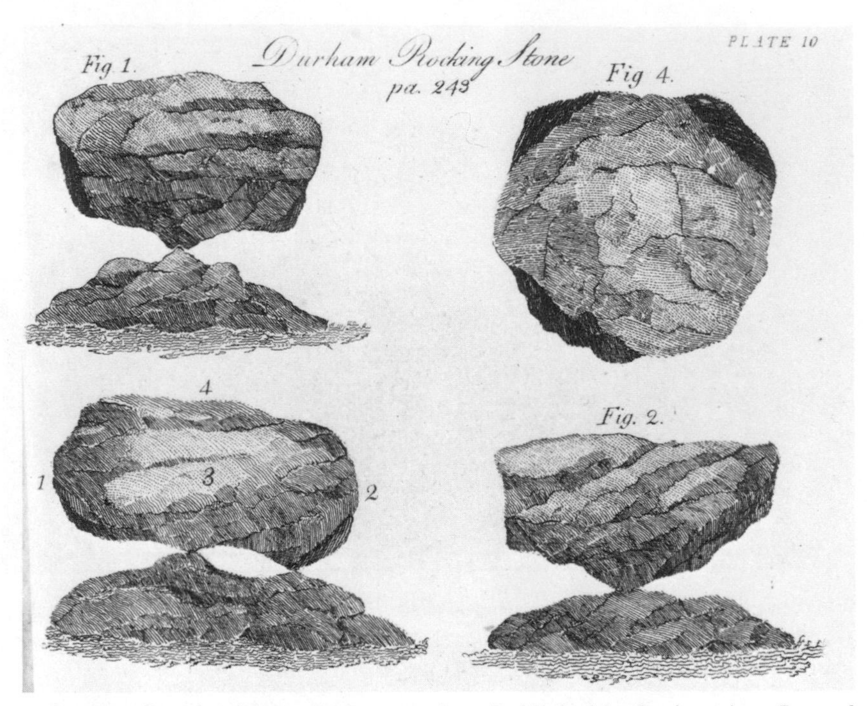

3. Durham Rocking Stone. *Line engraving. Published in the* American Journal of Science *(1823), 4: plate 10 (detail).*

4. Thomas Cole, American Scenery. *Engraving by G. B. Ellis. Published in* The Token; A Christmas and New Year's Present *(Boston, 1831), 4: opp. p. 55.*

5. Thomas Cole, Expulsion from the Garden of Eden, *1828. Oil on canvas, 39 × 54 inches. Courtesy of the Museum of Fine Arts, Boston; M. and M. Karolik Collection.*

6. William Orme, View of Mount Vesuvius, *1799. Aquatint ("From the Original Transparent Drawing"). Courtesy of the Trustees of the British Museum, London.*

7. Thomas Cole, The Deluge, *1829. Sketchbook drawing (after a mezzotint by John Martin). Cole Papers, New York State Library, Albany.*

8. Present State of the Temple of Serapis at Puzzuoli, *1830. Engraving by T. Bradley. Published in Sir Charles Lyell,* Principles of Geology *(London, 1830), 1: frontispiece. Photo courtesy of the U.S. Geological Survey Library, Reston, Virginia.*

9. Thomas Cole, Ruins, or The Effects of Time, *ca. 1832–1833. Pencil drawing. 9-1/2 × 15-3/16 inches. The Detroit Institute of Arts; Purchase, Founders Society, William H. Murphy Fund.*

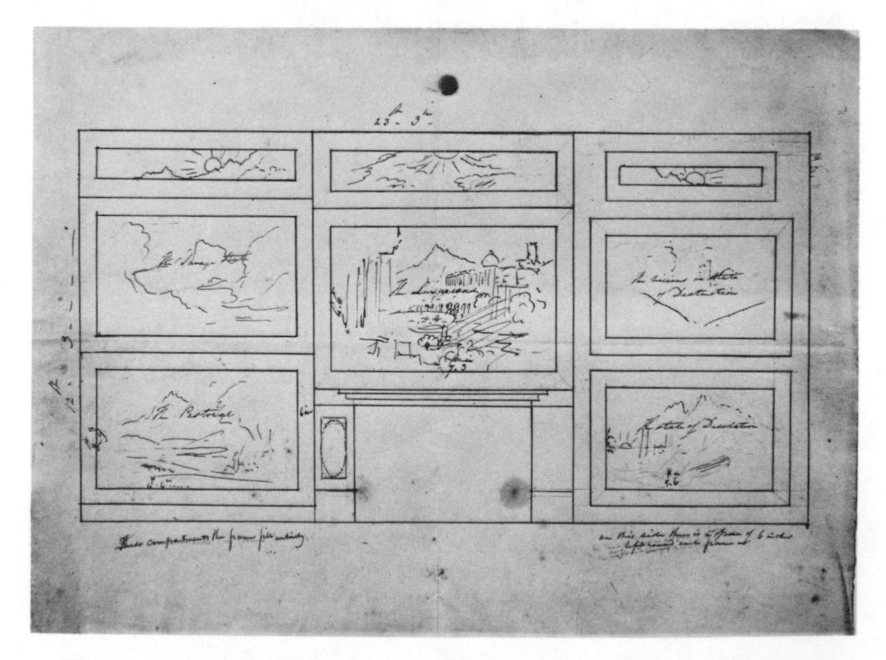

10. Thomas Cole, Installation Diagram for The Course of Empire, *September 1833. Pen and ink drawing. The Detroit Institute of Arts; Purchase, Founders Society, William H. Murphy Fund.*

11. Thomas Cole, The Course of Empire, *1833–1836. Five paintings reproduced to scale and arranged as the artist wanted them installed in Luman Reed's Gallery, 13 Greenwich Street, New York City. Courtesy of The New-York Historical Society.*

12. Detail of The Savage State, *showing the steep cliff and the singular, isolated rock on top of it which serves as a mariner's landmark as well as a fixed reference point for the viewer in all five paintings of the series. Courtesy of the New-York Historical Society.*

Whigs and Tories

Geology in 1776: Some Notes on the Character of an Incipient Science

Kenneth L. Taylor

What was the condition of geology 200 years ago? Perhaps the first point to establish is that if we take geology to be a certain form of inquiry and knowledge developing and changing in the hands of its most original practitioners, in 1776 this was still an Old World preserve. The point may need to be made as an apology for not looking, in this paper, at American geology, since our conference is devoted mainly to a bicentennial celebration and retrospective of geology in America. I think it will not be disputed, however, that in the late eighteenth century, whatever interest may attach to the activities of scientists elsewhere, Europe remained the locus of innovation and change in geology, as in the sciences generally.

One way of recounting what was going on in the world of geology in 1776 might consist of a record of the activities and statuses of important figures. For example, in 1776 Abraham Gottlob Werner had recently published his first book, *On the External Characters of Minerals*, and was completing his first year on the faculty of the Bergakademie Freiberg. James Hutton, aged 50, having long since given up scientific farming for the amenities of Edinburgh, was occupied with diverse commercial and intellectual interests, but the first publication of the theory of the earth he had been developing privately still lay nine years in the future. Horace Bénédict de Saussure was finishing up his two-year service as Rector of the Geneva Academy, and was already well embarked on the series of Alpine investigations that would be the substance of his *Voyages dans les Alpes*. That summer 200 years ago, in fact, Saussure took the diplomat and volcanic enthusiast William Hamilton with him on one of his two trips to Chamonix.

The deficiencies of this sort of heroic chronicle are obvious. Knowledge of such things is interesting, even essential, in our historical consideration, but it is far from sufficient. And the insufficiency does not stem merely from the incompleteness of the account. It will not suffice to continue down the roster of geologists, no matter how far we go past all the famous figures and in among the geologically obscure. To understand satisfactorily what was going on in geology in 1776, even the most exhaustive inventory of persons and their doings will not answer.

What is needed is to generalize appropriately about the problems, methods, assumptions, aspirations, accomplishments, and failures of the age's geologists. Like all disciplines, historical inquiry is a search for an order binding together a diversity of events, conferring on them a meaning they would not otherwise display. An article of the historian's faith is that a vision of such an order can be formulated which will give focus and significance to otherwise designless recitations of particulars.

There have been numerous scholarly attempts to provide parts of the historical vision I speak of.[1] They have given us the capability of making general historical formulations about the state of geology around 1776, some of which might be stated in this way:

(a) By 1776 a sense of geological time was being worked out. A bold few had directly challenged the Mosaic chronology, but perhaps at least as importantly a larger number of scientists were proceeding cautiously with the opinion that the evidence was soundly interpretable only through admission of a terrestrial age exceeding the Biblical allotment, perhaps by a large but as yet uncertain margin.[2]

(b) The organic origin of certain fossils having long since been widely acknowledged, a stratigraphic geology linking fossils to specific formations was being conceived.[3]

(c) The majority of writers on geological dynamics continued, as they had done for some time, to look on the aqueous processes of crystallization, sedimentation, and erosion as the chief agents by which the earth's features were fashioned. Nevertheless there had been and continued to be a smaller number who championed the dynamic role of a supposed internal heat.[4]

(d) In a movement on the whole distinct from the doctrinal concerns of the central-fire advocates, there was developing in 1776 an awareness of effects of volcanic action far broader than had before been suspected. This fostered an inquiry into the causes and effects of volcanoes, but did not at this time lead most participants of the movement to abandon a general preference for a neptunistic version of geological dynamics. The seeds of the basalt controversy had thus been sown, but they were only now beginning to germinate, and would not produce fruit for over a decade to come.[5]

(e) After a long period of relative neglect, geomorphology was again receiving the attention of imaginative thinkers. One consequence in the offing was Hutton's brilliant resolution of the denudation dilemma.[6]

(f) A long tradition of attempting to improve the classification of minerals was leading by 1776 to a keener sense of chemistry's relevance to a theory of the earth, and to greater use of a historical criterion to understand the place of minerals in the earth's economy.[7]

(g) The advent of new cartographic techniques for portrayal of geological phenomena was beginning to impinge on the geologist's sense of how he should approach his task, and was helping to give to geology a practical expressibility that enhanced its prospects for subsidy.[8]

These sketchy summaries of only a few among many generalizations already made available by scholars, although they represent a significant legacy, for which we are in debt, are not all we have. For example, many valuable investigations have been made, in historical context, of individual figures or works from eighteenth-century geology.[9] All the same, I think that none of those to whom the debt is held would claim that the story of which they have illuminated some part is ready to be fully told.

Yet it is not just historical information we need; what is also missing is a better perspective on how historical information should be sought and used. Probably more than in other areas of the history of science, terminalistic or Whiggish perspectives have continued to prevail. There has been a less thorough realization than elsewhere in the history of science of the disadvantages in presuming that the task is to show where modern scientific convictions come from, to identify and trace the sources of the present, to use the way things turned out as the criterion for judging what went on before. I think there has been in the history of geology an inadequate reformulation of historical tasks in terms of gaining an understanding of the past on its own ground. To be sure, in recent years some excellent scholarship has shown how the history of geology can avoid the terminal fallacy. But even in such cases there is a tendency to fall easily into the habit of using labels and categories of questionable validity. Recall the terms used or implied in the above generalizations: *geological time*, *stratigraphic geology*, *geological dynamics*, *volcanic geology*, *geomorphology*. Terms like these are often used with little historical discrimination, as if there were no problem in their intrusion into any historical period. Perhaps there is a problem. In what follows I want to discuss one way we may be able to throw a clearer light on the eighteenth-century developments that concern us: by considering with greater care the terms in which we approach them.

Let us return to the question with which we began: What was the condition of geology in 1776? A good first step, I suggest, is to recognize that, in a significant way, the question is wrongly put. Speaking strictly, there was *no* geology in 1776. The term itself had not yet been coined, at least not with the scientific meaning it was shortly to acquire. Not until 1778 did Jean André DeLuc perform that service, and only in the early nineteenth century did the word gain something approaching universal recognition and acceptance.[10]

The absence of the term geology from the lexicon in 1776 was not accidental. We should not allow ourselves to assume casually that geology was actually there, but had not yet been named. Words of all kinds, not excluding scientific ones, have a way of making their appearances in response to perceived needs. The need that DeLuc saw in 1778 had quite arguably been recognized, in different ways and degrees, by some few others before him. Nonetheless, the need had not been responded to in just this way, and the response itself became a historical act with its own consequences.

It would be excessive, of course, to argue that the concept of geology appeared only with the word. Ideas emerge slowly, by stages. This is likely to be especially true of the idea of a discipline, which encompasses purposes, methods, procedural rules. There may occur, in the evolution of a disciplinary concept, substitutions of symbolic terms in parallel with shifts in the idea. So we should be surprised if close relatives to the concept of geology were not to have existed before it. Indeed they did, frequently in association with organized knowledge about the objects to be found in the earth.

Consider mineralogy. The idea of dealing with mineral substances as a realm of natural history dates back to ancient times. The English *mineralogy*, however, may have come into use only in the seventeenth century; Boyle employed the term in 1690.[11] French use of *minéralogie* appeared no later than 1732;[12] Louis Bourguet referred to the studies of *minéralogues* in 1742;[13] and in 1753 Baron d'Holbach (Paul

Henri Thiry) suggested calling one learned in this field a *minéralogiste*.[14] As names for recognized avenues of knowledge, mineral natural history and mineralogy came in the eighteenth century to have widely acknowledged meaning—primarily as descriptive sciences but in some cases bearing a broader historical sense of the mineral realm as a totality.

Several somewhat more exotic suggestions were made during the eighteenth century for terms to designate knowledge of fossils (in the traditionally wide sense of things dug out of the earth), stones, and rocks. A work of Johann Jacob Scheuchzer, *Specimen Lithographiae Helveticae Curiosae* (1702), spurred the Welshman Edward Lhuyd to use the Anglicized *lithography* in 1708 for a description of stones and rocks.[15] *Lithology* was used in English in 1716,[16] and *lithologie* by the French as early as 1742, in a work by Antoine Joseph Dezallier d'Argenville.[17] *Oryctologie* (the natural history of fossils) was in the title of another of his books in 1755.[18] For a time both *lithologie* and *oryctologie* made a bid for fairly extensive use in France, together with their variants *lithographie* (1752) and *oryctographie*, whose suffixes emphasized the descriptive function of these disciplines.[19] The *Encyclopédie* included brief definitions and remarks on each of these words in 1765, calling *oryctologie* and *oryctographie* synonyms for *minéralogie*.[20] *Lithologue*, as a name for a practitioner of lithology, was proposed in 1752 and accepted by the Académie Française, together with *lithologie*, a decade later.[21] Jacques Christophe Valmont de Bomare referred to a *lithologiste* in 1762 as one "who possesses the science of stones, and teaches it,"[22] and Emanuel Mendes da Costa spoke of lithologists in 1746.[23] The Supplement to Chambers' *Cyclopaedia* (1753) defined oryctography as "that part of natural history wherein fossils are described" and oryctology as "that part of physics which treats of fossils."[24] An even more advanced neologism was devised by Johann Heinrich Pott, who offered *lithogeognosie* in the title of a 1746 book as a label for knowledge of the earth's parts.[25] In a 1776 book George Edwards put forward *fossilogy* as his preferred name for the science of nonextraneous mineral bodies, or minerals not of organic origin.[26] For Edwards a specialist in this field was a fossilist, a word used previously by Mendes da Costa and Thomas Pennant in reference to a student of extraneous or organic fossils.[27]

The descriptive nature of these earth sciences, especially as understood before the latest part of the eighteenth century, is striking. Lithology, oryctology, mineral geography, orography (knowledge of the physical arrangement of mountains), and many other terms were usually agreed to fall within the province of descriptive natural history. Mineralogy and physical geography were often treated in a similar manner, although in some cases they tended more easily to acquire developmental overtones. Also within the tradition of natural history lay the practice of recording the phenomena of springs, mineral ore deposits, earthquakes, volcanoes, and the like, all studied with the general assumption that their foremost characteristics are to be considered as fixed aspects of an essentially static world. Insofar as they conformed to this natural history outlook, the men who contributed their efforts to these endeavors were inclined to believe that they were dealing with "the immanent properties and processes of the physical earth and its constituents."[28]

The ancestry of geology is not exclusively in descriptive natural history. Relevant activity was taking place also in already existing analytical sciences, like chemistry. For example, many of the most prominent French students of a theory of the earth living in 1776 had received their first instruction during the 1750's and 1760's from the influential chemist Guillaume François Rouelle.[29] It is well known that chemists over a long period of time have commonly taken an interest in minerals and earths, and it is not rare to find chemists of the eighteenth century who took a generalized interest in an understanding of both the earth's constitution and its history. Since we all have an idea of what chemistry and geology are, it is tempting to suppose that such chemists were indulging their additional interest in mineralogy or geology. It is a temptation to resist, however, in the name of better history, and I would like to expand on the point with further comment on Rouelle.

The question is whether Rouelle, in imparting to his students rather definite and detailed ideas about the earth's constitution and past, was getting into something beyond the scope of his chemistry. I think not. In my opinion Rouelle did not think, as he dealt with the earth, its parts, and its historical development, that he was doing something apart from chemistry. On the contrary, he regarded this terrestrial inquiry as a significant dimension within the broad scope properly entailed by chemistry. To Rouelle, as to many chemists before him, chemistry was the understanding of the formation of material, not just in principle, but also in the specific instances the world affords. And what grander or more important instance was there than the composition and differentiation of the earth itself? Chemistry was knowledge of the material world's specific products as well as of chemical operations; it was knowledge of history as well as of process.

These assertions require historical demonstration, something I can do no more than indicate in rough outline here. Our knowledge of

Rouelle's ideas about the theory of the earth is derived almost entirely from what some of his students had to say about it, and these students created a record of two different kinds.[30] First, there exist numerous copies of detailed notes taken at Rouelle's chemistry course, a part of which he evidently devoted regularly to a discussion of how the earth's crustal parts should be classified and subjected to historical interpretation. Secondly, several of Rouelle's distinguished pupils who later became geologists or engaged in further geological investigations (notably Nicolas Desmarets, Antoine Laurent Lavoisier, Antoine Grimoald Monnet, and Jean Darcet) left commentaries on their teacher's views. This second type of evidence about Rouelle's opinions has been given close attention in trying to reconstruct Rouelle's ideas. The pupils who later made good in geology had more reason than others to give consideration to the value of what Rouelle had taught in an area that came to have special importance to them. But without impeaching the significance and usefulness of these sources, should we not also consider that they may reflect an evaluation of Rouelle's teachings in light of the fact that geology had subsequently become an identifiable field of investigation? If we wish to know how Rouelle categorized his discussion of the theory of the earth, it may well be better to rely on the first type of source, contemporary notes which cannot reflect the bias of a subsequent reorganization of disciplines.

As far as I can see, the notes of Rouelle's *Cours de chimie* do not justify any argument that his theory of the earth lies outside his chemistry. On the contrary, the organization of the course in parts corresponding to the three realms (vegetable, animal, mineral) reflects a broad, all-encompassing definition of the science. The course notes display a conception of chemistry that one observer (Jean Mayer) calls "très ambitieuse" (meaning both ambitious and pretentious). Mayer adds that "if [Rouelle] could have, he would have required of [chemistry] the explanation of the genesis of the world and the answer to the riddle of life."[31] While Mayer sounds condescending, historically he is close to the mark.

One more illustration of the breadth of historical linkages in science may assist our understanding of how eighteenth-century thinkers could have had notions we find strange on how to proceed in studying the earth. Recent scholarship in the history of Renaissance chemistry (or chemical philosophy) has revealed with greater clarity than ever before that the conception of chemistry held in the sixteenth and seventeenth centuries, especially by those influenced by traditions of alchemy, iatrochemistry, and Paracelsianism, was that of a global inquiry into the whole of nature.[32] Chemical philosophy tended to hold all of nature to be "a vast chemical laboratory," and chemistry, if properly cultivated, was thought capable of illuminating the order, structure, and processes of the terrestrial sphere, among other things. Now, the chemistry of the eighteenth century was obviously more than just the direct outgrowth of this Renaissance version. It involved extensive new developments, and overturned portions of its Renaissance antecedent. But for all that, Rouelle's conception of the scope and role of chemistry may have owed something to alchemical and Paracelsian influences. Furthermore, the chemical articles of the *Encyclopédie*, many of them written by Rouelle's pupil Gabriel François Venel, reveal an alchemical ideal of gaining knowledge of the whole of nature through chemistry.[33] Although innovative developments were in the process of transforming chemistry, an old tradition of the science's universality did not die easily.

A conclusion we can draw from this discussion of Rouelle's views on the place within science of a theory of the earth is that not much over 200 years ago some important ideas that we view as geological were nurtured as an integral part of an already existing science. Natural as it is for us to refer to these conceptions as having been developed by a chemist reaching beyond his field, that is not how Rouelle saw it. And if Rouelle's chemistry embraced a theory of the earth, other sciences of the time could have similarly included considerations that from our vantage point appear to have been geological.

Another distinct and long-standing branch of learning that encompassed serious investigation of the earth during the eighteenth century was geography. The importance of the activities and ideas of geographers in eighteenth-century developments concerning a science of the earth has been underestimated by most historians of geology. Opportunities for rectifying this situation are now enhanced by the appearance of two extensive studies of French geography.[34]

It seems clear that at least among French geographers of the eighteenth century a growing emphasis on physical geography was marked by some imaginative attempts to reduce empirical knowledge of the earth's surface features to general physical principles. Philippe Buache, for example, formulated a theory of an interlocking global network of mountain ranges that had important consequences not only for French traditions in physiography, but also for the ideas of such a scientist as Desmarets on the theory of earthquakes and the general organization of land masses. From 1730 until his death in 1773 Buache was *adjoint géographe* in the Royal Academy of Sciences of Paris, the only member who held a position explicitly identifiable with a science of the earth. Not until the 1785 reorganization, with the creation of a class for natural history and mineralogy, would it become possible for any Paris academician other than the *adjoint géographe* to have so official an identity with the scientific study of the earth.

There may have been other sciences in addition to chemistry and geography that possessed components later observers would feel inclined to separate out as geological. A possible example is botany. I suspect that for some of the eighteenth century's botanists who concerned themselves with earth science, this was a pursuit less thoroughly divorced from botany than has been supposed. The study of minerals and soils was relevant to knowledge of plant life, and the understanding of fossil plant forms in historical terms was accepted by some botanists as part of their work.

These comments on how some eighteenth-century scientists looked upon their own concern for scientific understanding of the earth can be summarized as a challenge to what I think is a rather widespread misconception about the organization and structure of eighteenth-century earth science. That misconception can be put like this: Throughout the eighteenth century, scientists who carried out investigations of a sort we would call geological generally assumed the existence of a scientific enterprise corresponding in nature and scope, though not in name, to the geology that came to be recognized around the end of the century. These scientists would have had little difficulty in grasping geology as a concept if it could have been presented to them. With or without an appropriate label, they shared a sense of the enterprise that was not significantly altered by the advent of the word geology.

Our brief inquiry into some facets of the organization of geological endeavor some 200 and more years ago suggests that the science of geology as it came to be conceived in the train of the term's invention was not a direct, lineal descendant of a more primitive eighteenth-century version of the same thing. Instead, the late eighteenth-century developments that set the stage for the emergence of geology as a new discipline included a reordering or reorganization of the sciences, making room in a new way for a distinct science of the earth.

I suggest that the devising and use of the new term geology in the late eighteenth century did mark a change, identifying a growth toward a common theoretical understanding of a scientific goal engaging thinkers of diverse types. We have seen that the interests of these diverse types had been focused by men who considered themselves to be students of chemistry, geography, mineralogy, and other sciences. If there was a genuine change involved in the production of a new name, it was mainly bound up in an increasing recognition that various groups of scientists practicing relatively distinct disciplines had come to be united by a set of ideas held in common. To some extent, geology came into existence as a reorientation of existing disciplines, or parts of disciplines, under a new umbrella. It was the fashioning of a new relationship among activities that had already been going on for some time.

It remains to identify the essential ingredients of the new, integrated relationship I am claiming was brought about late in the eighteenth century. I offer my opinion briefly as a conjecture in need of further study. It is that there was a growing acceptance of *development* or *history* as the framework within which the descriptively studied objects and events of the earth were to be explained in terms of *process*. Fossils, minerals, strata, ore deposits, and mountains, as well as earthquakes, volcanoes, floods, and shifting shorelines—these phenomena were increasingly looked upon as data to be understood through the processes that caused them, processes that were the mainsprings of terrestrial change. Geology came to denote a collection of activities or disciplines subjected to the unifying themes of law-like process and historical development. There was still ample room within this framework—there had to be—for major differences of opinion regarding the nature of the main historical events and their temporal extent, but in the main controversies of the day the disputants could generally operate as geologists because they agreed at least that the information they dealt with had to be rendered intelligible within a matrix of orderly events occurring in time. They expected their science to involve "the determination of configurational sequences, their explanation, and the testing of such sequences and explanations."[35]

It has been my central concern to argue that key terms used by scientists in specific historical conditions are reflections of distinctive scientific ideas, and that the historian of geology can get useful historical information from study of such terms. I have also argued that the pattern of development of words for earth-related disciplines supports the thesis that geology hardly existed as a distinct science until late in the eighteenth century. This is not to say that scientists before then made no valuable contributions to knowledge of the earth, or that they did not possess understanding that was in some way meaningful. Rather it means that they made their contribution or gained their understanding within a conceptual framework foreign to that of later geologists. They would not have recognized their work as it would later be fitted into an order understood by, say, Lyell. Lyell is, in fact, a figure with whom it is hard to juxtapose most of our eighteenth-century geologists. Several of the contributions to the Lyell symposium in 1975 made clearer than ever the degree to which conventional historical interpretations of geology's past have been governed by Lyell's use of history to advance his own geological ideas.[36] We are still in the process of releasing ourselves historically from a Lyellian bondage.

It should go without saying that if our purpose is to understand eighteenth-century science on its own terms, we shall have to make an effort to overcome the habit of imposing our categories (or Lyell's) on it after the fact. But even if our purpose is—less historically—to trace more fully and accurately the paths by which we have arrived at our present condition, we still need to make the same effort. A faithful account of how the past led to the present will be acquired best by acting as though the past had value only for itself. By excluding consideration of later developments as fully as possible, we can come closest to reconstructing a past sustained by the evidence, rather than by our expectations of it.

Acknowledgment

I am grateful to Rhoda Rappaport and Alexander M. Ospovat for useful comments on this paper, and to John H. Eddy, Jr., for research assistance.

Notes

1. For general discussion of eighteenth-century geology see Rhoda Rappaport, "Problems and Sources in the History of Geology, 1749–1810," *History of Science* (1964), 3:60–77; V. A. Eyles, "The History of Geology: Suggestions for Further Research," ibid. (1966), 5:77–86; and V. A. Eyles, "The Extent of Geological Knowledge in the Eighteenth Century, and the Methods by Which It Was Diffused," in *Toward a History of Geology*, ed. Cecil J. Schneer (Cambridge, Mass., and London, M.I.T. Press, 1969), 159–183. Also important are Jacques Roger's introduction and notes to his critical edition of Buffon's *Les Époques de la nature*, Mémoires du Muséum National d'Histoire Naturelle, Ser. C, "Sciences de la terre," Tome X, Paris, Éditions du Muséum (1962), and John C. Greene, *The Death of Adam: Evolution and Its Impact on Western Thought* (Ames, Iowa State University Press, 1959). Still

useful despite its historiographical limitations is Sir Archibald Geikie, *The Founders of Geology*, 2nd ed. (London and New York, Macmillan, 1905).

Mention must also be made of the following items published since this paper was written: Roy Porter and Kate Poulton, "Research in British Geology 1660–1800: A Survey and Thematic Bibliography," *Annals of Science* (1977), 34:33–42; W. R. Albury and D. R. Oldroyd, "From Renaissance Mineral Studies to Historical Geology, in the Light of Michel Foucault's *The Order of Things*," *British Journal for the History of Science* (1977), 10:187–215; and especially Roy Porter, *The Making of Geology: Earth Science in Britain 1660–1815* (Cambridge, Cambridge University Press, 1977).

Each of these sources has relevance to several of the points raised below in items *a* through *g*, and some of the works cited below in notes 2–8 treat general issues in addition to the ones which they annotate and should be consulted accordingly.

2. Francis C. Haber, *The Age of the World: Moses to Darwin* (Baltimore, Johns Hopkins Press, 1959). Also Stephen Toulmin and June Goodfield, *The Discovery of Time* (New York, Harper and Row, 1965). On two of the bolder advocates of a broad time scale, see Albert V. Carozzi, "De Maillet's Telliamed (1748): An Ultra-Neptunian Theory of the Earth," in Schneer, ed., *Toward a History of Geology*, 80–99, and Carozzi's introduction to his translation of *Telliamed, or Conversations between an Indian Philosopher and a French Missionary on the Diminution of the Sea* (Urbana, University of Illinois Press, 1968); and Jacques Roger, "Un Manuscrit inédit perdu et retrouvé: *Les Ancedotes de la nature*, de Nicolas-Antoine Boulanger," *Revue des sciences humaines* (1953), 71: 231–254.
3. Although it pays relatively little attention to the eighteenth century, the best point of departure is Martin J. S. Rudwick, *The Meaning of Fossils: Episodes in the History of Paleontology* (London, Macdonald, and New York, American Elsevier, 1972).
4. Rappaport, "Problems and Sources," 66–69; John G. Burke, "Romé de l'Isle and the Central Fire," XIIe Congrès International d'Histoire des Sciences, *Actes* (Paris, Albert Blanchard, 1971), Tome VII, "Histoire des sciences de la terre et de l'océanographie," 15–17.
5. Robert Michel, "Les Premiers recherches sur les volcans du Massif Central (18e–19e siècles) et leur influence sur l'essor de la géologie," *Symposium Jean Jung: Géologie, géomorphologie et struc-*

ture profonde du Massif Central français (Clermont-Ferrand, Plein Air Service Ed., 1971), 331–344; Kenneth L. Taylor, "Nicolas Desmarest and Geology in the Eighteenth Century," in *Toward a History of Geology*, 339–356; Otfried Wagenbreth, "Abraham Gottlob Werner und der Höhepunkt des Neptunistenstreites um 1790," in *Bergbau und Bergleute: Neue Beiträge zur Geschichte des Bergbaus und der Geologie*, Freiberger Forschungshefte, D 11 (Berlin, Akademie-Verlag, 1955), 183–241; Alexander M. Ospovat, "Abraham Gottlob Werner and His Influence on Mineralogy and Geology," Ph.D. dissertation, University of Oklahoma, Norman, 1960; Albert V. Carozzi, "Rudolf Erich Raspe and the Basalt Controversy," *Studies in Romanticism* (1969), 8:235–250.

6. Gordon L. Davies, *The Earth in Decay: A History of British Geomorphology, 1578–1878* (New York, American Elsevier, 1969).
7. Introduction and notes by Alexander M. Ospovat accompanying his translation of Abraham Gottlob Werner, *Short Classification and Description of the Various Rocks* (New York, Hafner, 1971); Charles Spencer St. Clair, "The Classification of Minerals: Some Representative Mineral Systems from Agricola to Werner," Ph.D. dissertation, University of Oklahoma, Norman, 1965; D. R. Oldroyd, "Mechanical Mineralogy," *Ambix* (1974), 21:157–178; D. R. Oldroyd, "Some Phlogistic Mineralogical Schemes, Illustrative of the Evolution of the Concept 'Earth' in the 17th and 18th Centuries," *Annals of Science* (1974), 31:269–305; Seymour H. Mauskopf, *Crystals and Compounds. Molecular Structure and Composition in Nineteenth-Century French Science*, Transactions of the American Philosophical Society, New Series, Vol. 66, Pt. 3 (Philadelphia, American Philosophical Society, 1976), esp. 7–16.
8. Rhoda Rappaport, "The Geological Atlas of Guettard, Lavoisier, and Monnet: Conflicting Views of the Nature of Geology," in *Toward a History of Geology*, 272–287; and Rappaport's Ph.D. dissertation, "Guettard, Lavoisier, and Monnet: Geologists in the Service of the French Monarchy," Cornell University, Ithaca, 1964. A stimulating study of the question of geology's links with practical motives for science in Britain is Roy Porter, "The Industrial Revolution and the Rise of the Science of Geology," in *Changing Perspectives in the History of Science: Essays in Honour of Joseph Needham*, ed. Mikuláš Teich and Robert Young (Dordrecht and Boston, D. Reidel, 1973), 320–343.
9. For example, Rhoda Rappaport on Lavoisier: "Lavoisier's Geologic Activities, 1763–1792," *Isis* (1967), 58:375–384; "The Early Disputes between Lavoisier and Monnet, 1777–1781," *British*

Journal for the History of Science (1969), 4:233–244; and "Lavoisier's Theory of the Earth," ibid. (1973), 6:247–260. Also Alexander M. Ospovat on Werner: in addition to works already cited, "Reflections on A. G. Werner's 'Kurze Klassifikation'," in *Toward a History of Geology*, 242–256, and "The Work and Influence of Abraham Gottlob Werner: A Reevaluation," *Proceedings of the XIIth International Congress of the History of Science* (Moscow, Editions Naouka, 1974), Sec. VIII, "The History of Earth Sciences," 123–130; as well as others. Also Jacques Roger on Buffon's *Les Époques de la nature* (above, note 1).

10. Jean André DeLuc, *Lettres physiques et morales, sur les montagnes et sur l'histoire de la terre et de l'homme* (En Suisse: Chez les Libraires Associés, 1778), [vii]-viii n. See also Frank Dawson Adams, "Earliest Use of the Term Geology," *Bulletin of the Geological Society of America* (1932), 43:121–123; "Further Notes on the Earliest Use of the Word 'Geology'," ibid. (1933), 44:821–826; *The Birth and Development of the Geological Sciences* (Baltimore, Williams and Wilkins, 1938), 165–166; and Arthur Birembaut, "La Minéralogie et la géologie," in *Histoire de la science*, ed. Maurice Daumas (Paris, Gallimard, 1957), 1104–1105, 1113. Birembaut calls attention to Diderot's use of "géologie" in 1751 as one of four subheadings of "cosmologie" (the others being "uranologie," "aerologie," and "hydrologie") in the "Système figuré des connoissances humaines" for the *Encyclopédie* (Vol. I, facing p. xlvii). The organizing principle here appears to be of cosmic regions rather than types of investigation, and this is substantiated by the appearance of "minéralogie" elsewhere in the scheme.
11. *Oxford English Dictionary*, VI, 467.
12. Ferdinand Brunot, *Histoire de la langue française des origines à nos jours*, 13 vols. (Paris, Armand Colin, 1966–72), Vol. VI: *Le XVIIIe Siècle*, 625; Adolphe Hatzfeld and Arsène Darmesteter, *Dictionnaire général de la langue française du commencement du XVIIe siècle jusqu'à nos jours, précédé d'un traité de la langue*, 2 vols., 6th ed. (Paris, Delagrave, 1920), II, 1524.
13. Louis Bourguet, *Traité des petrifications*, 2 pts. in 1 vol. (Paris, Briasson, 1742), I, 1.
14. Brunot, *Le XVIIIe Siècle*, 627.
15. *Philosophical Transactions*, 1708–1709 (1710), 26:161.
16. *Oxford English Dictionary*, VI, 347.
17. *L'Histoire naturelle éclaircie dans deux de ses parties principales, la lithologie et la conchyliologie, dont l'une traite des pierres et l'autre des coquillages* (Paris, De Bure l'Aîné, 1742).

18. *L'Histoire naturelle éclaircie dans une de ses parties principales, l'oryctologie, qui traite des terres, des pierres, des métaux, des minéraux, et autres fossiles* (Paris, De Bure l'Aîné, 1755).
19. Paul Robert, *Dictionnaire alphabétique et analogique de la langue française. Les Mots et les associations d'idées*, 6 vols. (Paris, Société du Nouveau Littré, 1954–1964), IV, 282; Hatzfeld and Darmesteter, *Dictionnaire général de la langue française*, II, 1413. Scheuchzer used "oryctographia" in his *Helvetiae historia naturalis*, 3 vols. (Zürich, In der Bodmerischen Truckerey, 1716–1718), III: *Meteorologia et oryctographia Helvetica.*
20. *Encyclopédie*, IX (1765), 587; XI (1765), 677.
21. Hatzfeld and Darmesteter, *Dictionnaire général de la langue française*, II, 1414.
22. Valmont de Bomare, *Minéralogie, ou nouvelle exposition du règne minéral*, 2 vols. (Paris, Vincent, 1762), II, 348.
23. Mendes da Costa, "A Dissertation On Those Fossil Figured Stones Called Belemnites," *Philosophical Transactions*, 1747 (1748), 44 (pt. 2):398, 406. The communication was dated 1746.
24. *A Supplement to Mr. Chambers's Cyclopaedia: Or, Universal Dictionary of Arts and Sciences*, 2 vols. (London, printed for W. Innys [*et al.*], 1753).
25. . . . *Chymische Untersuchungen welche fürnehmlich von der Lithogeognosia oder Erkäntniss und Bearbeitung der gemeinen einfacheren Stein und Erden ingleichen von Feuer und Licht handeln* (Potsdam, C. F. Voss, 1746).
26. Edwards, *Elements of Fossilogy: Or, An Arrangement of Fossils, into Classes, Orders, Genera, and Species; With Their Characters* (London, printed by B. White, 1776).
27. Mendes da Costa, "Dissertation," 406; Pennant, *British Zoology*, 4 vols. (London, printed for Benjamin White, 1768–1770), I, 41.
28. The words are those of George Gaylord Simpson, "Historical Science," in *The Fabric of Geology*, ed. Claude C. Albritton, Jr. (Reading, Mass., Addison-Wesley, 1963), 25.
29. Rhoda Rappaport, "G.-F. Rouelle: An Eighteenth-Century Chemist and Teacher," *Chymia* (1960), 6:68–101; "Rouelle and Stahl—The Phlogistic Revolution in France," ibid. (1961), 7:73–102; and Jean Mayer, "Portrait d'un chimiste: Guillaume-François Rouelle (1703–1770)," *Revue d'histoire des sciences* (1970), 23:305–332.
30. Extant original sources on Rouelle's courses are indicated in the articles by Rappaport and Mayer, and in Rappaport's article on Rouelle in *Dictionary of Scientific Biography* (1975), XI, 562–564.
31. Mayer, "Portrait d'un chimiste," 332.

32. See, for example, publications of Allen G. Debus, including *The English Paracelsians* (London, Oldbourne, 1965); "Renaissance Chemistry and the Work of Robert Fludd," *Ambix* (1967), 14: 42–59; "Edward Jorden and the Fermentation of the Metals: An Iatrochemical Study of Terrestrial Phenomena," in *Toward a History of Geology*, 100–121; "The Chemical Debates of the Seventeenth Century: The Reaction to Robert Fludd and Jean Baptiste van Helmont," in *Reason, Experiment, and Mysticism in the Scientific Revolution*, ed. M. L. Righini Bonelli and William R. Shea (New York, Science History Publications, 1975), 19–47; and *The Chemical Philosophy: Paracelsian Science and Medicine in the Sixteenth and Seventeenth Centuries*, 2 vols. (New York, Science History Publications, 1977).
33. Jean-Claude Guédon, "Le Lieu de la chimie dans l'*Encyclopédie* de Diderot," *Proceedings of the XIIIth International Congress of the History of Science* (1974), VII ("The History of Chemistry"), 80–86; also Guédon's doctoral dissertation, "The Still Life of a Transition: Chemistry in the *Encyclopédie*," University of Wisconsin, Madison, 1974.
34. Numa Broc, *Les Montagnes vues par les géographes et les naturalistes de langue française au XVIII*[e] *siècle: Contribution a l'histoire de la géographie* (Paris, Bibliothèque Nationale, 1969), and *La Géographie des philosophes: Géographes et voyageurs français au XVIII*[e] *siècle* (Paris, Éditions Ophrys, 1974?).
35. Again, the words of George Gaylord Simpson (above, note 28). It is understood, of course, that the conceptions of process and development were not created by scientists of the late eighteenth century. What is proposed here is the advent at that time of a more or less collective recognition of these concepts' parallel applicability to diverse terrestrial phenomena.
36. *British Journal for the History of Science*, Lyell Centenary Issue (1976), 9: Part 2. See especially Roy Porter, "Charles Lyell and the Principles of the History of Geology," 91–103, and Alexander M. Ospovat, "The Distortion of Werner in Lyell's *Principles of Geology*," 190–198.

Geology and the Industrial-Transportation Revolution in Early to Mid Nineteenth-Century Pennsylvania

William M. Jordan

Introduction

The distribution and character of landform and mineral resources is a well known determinant of any nation's historical development; this was especially so in the young republic of the United States. The influence of local geology on the history of Pennsylvania, taken as a microcosm of the new nation, has been described by Willard and others.[1] This paper briefly examines a corollary aspect of early national history; the nature of the geological-technological relationship that, through its various manifestations in the early to mid nineteenth-century industrial-transportation revolution, influenced the development of geology as a science in Pennsylvania.

The relationship of geology to the technological and engineering pursuits of a society, and the mutual interdependence of geology and industry, is clear enough for Legget to state on the occasion of the 75th anniversary of the Geological Society of America that "so intimate would appear to be the links between the study of the earth's crust and the daily activities of man that . . . a paper [such] as this . . . might at first sight appear to be superfluous."[2] In questioning the relatively delayed historical development of geology, Legget speculated on the influence of the industrial revolution and "the practical demands of modern engineering," with the symbiosis of geology and civil engineering in the person of William Smith (1769–1839) being an outstanding example.[3] This mutual relationship shows clearly in the title of Smith's county map atlas, issued between 1819 and 1824: "A New Geological Atlas of England and Wales [showing] the situation of the best materials for Building, Making of Roads, the Construction of Canals, and pointing out those places where Coal and other valuable Materials are likely to be found."[4]

This concept is elaborated by Tomkeieff who presented a graph showing the relation of the "Heroic Age of Geology" to curves of production of coal and pig iron in Britain.[5] An early period of steady mutual growth is bracketed by the dates of Watt's first practical steam engine (1775) and the first passenger railroad operated by locomotive (1825), with the production of industrial commodities and geology both expanding rapidly thereafter. Similarly Fuller, demonstrating the industrial basis of stratigraphy, used a bar graph that chronologically depicts, for this early Heroic period, the comparative frequency of Acts of Parliament relating to enclosures of land and to the construction of roads and canals.[6]

In Britain the natural advantages of abundant mineral resources, relative political and economic stability, and maritime dominance enhanced rapid industrial development during the early Industrial Revolution. The American colonies of the mid-eighteenth century had these same advantages but were stifled by a 1750 Act of Parliament prohibiting the rolling of iron, steelmaking, and related activities while permitting the exchange of pig iron for wool and British manufactures. Ironically this mercantilist system, which concurrently stifled American scientific as well as industrial development, was attacked by Adam Smith in *Wealth of Nations* the same year that the Declaration of Independence introduced decades of economic and political instability to North America. Only with the beginning of the nineteenth century did the large-scale development of interior commerce and the widespread exploitation of mineral resources create conditions necessary for the stimulation of geology as a science.

It is contended in this paper that the development of transportation systems (canal and railroad) and the exploitation of coal in Pennsylvania was, as in contemporary England, a major stimulus to the development of geology. In Pennsylvania economic considerations led directly, by 1836, to establishment of the First Geological Survey, supporting the pioneering work in structural geology of Henry Darwin Rogers. However, after six years of operation, adverse economic conditions as well as political-sociological factors resulted in the demise of the first or Rogers geological survey. Resumption of large-scale public support for geology in Pennsylvania came only with the renewed mineral-fueled industrial boom that followed the Civil War.

Early Commercial Developments in Pennsylvania

Internal commerce generally flourished in Pennsylvania during the late eighteenth century and throughout the first quarter of the nineteenth, despite the panic and subsequent depression of 1819. Overland transportation had developed early with the Philadelphia-Lancaster Turnpike, the first in the United States, opening in 1772, but the development of water routes was somewhat slower. Even though William Penn had proposed a Schuylkill to Susquehanna River canal as early as 1690, construction of Pennsylvania's first artifical waterway, the Union Canal, lasted from 1811 to completion in 1827.[7]

The Lehigh Canal was started in 1818 and when opened in 1829 provided the first economically feasible means of exploiting the anthracite coal regions. "Canal Fever" was further stimulated by the 1825 opening of the Erie Canal in New York State and accordingly, on the Susquehanna River, construction of the Main Line of the Pennsylvania Canal System was undertaken by the state government beginning in 1826, in part to aid in further development of the anthracite coal fields to the north.

Anthracite coal was known and being used as a fuel in the Wyoming Valley as early as 1769 at what was to become Wilkes-Barre, although tradition carries knowledge of anthracite in Pennsylvania back to about 1750. The first commercial shipment of coal from this area was by barge down the Susquehanna to Harrisburg in 1776. Anthracite was later discovered in 1791 at Summit Hill near Mauch Chunk on the Lehigh River and, in that portion of the anthracite basin drained by the Schuylkill River, coal was known in the Pottsville region by 1770. Major deposits of Pottsville anthracite were found in 1790, but a wagon load taken to Philadelphia in 1800 was not sold because it could not be made to burn properly. Major commercial development awaited invention of the anthracite fire-grate in 1808, as well as the development of more efficient transportation by canal and later by railroad.

Rather rapidly, horse- and gravity-powered "anthracite railroads" were developed for local transportation within the new coal fields, the oldest of these, the nine-mile-long Mauch Chunk Railroad, having been started in 1818 by the Lehigh Coal and Navigation Company. The first but short-lived use of a steam locomotive in America was in 1829, at Honesdale, where the anthracite railroad from the Wyoming Valley connected with the new Delaware and Hudson Canal. In this same period, illustrating the quickening development of the transportation revolution, the world's first steam-powered passenger-carrying railroad, the Stockton and Darlington, opened in Britain in 1825. Groundbreaking for America's first passenger railroad, the Baltimore and Ohio, did not occur until July 4, 1828. On that same day, in a similar ceremony, President John Quincy Adams initiated construction of the rival and ill-fated Chesapeake and Ohio Canal. Canals were eventually to be replaced by the new and more efficient railroad systems. What was to become the Pennsylvania Railroad was first privately chartered in 1823, but was constructed by the state government between 1829 and 1834 as the Philadelphia and Columbia Railroad to provide access to the growing but also ill-fated state-owned canal system on the Susquehanna. These transportation developments postdate the earliest practice of geology in Pennsylvania but, by greatly stimulating commerce and industry, laid the social, economic, and political groundwork for later governmental formalization of this activity.

Early Geology in Pennsylvania

Although Philadelphia was scientifically more cosmopolitan than colonial New England, where the constraints of mercantilism concentrated early scientific activity in areas such as mathematics, astronomy, meteorology and surveying, geologic activity in Pennsylvania was at first fitful and sporadic. The American Philosophical Society had been founded in 1743, its wide range of scientific interests represented by men such as Benjamin Franklin, David Rittenhouse, Benjamin Rush, and John Bartram.

The first known geologic paper concerned with a locale in Pennsylvania, Hutchins' geologic description of the falls of the Youghiogeny,[8] dates from 1786, although some results of more general investigations, such as in mineralogy, had been published earlier.[9] Willard characterizes these early applications of modern geology as "a few desultory papers of little significance," commenting with considerable *presentism* that "many of these articles seem quite naive today."[10] J. P. Lesley, the second state geologist of Pennsylvania, provides a more complete and objective summary of these early observations and papers in his 1876 *Historical Sketch of Geological Explorations in Pennsylvania and Other States.*[11]

After the Revolution scientific activity in general languished, perhaps as part of an overall romantic reaction to the enlightenment, and the first professionalization of science, including geology, was delayed until the second decade of the nineteenth century.[12] This period saw publication of the first of Maclure's *Observations on the United States of America* in 1809, its revision in 1817, and Silliman's founding of *The American Journal of Science and the Arts* in 1818, a year after *The Journal of the Philadelphia Academy of Natural Sciences* had commenced publication in 1817. Nevertheless, investigations of Pennsylvania geology proceeded slowly until encouraged by the establishment in 1832 of the Geological Society of Pennsylvania and the eventual legislative authorization of the First Geological Survey of Pennsylvania in 1836.

The First Geological Survey of Pennsylvania

The First Geological Survey of Pennsylvania arose out of the efforts of the Geological Society of Pennsylvania "for the scientific and practical study of the mineral resources of the Commonwealth" and with the express purpose of memorializing the legislature to authorize an official survey. The Society included among its members R. C. Taylor, William Maclure, and G. W. Featherstonhaugh. Its constitution reads in part that "the objects of this Society are declared to be, to ascertain as far as possible, the nature and structure of the rock formations of this State . . . *and particularly the uses to which they can be applied in the arts, and their subserviency to the comforts and conveniences of man*" (italics mine).[13]

Such "scientific and practical" concerns had already resulted in the establishment of state geological surveys in North Carolina (1823), Massachusetts (1830), Tennessee (1831), and Maryland (1834). In 1835 New Jersey, Connecticut, and Virginia (with W. B. Rogers as state geologist) also created surveys. The First Geological Survey of Pennsylvania was established by an act of the legislature on May 29, 1836, the same year as the beginnings of the Maine, New York, and Ohio surveys. Henry Darwin Rogers, brother of W. B. Rogers and at that time state geologist of New Jersey and Professor of Geology and Mineralogy at the University of Pennsylvania in Philadelphia, was appointed the first Pennsylvania state geologist.[14]

Rogers, with two assistants (James C. Booth and John F. Frazer) and a chemist (Robert E. Rogers), began the work of the survey with a five-year annual appropriation of $6,400. Staffing proved a major problem since, of the four assistant geologists of 1837, only one, James D. Whelpley, had any formal geological training, in this case under Silliman. Another assistant, Samuel S. Haldeman, later to become prominent as a paleontologist, was from a Lancaster County iron mining and smelting family. Their homestead and iron furnace was located near Chickies Rock, a prominent landmark overlooking the Susquehanna which was to become the type locality of the Chickies Quartzite and *Skolithos linearis* described by Haldeman in 1840.[15] Lesley notes that the workers of the first survey relied heavily on the observations of laymen. "Many a professor of geology is put to the blush by obscure farmers, hunters and miners, who keep their eyes open and their mouths shut, speak little and publish nothing, but do a deal of thinking and do it well . . . The first Geological Survey of Pennsylvania revealed a wonderful amount of actual science stowed away among the people."[16]

The early iron industry of Pennsylvania is a good illustration of the initial irrelevance of formal training in geology. As with William Smith, later geologic theory was a derivative of activities basically utilitarian in purpose since geology as a science followed mining and smelting as an economic activity. Iron had been produced in Pennsylvania, using charcoal as a fuel and reducing agent, from the earliest eighteenth and even the late seventeenth century.[17] It did not require trained geologists, then or later, to discover either "bog" or the Appalachian-type residual iron ores. At what was to become Ironton in Lehigh County "it lay in lumps upon the surface, some large boulders weighing several tons, and found in such profusion that its presence was a serious impediment to the prosecution of agriculture,"[18] Early discoveries of coal were no less a matter of serendipity. At Summit Hill, near Mauch Chunk, in 1791 a hunter, Philip Ginter, found fragments of coal lodged in the roots of an upturned tree and similarly, near Pottsville in 1790 another hunter, Necho Allen, accidentally discovered coal by igniting it with his campfire.[19]

By the time of the first survey, however, technical expertise was becoming more of a necessity in coal prospecting, as had been recognized by those who advocated the survey's establishment. This was not always fully appreciated by the locals, who, to that point, had generally been successful without outside advice. In his economic and social history of Schuylkill County, Yearley notes that the vigorous frontier society of that region, despite increasing difficulty in extending the mines, rejected outside geologic advice without serious examination since, according to Yearley, geologic theory of the day was too "prolix, academic, speculative, sometimes undeniably incorrect, and, worst of all in the eyes of local wise men, time-consuming and impractical."[20]

Perhaps this local reliance on intuition and practical knowledge, when successful, led as much to the demise of the first survey as "the financial embarassment of the Commonwealth," the official reason given for the failure of the legislature to renew appropriations in 1842. An overly ambitious system of internal improvements, in particular the state built and operated Pennsylvania Canal System, as well as the Panic of 1837 and the subsequent depression lasting until 1842, were responsible for this financial condition. Nevertheless, Lesley, an employee of the first survey, still shows considerable bitterness in 1876 when he asserts, at length, that:

> the mines of the State were (with some most honorable exceptions) bossed by the commonest miners from foreign and quite different geological regions, who had suddenly exchanged the character and position of hewers of coal and pumpers of water at home, for the character and position of mining engineers in America . . . they were as unwilling to accept as they were unable to acquire a correct knowledge of our geology, so different from their own, and hated professional geologists because these had never lived in childhood, pick in hand, under ground,—because they taught new things hard to comprehend,—and because they denied the propriety of mining the coal of Schuylkill County on the plan of the colleries of South Wales, or of employing the ancient methods of the Cornish tinworks to the brown hematite banks of the Lebanon Valley. The jealousy of professional and 'theoretical' interference with traditional and 'practical' usages . . . was in 1842 in all its vigor; and was shared by the landed proprietors, the directors of the companies and the general superintendents of colleries and mines. A wave of suspicion and dislike, pushed before it by the First Geological Survey through its whole progress, brought it at last to a dead stop.[21]

Although the foregoing is undoubtedly not an objective assessment and shows strongly felt anti-immigrant sentiment, Lesley is willing at least to imply that some of the blame for the collapse of state-funded geologic activity was also due to shortcomings of the geologists themselves: "The language of science was then [1842] an unknown tongue, and sounded in the ears of the people like the chattering of animals or idiots. The disputes of geologists respecting doubtful points, if listened to at all, were regarded as good evidence for the worthlessness of all their theories; and the truths in which they agreed seemed . . . the insanities of an exalted imagination, or the impious utterances of an irreligious temper."

Geologic Accomplishments

Despite the eventual denouement of the first survey, the work it accomplished, both practical and theoretical, was enormous. This was made possible by the more rapid and convenient transportation then available, as well as by the many new rock exposures being made by canals, railroads, and mining activity. Even before the survey, the opportunities thus provided were exploited by Benjamin Silliman, who, about the time of the opening of the Lehigh Canal in 1829, had visited the mines near Mauch Chunk and the anthracite district farther west in the Wyoming Valley. In 1832 R. C. Taylor, whose major reputation was based on his studies of coal, made use of excavations for the Pennsylvania Canal, in the narrows of the Juniata River below Lewistown, in his investigation of the stratigraphic position of *Fucoides alleghaniensis*, first described by Harlan the previous year. Schuchert describes Taylor as "a relative, pupil, and business associate of William Smith," an immigrant to Pennsylvania in 1830 who "stood preeminent in economic geology and in mining engineering."[22] Similarly, Edward Miller, the civil engineer in charge of constructing the Allegheny Portage Railroad, published in 1835 a geologic cross section of the "Old Red sandstone system" based upon exposures in the railroad cuttings on the face of the Allegheny escarpment.

The new survey itself made maximum utilization of artificial exposures in its investigations, which were primarily of a reconnaissance and resource inventory nature. The basic stratigraphy and structure of the Pennsylvania folded Appalachians were recognized in the first year. In particular, detailed work in the anthracite basins brought major results. The "Rule of the Steep North Dip" (based on the observation that north dipping beds are almost always steeper than south dipping beds) upon which H. D. Rogers built his wave theory, was formulated in 1837. Rogers' theory likened the alternating anticlines and synclines of the folded Appalachians to the billows of the sea, in this instance the waves being produced by undulations in a liquid stratum (fluid lava) lying beneath a flexible crust, the sudden and violent escape of elastic vapors and gases being responsible for the motion.

The field work of that second year was confined to the northeast quadrant of the state but the survey's first reconnaissance west of the Alleghenys began in 1838. This resulted in recognition of the threefold subdivision of the bituminous coal measures and construction of a 150-mile geologic section from New York to the Virginia (now West Virginia) border. The use of topography as a clue to the geologic structure—later perfected and called topographical geology by Lesley—was also recognized in these early years. By 1842, based upon the work of their separate surveys and their joint experience, the Rogers brothers were ready to present their classic joint paper on Appalachian structure.[23]

Despite the accomplishments of the survey and H. D. Rogers' scientific achievements, major public support for geology in Pennsylvania ended in 1842, although industry problems in the anthracite region did not. According to Yearley, considerable losses of time, money, and energy resulted from the breakdown in communication between "those who knew and those who needed to know."[24] This was only one result of the long and involved delay between Rogers' transmission of his final report manuscript to the state in 1847, partial revival of the survey in 1851 and 1852 for revisions, and final publication of the report in Edinburgh as late as 1858.[25]

Mid-century Developments

Even though the First Geological Survey of Pennsylvania had come to an end, the decade of the 1840's was one of heightened geological activity nationally. This included organization in 1840 of The Association of American Geologists and Naturalists, and subsequently, in 1848, the American Association for the Advancement of Science. The decade also saw Lyell's visits to America in 1841 and 1845 and Agassiz' arrival in 1846. The Smithsonian was founded in 1846, marking the first permanent institutionalization of federal support for science, and by 1850 twenty states had initiated geological surveys.[26]

Developments of the 1850's presaged the changed industrial character of the nation following the Civil War and the resulting increased economic, technologic, and social influences on geology that are beyond the scope of this paper. For example, the first recorded miner's strike in the anthracite region occurred in 1849, and the use of coal breakers, with their attendant unpleasant child labor practices, was a new development of the early 1850's.[27] By 1852 the Pennsylvania Railroad reached Pittsburgh, and the improved transportation it provided sped a westward shift in the location of the iron and coal industries.

J. P. Lesley, who was to become director of the Second Geological Survey of Pennsylvania upon its establishment in 1874, reemerged into the world of science in the 1850's after a decade of concern with religion and part-time assistance to Rogers, helping with the draft of the first survey's "final report." Resigning the pastorate of a Congregational church in 1851, Lesley became secretary of both the Iron and Steel Association and the American Philosophical Society in Philadelphia. His *Manual of Coal and Its Topography* appeared in 1856 and his *Iron Manufacturer's Guide* in 1859.[28] Finally, successful completion of the Drake Well at Titusville, also in 1859, and the Civil War itself initiated a new set of circumstances under which geology, industry, and society were to operate until well into the twentieth century.

Conclusions

This paper has been concerned not with the state of geological knowledge in Pennsylvania *per se* but rather with outlining the conditions and circumstances under which this knowledge developed through the first part of the nineteenth century. The discovery and communication of new knowledge from other regions has not been considered, nor have the totally personal characteristics of the individual investigators themselves. Instead, it has been shown that, as in western Europe, the rapid development of transportation, mining, and the iron and steel industry strongly paralleled and influenced the growth of geology. Pennsylvania's financial difficulties, due in part to overexpansion of the state canal system, as well as ineffective communication of the first survey's geologic results to those in industry and mining most concerned, brought an unfortunate halt of nearly thirty years to public supported geology in the Commonwealth. Only with the new economic boom and changed conditions following the Civil War did Pennsylvania geology again prosper as it had in the days of the Rogers survey.

Notes

1. Bradford Willard, "Cultural Influences of Pennsylvania's Mountain Gaps: I, Early Adaptations of Natural Routes," *The Scientific Monthly* (1943), 57:33–43; "II, Improving the Routes," ibid. (1943), 57:132–144. A related paper is Bradford Willard, "Geology and Wars," *Pennsylvania History* (1963), 30:393–419.
2. Robert F. Legget, "Geology in the Service of Man," in Claude C.

Albritton, Jr., ed., *The Fabric of Geology* (Stanford, Freeman, Cooper, 1963), 242, 246.

3. Recent treatments of this relationship include W. B. N. Berry, *Growth of a Prehistoric Time Scale Based on Organic Evolution* (San Francisco, W. H. Freeman, 1968), 53–59; J. M. Eyles, "William Smith: Some Aspects of His Life and Work," in Cecil J. Schneer, ed., *Toward a History of Geology* (Cambridge, Mass., MIT Press, 1969), 142–158; and John G. C. M. Fuller, "The Industrial Basis of Stratigraphy: John Strachey, 1671–1743, and William Smith, 1769–1839," *Bulletin of the American Association of Petroleum Geologists* (1969), 53:2256–73.
4. *William Smith's Geological Maps* (London, British Museum [Natural History] (1975), 4 pp. There is also a William Smith anthology prepared by Douglas A. Bassett, "William Smith, the Father of English Geology and Stratigraphy: An Anthology," *Geology: The Journal of the Association of Teachers of Geology*, (1969), 1:38–51.
5. S. I. Tomkeieff, "James Hutton and the Philosophy of Geology," *Proceedings of the Royal Society Edinburgh, Section B* (1948–49), 63:389.
6. Fuller, 2258. Fuller presents the thesis that "geological activity takes place mainly in response to industrial and social pressures": "The Geological Attitude," *Bulletin of the American Association of Petroleum Geologists* (1971), 55:1927–38. This theme is also examined by Roy Porter, "The Industrial Revolution and the Rise of the Science of Geology," in *Changing Perspectives in the History of Science* ed. M. Teich and R. Young (Boston, D. Reidel, 1973), 320–343. See also D. H. Hall, *History of the Earth Sciences During the Scientific and Industrial Revolution* (New York, Elsevier, 1976).
7. This and following data is drawn from a number of sources, including William H. Shank, *The Amazing Pennsylvania Canals* (York, Pa., Historical Society of York County, Pa., 1965); George Swetnam, *Pennsylvania Transportation* (2nd ed., Gettysburg, Pa., Pennsylvania Historical Association, 1968); Louis Poliniak, *When Coal Was King: Mining Pennsylvania's Anthracite* (Lebanon, Pa., Applied Arts Publishers, 1970); Frederick M. Binder, *Coal Age Empire: Pennsylvania Coal and Its Utilization to 1860* (Harrisburg, Pa., Pennsylvania Historical and Museum Commission, 1974); and Hudson Coal Company, *The Story of Anthracite* (New York, privately printed, 1932). Also relevant is James L. Livingood, *The Philadelphia-Baltimore Trade Rivalry: 1780–1860*

(Harrisburg, Pa., Pennsylvania Historical and Museum Commission, 1947).

8. T. Hutchins, "Description of a Remarkable Rock and Cascade Near the Western Side of the Youghiogeny River," *Transactions of the American Philosophical Society* (1786), 2:50–51.
9. John C. Greene, "The Development of Mineralogy in Philadelphia, 1780–1820," *Proceedings of the American Philosophical Society* (1969), 113:283–295.
10. Bradford Willard, "Pioneer Geologic Investigations in Pennsylvania," *Pennsylvania History* (1965), 32:243.
11. J. P. Lesley, *Historical Sketch of Geological Explorations in Pennsylvania and Other States* (Harrisburg, Pa., Second Geological Survey of Pennsylvania, 1876), 3.
12. Kenneth A. Aalto, "Specialization and Professionalization of Pre-Civil War North American Geology," *Journal of Geological Education* (1969), 17:91–94. John C. Burnham, *Science in America: Historical Selections* (New York, Holt, Rinehart and Winston, 1971), 71. George H. Daniels, *American Science in the Age of Jackson* (New York, Columbia University Press, 1968).
13. Willard, "Pioneer Geologic Investigations," 246. Lesley, 30–31. Anne Millbrooke, "The Geological Society of Pennsylvania 1832–1836," *Pennsylvania Geology* (1976), 7, no. 6:7–11; (1977), *8*, no. 2:12–16.
14. George P. Merrill, *The First One Hundred Years of American Geology* (New Haven, Conn., Yale University Press, 1924), 188.
15. W. Hantzschel, "Trace Fossils and Problematica," *Treatise on Invertebrate Paleontology, Part W: Miscellanea* (New York, Geological Society of America, 1962), 215.
16. Lesley, 15.
17. Arthur Cecil Bining, *Pennsylvania Iron Manufacture in the Eighteenth Century* (2nd edition, Harrisburg, Pa., Pennsylvania. Historical and Museum Commission, 1973).
18. C. R. Roberts, et al., *History of Lehigh County, Pennsylvania* (Allentown, Pa., Lehigh Valley Publishing Co., 1914).
19. Poliniak, 2, and inscription on monument at the discovery site, Summit Hill, Pa. Hudson Coal Company, 29, 32.
20. C. K. Yearley, Jr., *Enterprise and Anthracite: Economics and Democracy in Schuylkill County, 1820–1875* (Baltimore, The Johns Hopkins Press, 1961), 98.
21. Lesley, 111–112.
22. Charles Schuchert, *Stratigraphy of the Eastern and Central United States* (New York, Wiley, 1943), 126.

23. P. A. Gerstner, "A Dynamic Theory of Mountain Building: Henry Darwin Rogers, 1842," *Isis* (1975), 66:26–37. See also Patsy Gerstner, below, p. 175. Lesley, 54–55, 79.
24. Yearley, 103.
25. Lesley provides a long and detailed account, including his own participation in the travail of that period. Merrill, 373–379, provides a briefer account. Henry D. Rogers, *The Geology of Pennsylvania: A Government Survey with a General View of the Geology of the United States*, (4 vols., Edinburgh, Blackwood 1858).
26. Walter B. Hendrickson, "Nineteenth-Century State Geological Surveys: Government Support of Science," *Isis* (1961), 52:357–371.
27. Hudson Coal Co., 28. Polinak, 3.
28. J. P. Lesley, *Manual of Coal and Its Topography* (Philadelphia, Lippincott, 1856). *Iron Manufacturer's Guide* (New York, Wiley, 1859).

The Image and the Idea: Two Artist Naturalists

10

Geological Theory and Practice in the Career of Benjamin Henry Latrobe

Stephen F. Lintner and
Darwin H. Stapleton

Introduction

Benjamin Henry Latrobe was one of the great American architects and engineers. Born in England in 1764, he was educated in Moravian schools in England and Germany. While living in London, he studied engineering under John Smeaton and architecture under Samuel Cockerell, eminent men in their professions. Latrobe came to the United States in 1796 and remained there until his death in 1820. During those twenty-four years he was a major figure in the introduction of Neoclassical architecture, directed the construction of the United States Capitol and the White House, built the first comprehensive steam-powered water system at Philadelphia, and participated in numerous other architectural and engineering projects.

He was in addition a keen scientific observer, and one of the things that excited his interest in his new country was its geology. Latrobe acquired his knowledge of natural history from field observations, the current general literature, and involvement in scientific groups such as the American Philosophical Society.[1] He recognized the engineering applications of information derived from natural history, particularly with regard to hydraulic engineering, and his observations included the economic and utilitarian use of the environment.

Geological Observation and Theory

In the development of his geological thinking Latrobe was greatly influenced by his personal association with William Maclure (1763–1840) and Constantin F. C. Volney (1757–1820).[2] Latrobe met Volney while they both lived in Richmond, Virginia, and developed a deep admiration for his abilities. Because of Volney's influence, Latrobe embraced the Neptunist views of the preeminent German geologist Abraham Gottlob Werner (1750–1817). Latrobe felt from his field observations that Werner's ideas of rock genesis through precipitation and sedimentation from an ancient sea could explain the geology of the Atlantic coast states, and he followed the Wernerian system of classification of rock formations.[3] His support for this theory was not uncritical, however, and his "Memoir on the Sand Hills of Cape Henry in Virginia" (1799) stated that contemporary coastal processes were creating a geological formation that could be erroneously interpreted by some future philosopher using Wernerian concepts.[4] Besides debating the relative roles of the agencies of water and fire in earth processes, geologists debated the extent to which past catastrophic conditions were required to interpret the earth record.

Latrobe believed that many topographic features were formed by catastrophic events of an actualistic nature—that is, "the causes of geological changes in the past *differ not in kind*, though they may sometimes *differ in energy*, from those now in operation."[5] In describing the formation of the marble being quarried for the columns of the United States House of Representatives, he conjectured:

> Imagine these pebbles rounded and mingled by attrition for ages, and then to have been left, and cemented by some matter filling all interstices . . . so as to become a solid mass. Suppose then that the valley became the bed of a mighty torrent running from S.W. to N.E. over this cemented mass, wearing it down in the direction of its current unequally, according to the velocity of its veins; and employing, (as in all our rivers) the agency of loose stones to whirl deep basins into the solid mass, and thus giving to the rocks, now separated into distinct masses, that specific character, which the

> rocks of all of our rapids acquire by the action of water, *and which character cannot possibly be mistaken, or derived from any other known agency*. Imagine then that this torrent cease, leaving its bed dry, and the rocks bare, but covered in its lower parts with alluvial soil.[6]

In a single instance Latrobe adopted a nonactualistic catastrophic viewpoint—that is, "The causes of some geological changes of the past differ in kind and energy from those now in operation."[7] He hypothesized that the moon was derived from the earth and perhaps from the basin of the Pacific Ocean. Expanding upon this idea, he proposed that the North American continent was tilted westward because of uplift associated with the expulsion of the moon from the Pacific, thus accounting for the elevated marine deposits he had observed.[8] But of such speculation Latrobe wrote, "I do not pretend that this hypothesis is worth half a farthing.—I am sick of pursuing it. I hate hypothesis making . . ."[9]

Latrobe recognized the processes of contemporary river systems in the development and modification of landscapes, but he saw them as factors acting to refine and finish features rather than as primary agents. It is probable that like many contemporary observers he thought that the age of the earth was insufficient for currently observable processes to have modified more than a minute portion of the surface.[10] Nonetheless, he frequently discussed the impact of fluvial processes on the landscape. For example, he credited the Susquehanna River with cutting its channel through the mountain ridges.

> Another narrow ridge of granite hills crosses the river immediately below Columbia [Lancaster County, Pennsylvania], over which the river falls rapidly, and then enters the wider limestone valley known by the name of the Jochara valley. The river spreads here to the width of three miles [4.8 kilometers], its stream is gentle though rapid, and it abounds in beautiful and fertile islands. It then suddenly contracts and is received into the narrow ravine which it has *sawed* down into the granite hill called Turkey hill.[11]

He also attributed the development of minor valleys on the Delmarva Peninsula to gradual erosion,[12] and recognized a multicycle development of valleys in the Coastal Plain of Virginia.[13] In a strikingly modern fashion he explained the causes and impact of culturally accelerated erosion and sedimentation in the rivers of the Middle Atlantic states. In Latrobe's words:

> Since the clearing of the country, the heads of navigation have gradually become choaked with alluvion of the upper country deposited at the place where the velocity of the upper water is checked by the opposition of the tide, and by the expansion of the stream in the wide water below the rocky ridge of the falls.
>
> Thus by degrees, and without exception, I believe, have the ports immediately below the falls of all our rivers, gradually lost their shipping business, which has retired to the first spot at which deep water was found to be permanent—that is, below the expanse and distance required for deposition of the alluvion of the land water.[14]

Latrobe had an awareness of general rock types and successfully constructed an accurate map of the basic geology of the lower Susquehanna Valley.[15] His classification of rocks proceeded from an understanding and recognition of their mineralogy. In one location he commented:

> The rocks at this spot are very hard grey granite in large rhombodial masses regular;—& remarkably parallel in the situation of their planes of chrystallization. They abound in mica, & being rounded on their upper surfaces by the attrition of ice & water, have a remarkable silvery grey color, forming a singularly beautiful contrast with the rest of the landscape. These rocks abound in beautiful garnets interspersed among the feldspar of the composition & also with cubical pyrites of remarkably hard consistence, & correct shape.[16]

He recognized the variable resistance to weathering within and between rock types, which he attributed principally to the mechanical forces of water and ice. Latrobe probably underestimated the magnitude and extent of chemical weathering, although he was doubtless familiar with the varying environmental stability of such common building stones as granite, limestone, marble, slate, and sandstone.[17] He recognized soils (including saprolite) as the product of rock weathering rather than as materials aggregating to form rock.[18] Continuing from his earlier observation on the weathering of granitic rocks of the Piedmont Plateau of Virginia,[19] he characterized the relief relationships of rock and water along the lower Susquehanna River by observing:

> Wherever the river crosses a valley of limestone or slate, the rocks are worn into a smoother and wider bed: but when it has to

> cross a ridge of granite, its course is immediately broken by irregular masses and ranges of rocks; its bed is narrow and enclosed by precipices, and its torrent furious and winding.[20]

Hydrology was another area in which Latrobe exhibited some special knowledge and interest. He knew something of hydraulic engineering by an early informal study under a German engineer and his work under Smeaton, as well as from technical treatises. He was strongly influenced by the works of Bernard Belidor, who was regarded by contemporaries as the foremost authority in hydraulic engineering. On several occasions Latrobe cited Belidor to support his arguments, though he mentioned others.[21]

Latrobe had a keen awareness of the dynamic relationship between hydraulic factors and channel morphology, and sought to use natural forces rather than engineering works in the maintenance or modification of channels. In the improvement of the Susquehanna he employed this principle by attempting to create a channel which would clear itself of silt and debris.[22] He knew that trees and understory vegetation are involved in the retention of soil, stabilization of river banks, and obstruction of water flow, so he ordered them removed from selected areas in order to open and maintain channels. He also had some notion of groundwater hydrology, and his observations on subsurface conditions of Philadelphia illustrated his understanding of how subterraneous and surficial geology are interrelated with water supply, sewage disposal, and contamination.[23]

Application to Architecture and Engineering Practice

In considering Latrobe's application of his geological knowledge, we are dealing with his work in the fields of architecture and civil engineering, particularly the latter. Latrobe was the engineer of two canals (Chesapeake and Delaware, and Washington) and two urban waterworks (Philadelphia and New Orleans) and the architect of important public buildings in Philadelphia and Washington, and was consulted on a number of other projects. In many instances his geological knowledge was vital to the successful execution of his work.

Within a few years of his arrival in America, Latrobe established in his mind a map of the geology of the Atlantic coast states and the natural forces at work there. For example, he thought of the Fall Line as "the great granite ridge," and of the South Mountain as rich in copper ore. He recognized at an early point that the silting of the estuaries of the coastal rivers was a recent phenomenon caused by the agricultural exploitation of the hinterland by European agricultural technology. When he was consulted at different times on the problems caused by silting of ship channels of the Delaware and Potomac, he offered little hope and referred to causes beyond remedy of the engineer's art.[24]

This mental map of America was useful when Latrobe made surveys for canal routes. To be economical, canals must take the shortest line possible, have the least possible change in elevation and have few costly excavations or aqueducts. In an era without standard topographical maps, the engineer who was able to visualize accurately the lay of the land prior to the survey could reduce his options and find the best route more quickly. Latrobe explained this approach in a report on the survey of the Chesapeake and Delaware Canal, which was to connect the Chesapeake and Delaware bays.

> It will be easy to comprehend the present state of the Land between the two Bays, if we suppose the whole peninsula to have been once a plain composed of soft alluvial soil extending from the foot of the Granite hills [that is, the Piedmont] to the ocean, and gently declining from the North west, to the South east. In the course of many centuries the water falling upon the surface and discharging itself into either Bay as accident or some unknown cause directed would wash this plain into vallies.
>
> Between these vallies ridges of small width would naturally retain the level and inclination of the original plain, while the vallies would become deeper and wider, the water within them collecting from the heavens, and from the springs starting from their sides . . . And soon after the commencement of the operation of the survey, these conjectures proved so exactly coincident with the facts, as to be of great service in chusing the ground over which to carry the level. For in running along the ridges higher ground was invariably to be found to the northwest, and lower by running more the eastward of South.[25]

Fifteen years later he took a similar approach when consulted on a proposed canal in North Carolina. He stated that the "natural formation of our continent is so exactly similar from the state of New York to the Roanoke [River]" and that circumstance allowed him to suggest the general procedures to be followed in laying out the canal, even though he had never been to North Carolina.[26]

Once Latrobe actually began a survey, he carefully examined the geological formations he encountered. Undoubtedly most of his knowledge came from examination of exposures of rock and soil, and did not involve taking samples. But he did dig pits or bore into the ground several times to determine the nature of the surface and subsurface conditions.[27] He tended to identify what he found qualitatively, rather than generically.

Latrobe's geological observations helped him to decide what means he could use to solve an engineering problem. At the Philadelphia Waterworks he found that a tough granite rock stratum underlying the city made it possible to excavate an unreinforced tunnel from near the water intake at the river to the well of the pump house. Similarly, in his proposal for a stream by-pass project in Baltimore, he noted that the planned diversion tunnel would have "strata . . . very favorable to the work, for the roof will fall in a tough substance, consisting of decomposed Rock." And his observations on the route of the Washington Canal convinced him that "the whole of it passes thro' clay and gravel, without any rock," indicating that excavation should be easy.[28]

He was sometimes badly deceived. On the Washington Canal there was no rock, but a significant portion of the canal went through a fluid substratum which made it difficult to establish stable banks. A part of the Chesapeake and Delaware Canal passed through a tongue of land which Latrobe thought would be "easy cutting," but turned out to contain a tough clay.[29]

Latrobe's geological observations in the neighborhood of his works also focused on locating suitable stone for construction. Stone was Latrobe's favorite building material, because he believed it to be the most enduring as well as aesthetically pleasing of any material available to engineers of his era. Brick was an acceptable substitute, iron could be safely used in only a few instances, and wood was the most temporary. In his preference for stone Latrobe followed the principles of British engineering in which he was trained, not the American tradition. Though British engineers believed in building for the ages, American engineers had different values. They knew that canal, turnpike, and bridge companies were often underfinanced, and that the cheapest construction would be most likely to ensure a project's completion. Moreover, temporary construction made more sense when population and economic growth were both rapid and unpredictable. Who could tell where new cities or major routes of commerce would be in another decade? Latrobe, however, never accommodated himself to this viewpoint.

In his search for stone, Latrobe was constrained by the need for quarries to be near excellent transportation, preferably water. For his works in Philadelphia and Washington he depended mostly upon previously opened quarries along major rivers, and found them satisfactory.[30] But when he wanted something to replace the columns destroyed by the fire in the House of Representatives during the War of 1812, he felt that he had to find a new stone befitting a national monument. He knew from twenty years' observation that a certain ridge in Virginia, Maryland, and Pennsylvania, contained a beautiful conglomerate marble. Since there was no opened vein, he and his son spent days in the Virginia countryside searching for a place where the marble could be quarried. Finally they located a workable outcropping on the upper Potomac River, and had samples cut and polished. It turned out to be as rich and attractive a material as Latrobe had anticipated, and the columns stand in the old House of Representatives (known now as Statuary Hall).[31]

On a smaller and more mundane scale, Latrobe searched for stone near the feeder of the Chesapeake and Delaware Canal. The feeder, a navigable branch of the main canal which was designed to bring additional water to it, was the first part of the work to be attempted. Latrobe planned that the feeder's several structures—aqueducts, culverts, and bridges—should be masonry, and in the company of an experienced stone mason he carefully examined the surrounding area for a suitable stone. What they found and quarried was a variety of granitic gneiss. Several of the feeder's surviving structures testify to his sound judgment in selection of a durable building material.[32] The same quarry is currently (1976) being commercially exploited.

Conclusions

Although geology was not a field in which early nineteenth-century engineers had to be especially competent in order to do their job, Latrobe's career suggests why geology was helpful. It contributed to his development of generalized geological schema of the areas which he had to survey for canals, and made possible the early selection of potential routes. It led him to sample rock and soil to help develop the best approach to several of his engineering problems. And his interest in geology was interrelated with his commitment to masonry construction. Benjamin Henry Latrobe would have been a vastly different engineer had he not also been something of a geologist.

Notes

All Latrobe journals and letters are in the collection of the Papers of Benjamin Henry Latrobe (PBHL), Maryland Historical Society, 201 West Monument Street, Baltimore, Maryland. A microfiche edition of the Latrobe Papers has been published (1976) by James T. White and Company, Clifton, N.J. Under Edward C. Carter II (Editor in Chief), a selected and annotated letterpress edition of Latrobe's Papers will be published by Yale University Press. Portions of this article appeared in Stephen F. Lintner's essay on the Susquehanna River map in Darwin H. Stapleton, ed., The Engineering Drawings of Benjamin Henry Latrobe (New Haven: Yale University Press, 1979), copyright © 1979 by Yale University.

1. See Gordon L. Davies, *The Earth in Decay: A History of British Geomorphology, 1578–1878* (London, Macdonald, 1969), 126; V. A. Eyles, "The Extent of Geological Knowledge in the Eighteenth Century, and the Methods by Which It Was Diffused," in Cecil J. Schneer, ed., *Toward a History of Geology* (Cambridge, Mass., M.I.T. Press, 1969), pp. 159–183. In his writings Latrobe refers to such works as Erasmus Darwin's *Zoonomia* (1796–1797); Andre Michaux's *Flora Boreali-Americana* (1803); Lazzaro Spallanzani's *Viaggi Vulcanici* (1778); and Alexander Wilson's *American Ornithology* (1808–1814).
2. Talbot Hamlin, Benjamin Henry Latrobe (New York, Oxford University Press, 1955), pp. 79–80; Constantin F. C. Volney, *Tableau du climat et du sol des Etats-Unis d'Amerique*, 2 vols. (Paris, Courcier, 1803); "William Maclure," *Dictionary of Scientific Biography*, ed. Charles Gillispie (New York, Charles Scribner's Sons, 1970–1976), VIII, 615–617; Latrobe, "An Account of the Freestone Quarries on the Potomac and Rappahannoc Rivers," *Transactions of the American Philosophical Society* (1809), VI, 283; Latrobe to Maclure, 15 June 1809, PBHL.
3. Latrobe, Journals, 6 February 1798, PBHL; Latrobe, "Memoir on the Sand Hills of Cape Henry in Virginia," *Transactions of the American Philosophical Society* (1799), IV, 439–444; Latrobe, "An Account of the Freestone Quarries," 283–293.
4. Davies, *Earth in Decay*; Latrobe, "Sand Hills of Cape Henry"; Alexander M. Ospovat, "Reflections on A. G. Werner's 'Kurze Klassification,'" in Schneer, ed., *Toward a History of Geology*, 242–256.
5. Reijer Hooykaas, "Catastrophism in Geology: Its Scientific Char-

acter in Relation to Actualism and Uniformitarianism," *Koninklijke Nederlandse Akademie von Weten Schappen* (1970), Vol. XXXIII, No. 7, 273.

6. Latrobe to Gales and Seaton, 18 January 1817, PBHL.
7. Hooykaas, "Catastrophism."
8. Latrobe, Journals, 6 February 1798, PBHL.
9. Ibid.
10. Ibid. Latrobe notes: "The idea that this continent [North America] is not so old as the Eastern Hemisphere, is by no means new.—It would require a very accurate, laborious examination during the course of many years of collect and arrange all the circumstances and appearances in the state of geology & natural history of North America that favor this opinion, in order to be able to treat the subject conclusively. It would be necessary to compare the whole *map* of the country with an equal map of the other side of the Atlantic. —I am very little qualified to say anything definite upon the Subject. My own knowledge even of Europe is very superficial compared with the immensity of the subject.—"
11. Thomas C. Cochran, ed., *The New American State Papers. Transportation*. (Wilmington, Del., Scholarly Resources, 1972), I, 266. Maclure and Volney believed that the major gaps in the Appalachians were the result of a vast transmontane lake breaking through the Appalachians in an earlier geologic period.
12. 21 October 1803, Reports of the Engineer, Chesapeake and Delaware Canal Company Papers, Historical Society of Delaware, Wilmington, Delaware. See also Latrobe, Journals, 6 February 1798, PBHL.
13. Latrobe, Journals, 6 February 1798, PBHL.
14. [*Papers Relative to the Completed Bridge Across the Potomac*] (Washington, Duane and Son, 1807), 7. Cf. Stanley W. Trimble, *Man-Induced Soil Erosion on the Southern Piedmont, 1700–1970* (Ames, Iowa, Soil Conservation Society of America, 1974).
15. Stephen F. Lintner, "The Susquehanna Map," in *The Engineering Drawings of Benjamin Henry Latrobe*, edited and with an essay by Darwin H. Stapleton, The Papers of Benjamin Henry Latrobe. Series III, Vol. I (New Haven, Conn., Yale University Press, 1979).
16. B. H. Latrobe, Sketchbook VIII, no. 29a, PBHL. See also Latrobe, Journals, 6 February 1798, PBHL; Lintner, "Susquehanna Map," map annotations.
17. B. H. Latrobe to Gales and Seaton, 18 January 1818, PBHL.
18. Ibid. Latrobe, Journals, 30 April 1798, PBHL.

19. Latrobe, ibid.
20. Cochran, ed., *The New American State Papers. Transportation*, I, 265.
21. Latrobe to the Mayor and Council of New Orleans, 10 June 1816, Mrs. Gamble Latrobe Collection, PBHL; Latrobe to More, 20 January 1811, PBHL; Asit K. Biswas, *History of Hydrology* (New York, American Elsevier, 1970), 261, 262, 279; Latrobe, *Remarks on the Address of the Committee of the Delaware and Schuykill Canal Company* (Philadelphia, Zachariah Poulson, Jr., 1799), 10; Latrobe, *Opinion on a Project for Removing Obstructions to a Ship Navigation to Georgetown, Col.* (Washington, W. Cooper, 1812), 17, 21, 24–25.
22. Luna B. Leopold, M. Gordon Wolman, and John P. Miller, *Fluvial Processes in Geomorphology* (San Francisco, W. H. Freeman, 1964); Latrobe to Fitzsimmons, 25 May 1807, PBHL; Latrobe to Bates, 21 November 1809, PBHL.
23. Latrobe, Journals, 29 April 1798, PBHL; Latrobe, "Designs of Buildings . . . in Philadelphia," 1799, Historical Society of Pennsylvania, Philadelphia, Pa.
24. Latrobe to Thomas, 5 October 1810, PBHL; Latrobe to Gales, 16 April 1818, PBHL; Latrobe to Greenleaf, 4 January 1813, PBHL; Latrobe to Fitzsimmons, 25 May 1807, PBHL; Latrobe, *Opinion on a Project*. Latrobe, "Designs of Buildings . . . in Philadelphia."
25. Reports of the Engineer, Chesapeake and Delaware Canal Company Papers, Historical Society of Delaware, Wilmington, Del., 21 October 1803.
26. Latrobe to Gales, 16 April 1818, PBHL.
27. Latrobe to Poulson, 24 October 1801, in *Poulson's American Daily Advertiser*, 27 October 1801; Latrobe to Harper, 9 February 1818, PBHL; Washington Canal Co. to Latrobe, Account, 25 September 1810, PBHL; Latrobe to Cochran, 30 March 1810, PBHL; Latrobe to Henry Latrobe, 16 August 1812, PBHL.
28. Latrobe to Cochran, 30 March 1810, PBHL; Latrobe, "Report on the Improvement of Jones' Falls, Baltimore," 31 May 1818, PBHL; Latrobe, "Designs of Buildings . . . in Philadelphia."
29. Latrobe to Fulton, 31 July 1811, PBHL; Latrobe to C. I. Latrobe, 5 June 1805, PBHL.
30. Hamlin, *Benjamin Henry Latrobe*, 264; Thomas Wilson, ed., *The Philadelphia Directory and Stranger's Guide for 1825* (Philadelphia, John Bioren, 1825), xxiii.
31. Latrobe to Gales and Seaton, 18 January 1817, PBHL; John E. Semmes, *John H. B. Latrobe and His Times, 1803–1891* (Balti-

more, Norman, Remington [1917]), 64–65; United States Department of the Interior, Geological Survey, *Building Stones of Our Nation's Capitol* (Washington, Government Printing Office, 1975), 11–12.

32. Latrobe to Vickers, 10 February 1804, PBHL; Joshua Gilpin, *A Memoir on the Rise, Progress, and Present State of the Chesapeake and Delaware Canal* (Wilmington, Del., Robert Porter, 1821), appendix, 27; Darwin H. Stapleton and Thomas C. Guider, "The Transfer and Diffusion of British Technology: Benjamin Henry Latrobe and The Chesapeake and Delaware Canal," *Delaware History* (1976), 17:127–138.

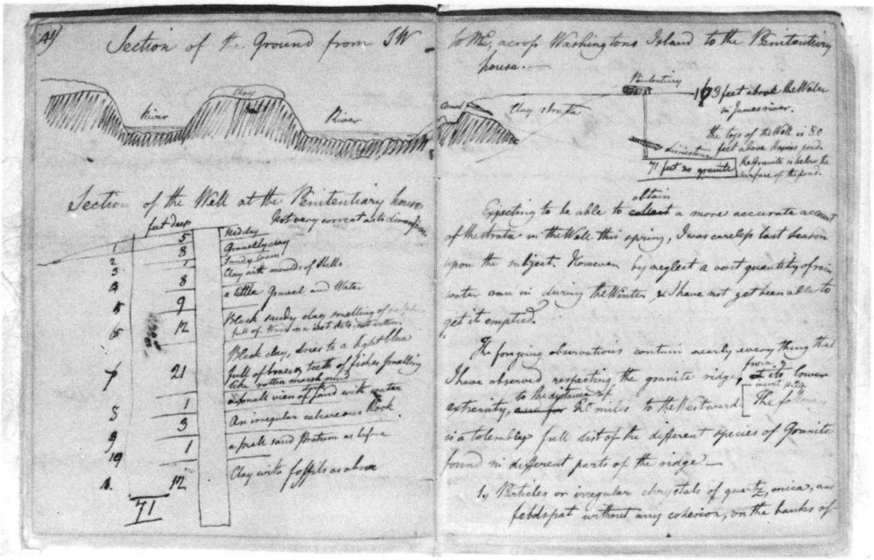

1. Two pages from Latrobe's journals, dated May 4, 1798. Across the top is a section from south (left) to north (right) of Richmond, Virginia, and the adjacent James River. The Penitentiary (of Latrobe's design) and its well are on the right, and the strata of the well are sketched on the lower left-hand page. Latrobe, Journals, May 4, 1798, The Papers of Benjamin Henry Latrobe, Maryland Historical Society, Baltimore, Maryland. Courtesy of the Papers of Benjamin Henry Latrobe.

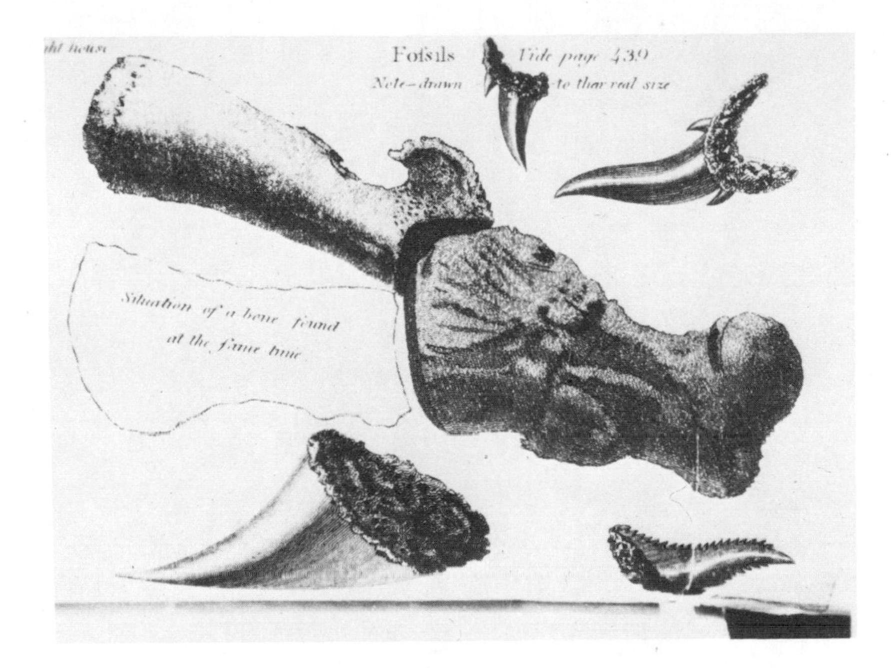

2. An engraving of Latrobe's drawing of teeth and bones removed from the Richmond Penitentiary well. Latrobe identified them as shark's teeth and the paw of an animal. Transactions of the American Philosophical Society *(1799), 4: plate following p. 443. Courtesy of the Papers of Benjamin Henry Latrobe.*

11

Le Crâne de Bone Bank (Indiana) et les fossiles de Walnut Hills, par Charles-Alexandre Lesueur, 1778–1846

Marie-Louise Hemphill

Introduction

Dans le compte-rendu de la séance du lundi, 5 Février 1844, de l'Académie des Sciences, François Arago, dans son rapport, présente la planche lithographiée[1] des "Vues et coupes du Cap de la Hève" de Lesueur, disant que ce dessin doit intéresser: "*le géologue* à raison des terrains divers qui le composent, *le naturaliste* à raison de l'étude des nombreuses espèces de fossiles qui s'y trouvent," et il ajoute: "C'est à ce double titre que l'auteur de ce tableau, auquel il conviendrait également celui d'artiste exercé, s'est attaché à faire de ce cap une étude spéciale et très approfondie."[2]

En effet, Charles-Alexandre Lesueur (1778–1846), lorsqu'il revint d'Amérique où, sur le conseil de Cuvier, il était parti en 1815, accompagnant le géologue américain William Maclure, s'adonnait à la géologie. Avec un pouvoir d'adaptation caractéristique, déjà en 1804, au retour de l'Expédition Baudin aux terres australes, subissant l'influence du savant François Péron, ses vélins d'animaux, de méduses, conservés au Muséum d'Histoire Naturelle du Havre, révèlent, non seulement ses qualités d'artiste, mais aussi celles de naturaliste accompli. Chaque planche montre à côté de la vue d'ensemble de l'animal, les détails d'organes ou autres susceptibles d'intéresser l'homme de science.

En 1815, Lesueur part donc pour les Etats-Unis. Après neuf années passées à Philadelphie où il se fixa après avoir parcouru un certain nombre de régions avec Maclure, il y acquit une véritable renommée en tant que professeur de dessin; qui plus est, ses travaux, dont le plan méthodique d'ichtyologie fut présenté à la nouvelle académie des Sciences Naturelles.[3] Ses lithographies sont jugées dès 1822 "tout à fait remarquables."[4] Pourtant, en 1825, Lesueur se laisse persuader par Maclure de quitter cette ville où il est heureux et honoré. Avec un groupe d'amis, il l'accompagne à New-Harmony, petite ville sise sur les bords du Wabash dans l'état de l'Indiana. Le départ a lieu à bord d'un de ces "flat-boats" sorte de chalands selon la coutume de la navigation fluviale en Amérique à l'époque. Celui-ci le "Philanthropist" évoque par son nom les buts du voyage.[5]

Le riche et philanthrope Maclure veut en effet se joindre à Robert Owen et fonder à New Harmony, une société communautaire en laquelle Lesueur n'a personnellement aucune foi. Sa réputation en fait, fit tout autant pour attirer les visiteurs européens de marque, en cet endroit, que l'entreprise de Maclure elle-même qui finit d'ailleurs par péricliter.

De New Harmony, où Lesueur resta jusqu'à son retour en France, avec celui qu'il appela "l'Ami Troost" minéralogiste et élève de l'Abbé René Haüy du Museum de Paris, il explora la région. Les tombeaux l'attirent. Il fait des découvertes à New Harmony même trouvant des vestiges et un certain nombre d'objets, poteries, pipes, calumets. Puis en 1828, un plus long voyage est projeté. Lesueur doit en réalité obtenir d'un consul de France un certificat de vie, sans lequel il ne peut percevoir le montant de la faible pension qui lui est envoyé de Paris. Il décide donc de partir pour la Nouvelle-Orléans. Plusieurs lettres et quelques écrits de la main même de Lesueur sont conservés aux Archives Nationales, au Museum de Paris et à celui du Havre, relatifs à ce voyage. L'une d'elles, de Natchez adressée à M. Bosc, professeur "d'agriculture" (sic) au Museum, datée du 17 mai 1828, nous renseigne. Il profite du départ de M. Gratz qui se rend avec sa femme et ses trois enfants à la Nouvelle-Orléans, pour partir avec eux. Mais c'est surtout dans les volumes 5 et 6 du "Voyages en Amérique" conservés au

Havre, que nous avons puisé des détails ayant trait à ce voyage qu'il renouvelle plusieurs fois et en particulier l'année suivante, c'est à dire en 1829, précisément à la même date. Fait qui a son importance à cause des variations atmosphériques rencontrées d'une année sur l'autre. Ces volumes reliés par les soins de Gustave Lennier sont une sorte de Journal composés surtout de nombreux dessins, de quelques écrits dont le brouillon d'une lettre à quelque amie. Ils abondent en précieux détails.

Ainsi donc, au cours de ces quelques pages, nous allons d'abord parler de Bone Bank où Lesueur trouve, sur les bords du Wabash, un crâne qu'il envoie au Museum de Paris. La courte description donnée par le Dr. Hamy en 1904, a été complétée par Paule Reichlen en 1976.[6] Nous verrons ensuite, comment nos explorateurs arrivant bientôt à Walnut Hills (Mississipi), Lesueur décrit les lieux tout en les dessinant. Ces planches et ces descriptions, en français et en anglais, publiées seulement en partie, nous les avons retrouvées dans les Archives du Museum, aux Archives Nationales.

De retour en France, Lesueur ayant acquis des connaissances de géologie auprès de Maclure, nous l'avons dit, révèle tout l'intérêt du Cap de la Hève et de ses fossiles, à la suite du glissement de 1842. Depuis le congrès de Durham, tant à Washington qu'à Philadelphie, Baltimore ou New-Harmony (Indiana), nous avons pu nous rendre compte que le nom de Lesueur n'est nullement oublié. A Philadelphie, nous avons admiré son portrait par Charles Willson Peale dans la Bibliothèque de l'Académie des Sciences Naturelles dont il fut élu membre dès 1818, en même temps que Deleuze, bibliothécaire du Museum de Paris.[7]

Nous ne saurions oublier de faire mention des fossiles: *Lesueurella* (err. pro. *Lesueurilla* ou *Lesueuriella*) dont Koken a parlé en 1898. Alors que dès 1841, Milne Edwards mentionnait les *Lesueuria*. Ces détails ont été donnés par Ellis Yochelsen (U.S.G.S.) géologue de la Smithsonian Institution, évidence de l'importance du rôle de Lesueur dans l'histoire des sciences au XIXème siècle.[8]

I. Le crâne de Bone Bank

Par sa lettre datée du samedi 29 mars 1828, de Natchez, nous apprenons que Lesueur et ses amis quittent New Harmony pour la Nouvelle Orléans "de grand matin, alors que souffle une petite bise. La nuit avait été calme," dit-il.

A bord du flat-boat, ressemblant au "Philanthropist" d'antan, mais plus petit celui-ci, on s'entasse "avec quelques sacs de farine." La première escale sera "Bone Bank." Laissons ici Lesueur nous confier ses impressions:

> bientôt donc, nous appareillâmes ou pour mieux dire nous poussâmes notre bateau, carré comme une boite . . . Les bords du Wabash, sans beaucoup de diversité, ni d'inégalité dans ses rives, sont à peu près uniformes et couvertes d'arbres. On y rencontre le sycomore, le noyer noir, le chêne d'Amérique, le cotonnier, le peuplier, le saule et quelques autres qui se distinguent par leurs troncs très noirs tandis que les platanes ont la tête blanche . . . La berge la plus élevée (15 pieds) a été couverte par les dernières crues, partout, on en voit les traces sur les arbres dont l'écorce avait été enlevée par les troncs d'arbres déracinés et charriés par les eaux, certains étant restés suspendus et accrochés aux branches. L'autre côté était lavé et mangé (rongé) par les eaux et par les arbres renversés. Ailleurs, des bancs de sable s'étaient formés a leur dépens, de nombreux amas d'arbres couvrant d'autres points où le vent les avait poussé. Nous passâmes devant plusieurs habitations qui avaient souffert de ces crues et dont les habitants étaient revenus. Le vent nous forçait à aller vers le large et nous arrivâmes à une berge que l'on désigne sous le nom de "Bone Bank" (berge des os). Cette berge est ainsi nommée à cause de la quantité d'os humains que l'on y trouve à un, deux, ou trois pieds au plus de la surface des eaux.
>
> C'est dans le terreau noir de cette surface que se sont dégagés les corps ou squelettes, têtes tournées vers l'est et les pieds vers l'ouest, certains dépassant de la berge écroulée. Nous y trouvâmes deux têtes aussi sur la même ligne. Elles étaient dans un tel état de décomposition que l'une d'elles tomba en miettes lorsque nous voulumes la prendre. L'autre avait la mâchoire supérieure à découvert et toutes les dents en vue. Sur la tranche de la berge d'où nous l'enlevâmes avec trop de précipitation, la mâchoire inférieure se détacha, se désintégrant en tombant et perdant toutes ses dents. Le crâne était intact, mais je crains qu'en séchant il ne se brise. [Pl. I.] Comme la berge venait de s'écrouler sous l'effet des crues extraordinaires de cette année, beaucoup d'arbres qui la couvraient se trouvaient déracinés et leur chute avait entrainé avec elles un squelette entier que les racines retenaient. Avec ce squelette, nous trouvâmes un grand nombre de débris, de

> poteries provenant de tombeaux. Il y en avait tellement qu'on serait tenté de croire que cet endroit était le lieu où elles furent fabriquées.

Et plus loin, Lesueur poursuit sa description en parlant de pipes et calumets découverts en même temps que le crâne de Bone Bank. Il les dessine ainsi que quelques autres objets, sans oublier la berge, mais il omet de nous laisser un dessin du crâne. Qu'importe d'ailleurs puisqu'il a été retrouvé:

> Ces pipes, assez simples, prennent quelquefois la forme d'un bec d'oiseau plus ou moins grossièrement taillé. D'autres modèles ont la forme de grenouilles ou autres animaux, ce qui prouve que les indiens tirent leurs modèles de la nature . . .

A ce point, en véritable géologue, Lesueur prévoit que cette partie de butte est rongée par la rivière qui ne cesse d'en enlever, tous les jours, des parties considérables; elle est donc destinée à disparaitre.[9] Il parle de ces eaux du Wabash qui n'étaient pas encore rentrées dans leur lit ordinaire lors de son passage:

> les arbres sont dans l'eau plus ou moins enfoncés selon la pente de la berge. Des tourbillons nous faisaient pirouetter toutes les fois que nous doublions un cap. Nous n'avions point de boussole, aussi, Madame, ne faut-il pas vous attendre à entendre parler ici des points du compas. Nous n'étions pas pourvu de loch pour mesurer notre vitesse. Elle était réglée selon la force du vent favorable ou défavorable à la force du courant.

Lesueur achève sa lettre, à cette inconnue, en disant: "Nous sortimes du Wabash à douze milles de Bone Bank pour entrer dans l'Ohio."

Cette dernière remarque étant importante puisqu'elle nous a permis, en octobre 1976, de repérer l'endroit où se situait "Bone Bank" qui hélas n'existe plus aujourd'hui, comme le prévoyait Lesueur, en son temps.

Le crâne de Bone Bank fut envoyé par Lesueur, en 1829, au Museum d'Histoire Naturelle de Paris. Retrouvé en 1975, grâce à l'article du Dr. E. T. Hamy qui, en 1904, s'il ne le décrivit que très succinctement prit toutefois le soin d'en noter le numéro: 1255.[10] Il nous a ainsi été possible de le retrouver au Musée de l'Homme. Il n'appartenait pas, ainsi que nous l'avions craint tout d'abord, au Museum du Havre où il aurait été irrémédiablement perdu comme le reste des collections Lesueur, lors de l'incendie de 1944.[11] Si Lesueur l'avait décrit ou

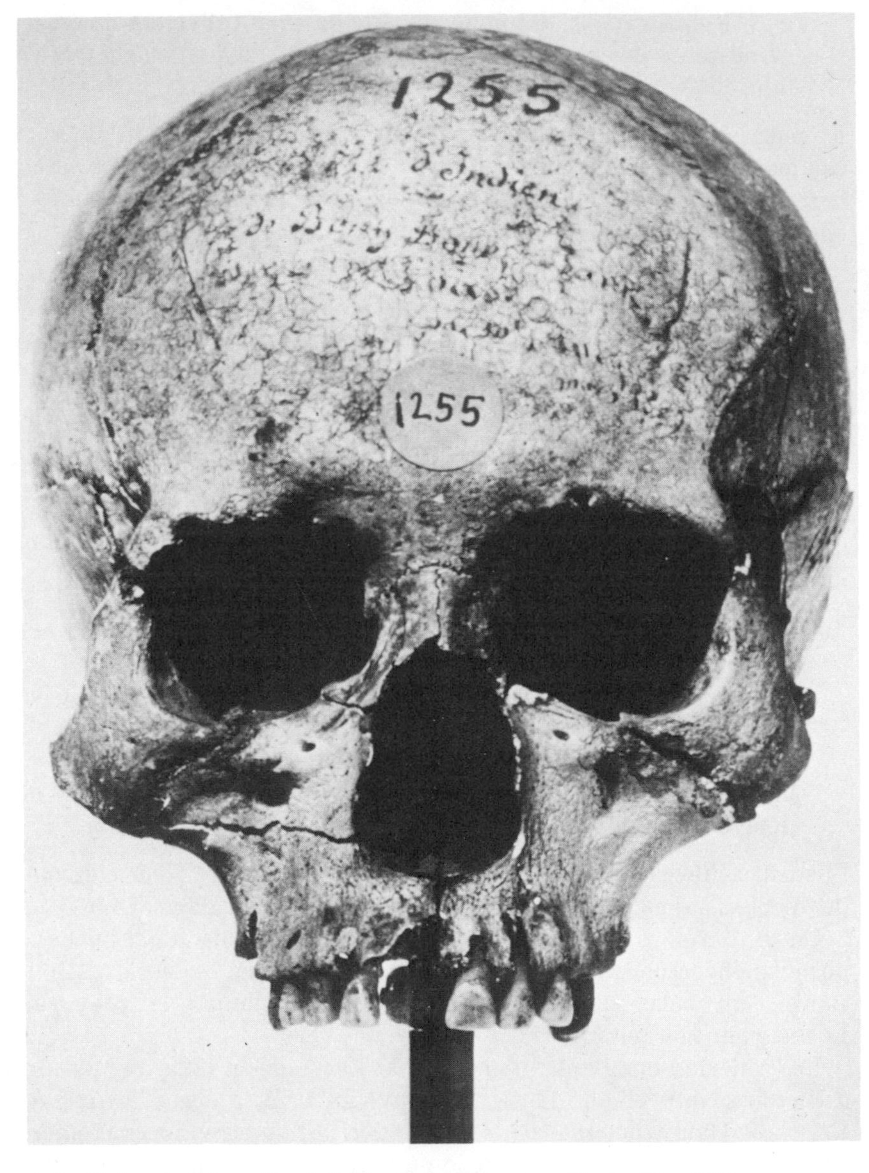

1. Le Crâne de "Bone Bank," Musée de l'Homme, Paris, No. 1255.

même dessiné en le trouvant, il eut précédé Squier et Davis dans leur description du Scioto Skull et le crâne de Bone Bank aurait eu sa place dans l'histoire des sciences et pris une importance considérable surtout pour les Américains.[12]

II. Les Coquillages Fossiles de Walnut-Hills

Dans sa lettre du 4 mai 1830, adressée aux professeurs du Museum de Paris, Lesueur annonce l'envoi de coquilles fossiles qui seront accompagnés de planches avec leurs descriptions.

Dans les documents retrouvés, le nom des fossiles donné par Lesueur (selon ce qu'il appelle "le système Lamarck"(?) ne correspond plus aux noms actuels. De plus, on ignore, en France, ce que signifie ce système. Il nous a paru inutile de faire perdre un temps précieux à des chercheurs, pour mettre à jour cette liste. La coupe de Walnut Hills ayant déjà été reproduite, nous nous abstiendrons de le faire ici, il nous suffit de dire qu'il en a montré les 14 sections ou couches différentes de terrain, terminant ainsi ses observations sur les rives du Mississipi.

Toutefois, Lesueur ayant décrit chacune de ces couches en détails il émet des idées qui semblent mériter d'être signalées puisqu'elles prouvent qu'il serait peut-être un précurseur.

Les divers fossiles qu'il rencontre l'incitent à expliquer tout l'intérêt que présentent les conditions locales. "Assumant," dit-il, "qu'une communication existait autrefois entre les divers états de l'Alabame, de Géorgie, et du Sud des montagnes de l'Alleghany, et de là traversant le nord de l'état de Carolina vers la Virginie et le Maryland, de nombreux fossiles, de part et d'autre, furent trouvés par M. Stephen Elliott, J. Finch et décrits par M. Say et d'autres par M. Morton." En examinant ces descriptions, Lesueur conclue en disant que les rapports existant entre ces divers dépôts marins pourraient ainsi s'expliquer. Mais pour l'affirmer, ajoute t-il, il faudrait faire de plus amples recherches. Et Lesueur poursuit par la longue description des 14 couches qu'il a pu observer, à la même époque précisément, en 1828 et 1829.

Le livre de Thomas Say illustré par sa femme ne comporte que deux planches de Lesueur, L et LI, et encore dans l'une seulement des deux éditions.[13] D'autre part, si quelques articles ont été publiés, nous savons que la majeure partie de ces feuillets restent inédits jusqu'à ce jour; il nous a donc semblé interessant d'en parler ici.[14]

Acknowledgments

[Dr. Hemphill expresses her thanks and gratefully acknowledges all those on both sides of the Atlantic who have generously assisted her in the preparation and presentation of this article.]

Notes

1. Ce rapport est conservé aux Archives de l'Académie des Sciences où Lesueur a l'honneur d'avoir un dossier.
2. Lors d'un récent voyage aux Etats-Unis (octobre-novembre 1976) nous avons lu la lettre dans laquelle Lesueur annonce à l'Académie des Sciences Naturelles de Philadelphie l'envoi de fossiles du Cap de la Hève (Bibliothèque). Coll.136A. Lettre datée du Havre, le 21 février 1845. Il a donc été jugé opportun d'envoyer à cette importante Académie américaine la photographie de la lithographie du dessin de Lesueur: "fossiles du Cap de la Hève" tout comme nous l'avons fait pour l'Académie des Sciences de Paris, où se trouve le rapport d'Arago.
3. A défaut de fossiles, un spécimen préparé par Lesueur, est précieusement conservé au Museum de l'Académie des Sciences Naturelles de Philadelphie, il s'agit d'un esturgeon, *Acipenser brevirostris*.
4. A propos de deux lithographies, *Cichla aena*, et *Sciaena oscula*, publiées dans le *Journal of the Academy of Natural Sciences of Philadelphia*, 2 (1822), June and July, Mr. Joseph Jackson écrivit que ce sont là les premières "bonnes" lithographies publiées en Amérique.
5. Voir le Catalogue de l'Exp. au Grd. Palais, Paris (Sept. 1976, Janv. 1977), "l'Amérique vue par l'Europe," n° 322: dessin appartenant au Mus. d'Hist. Natur. du Havre, crayon, 1826, 15/22.7cm flatboat sur lequel on aperçoit Lesueur et ses amis "bateau sur un fleuve, 'Le Philantropiste.'"
6. Hemphill-Reichlen, communication, congrès des Américanistes, Paris, Sept. 1976.
7. Dossier 20, Amer. Philos. Soc. Libr. Philad. 15 nov. 1816, lettre signée M. Vaughan, annonçant l'élection du "zoologiste, Lesueur" en même temps que celle de Deleuze, de Paris.
8. *Nomenclator Zoologicus* ed. Sheffield Avrey Neave (London 1939), p. 147; *Lesueurella* ou *Lesueurina* Fowler 1908, *Lesueuria*

Duncker 1928 (see *Lesueurigobius* Whitley 1950). *Proc. R. Zool. Soc.* N.S.W. (1948–49), Pisces.

p. 928; *Lesueurella* (err. pro. *Lesueurilla* Koken 1898) Perner 1907 in Barrande, Syst. Silur. Bohême, Rech. Pal. 4(2) IV Moll.

Lesueurella (emend. pro.—*rilla* Koken 1898) Cossmann 1915—Essais Paléon. Conch. Comp. 10,181 (also as *Lesueurella*). Moll.

Lesueuria, Milne-Edwards, 1841, Ann. Sci. Nat. (2) (Zool), 16.199—Coel.

Lesueuria Duncker 1928, Tierwelt Nord-u-Ostsee. 12 (g), 124 Pisces.

Lesueuriella, see Lesueurina.

Lesueurilla Koken 1898, N. Jahrb. Min. Geol. Pal. 1898 (1) 22-Moll.

Lesueurina Fowler 1908, Proc. Acad. Nat. Sci. Philad. 59,440 (also on *Lesueuriella*) Pisces.

Index Animalium A Carolo Davies Sherborn confectus, 1801, 1850, London. p. 3500, sous 47 dénominations, 47 articles sont cités.

9. Cox, A. T., *Geological Survey of Indiana* (1878), 125–126. Le géologue américain affirmait dès 1873 que à moins d'un changement dans le cours du Wabash, Bone Bank était appelée à disparaitre. En nous rendant sur les lieux mêmes en 1976, nous avons pu constater qu'il en était bien ainsi. Le fleuve a rongé sa rive et Bone Bank a disparu. Il faudrait cependant remonter le fleuve depuis l'Ohio pour pouvoir l'affirmer.
10. Dr. E. T. Hamy, *Bulletin des Américanistes de Paris* (1904), article qui a été traduit en 1975.
11. A la fin de la seconde guerre mondiale, un incendie a privé Le Havre de son Museum. Aujourd'hui reconstruit, il est ouvert au public depuis 1970. Les dessins de Lesueur avaient été mis à l'abri: ils ont donc été épargnés.
12. Squier, E. G. and Davis, E. H., "Ancient Monument of the Mississippi Valley," *Smithsonian Contribution to Knowledge* (Washington, D.C., 1848), I:288–292.
13. See Catalog, Lilly Library, Aug./Sept. 1970, Indiana University, Bloomington; *American Conchology* (New Harmony, The School Press, 1830–34): "All but two of the plates were drawn by Mrs. Say, and were hand-colored by pupils of the New-Harmony school."
14. Gardner, Julia, *Journal of Paleontology*, 12 (1919), "Lesueur's Walnut Hills Fossil Shells." The fact that Lesueur returned twice in 1828 & 1829 seems to have escaped Dall and this author.

From the State Surveys to a Continental Science

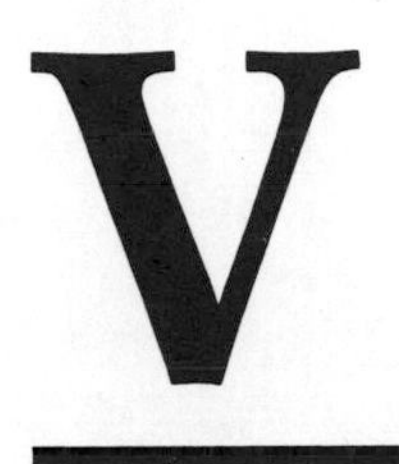

The Institutions

12

American State Geological Surveys, 1820–1845

Michele L. Aldrich

State surveys began in the South in the 1820's and spread quickly throughout the rest of the country, becoming important institutions for geological research within two decades. Economics played a part in determining what practical questions would be studied, as documented by Walter Hendrickson,[1] but early state geologists also investigated topics of purely intellectual interest, and geological theories imported from Europe or produced in the United States directed inquiry and stimulated field work. In this paper the accomplishments of the earliest surveys will be summarized as interesting in themselves and as a backdrop for studies of state efforts in later decades.

Surveys during 1820–45 covered most of what then constituted the United States. The Northeast would have been done entirely except for the slowness of Vermont, which did not begin its survey until 1844. More serious than the gap caused by Vermont was the uneven detail with which surveys were prosecuted, especially in the South and Old Northwest. The work in several states was essentially reconnaissance. Georgia's geologist issued no reports during his tenure of 1836–40, while others published a steady stream of annual reports but no final synthesis before 1845.[2] In Michigan and Tennessee the death of the geologist in charge prevented a final summary. Maryland, Pennsylvania, and Michigan repudiated all state debts during the financial contraction of 1841–42 and could not justify putting out expensive scientific works. The insensitivity of geologists to economic power groups in some states may have hurt the chances of final reports. Charles T. Jackson criticized the lumbering industry as a poor economic base for the state of Maine. The Rogers brothers—Henry Darwin in Pennsylvania and William Barton in Virginia—conscientiously worked on economic geology in their states, but the places on which they concentrated were those slated for future development, not where settlement was already heavy (the regions they favored also happened to be more geologically interesting). William Mather devoted much of his Kentucky reconnaissance to industry and little to agriculture. James C. Booth's behavior in Delaware presented an instructive contrast: faced with a legislative fiat requiring equal attention to all counties, Booth wrote about geology in the scientifically interesting regions and about agriculture in the geologically dull ones.[3]

The states that completed relatively detailed surveys and issued final reports (Massachusetts, Delaware, New York, New Hampshire, and Rhode Island) did so in part because of shrewd behavior by the scientists and solicitous concern for the surveys by strong governors. In New York party ideology, Whig or Democrat, seems to have had little influence on the decision to support or cut off the survey. Pragmatic politics—power plays, personalities, and effective lobbying—were the important variables.[4] Successful surveys courted the voters partly by discoveries in economic geology; two other favorite tactics were publishing a glossary of scientific terms with the annual reports to make them comprehensible to readers, and lavishly acknowledging the help of local residents in survey publications.[5]

The economic concerns of the early surveys, industry and agriculture, were manifested in later eras but with a focus on different problems and the adoption of different methods of analysis. Industry for the first surveys meant transportation (feasible canal and railroad routes) and mining; the geologists paid little attention to manufacturing. The materials that especially concerned surveyors were coal, metals (notably iron), salt, and limestone (for mortar, building stones, hydraulic cement in canal walls, and soil nutrition). Coal preoccupied the Pennsylvania geologist, Henry D. Rogers, while the New York, Indiana, and Michigan surveys devoted attention to salt. William B. Rogers' laboratory demonstration that magnesium in hydraulic cement increased its ability to set under water made his Virginia report on the subject scientifically interesting as well as economically valuable.[6]

Early surveys shared a close association with the transportation revolution. The surveys of Maryland and Virginia were directed by the state agency that oversaw public transportation lines; the Michigan survey appropriations came from the internal improvements fund. The surveys of Connecticut, Maryland, Indiana, and New York were tied in the legislative mind to canals and railroads.[7] These connections were useful as long as public transportation was no financial drain, but became dangerous when state funds for the systems were slashed during the economic downturns of the 1830's and 1840's. The Virginia survey in particular suffered from the correlation, and New York's Governor William H. Seward saved its survey by stressing its intellectual justification and contributions to mining rather than its relevance for transportation lines.[8]

Only five states (Georgia, Massachusetts, Michigan, and the two Carolinas) wanted detailed agricultural studies by the geologists, although most survey laws mentioned in passing the desirability of soil analyses.[9] The geologists of other surveys, wishing to justify their salaries and expenses by immediate economic benefits, included agricultural investigations; Charles T. Jackson was notable in this regard in his work for Rhode Island, New Hampshire, and Maine. Farming was viewed as a family enterprise, not as an industry, and the geologists attached no special romanticism to the calling. The state surveyors pointed out marl beds and urged farmers to use them to supplement organic manures. In going beyond practical hints and on to science, however, the surveys were mired in the intellectual swamp that was agricultural chemistry in the period.[10] Humphry Davy's *Elements of Agricultural Chemistry* (1813) was the only book that pulled the subject together, and its simple-minded soil analyses had been discredited by the time of the first surveys. Individual chemists had made isolated discoveries important for the discipline; most of them were European, but Henry D. Rogers and James C. Booth of the American surveys should be noted for discovering that potassium, not calcium, accounted for the corrective nature of green sand (glauconite).[11] In 1840, Justus Liebig finally assembled these bits of knowledge into a new synthesis. In his first work he subordinated the role of minerals to the part they played in his nitrogen cycle, but within three years Liebig redressed that balance and endorsed the importance of exacting, laboratory-controlled analyses of the mineral contents of soils.[12] His ideas were to influence considerably the surveys of the 1850's.

In addition to the patterns of geographical coverage of the early surveys, a major influence on their scientific achievement was the state of geological knowledge and practice at the time. The ancillary sciences of cartography, geography, and topography had not fully explored cis-Mississippi America and every state geologist complained of inadequate maps. In wilderness areas, state geologists found themselves reporting new streams, lakes, and mountains and correcting geographical errors. With the exception of Maryland,[13] the maps produced and used by early state geologists were hachured, not contoured, which meant that the geology colored over them could not yield quantitative data. Consequently, geological sections were crucial to state geologists for conveying scientific information, and they issued them in profusion and detail.

Another intellectual heritage of the state surveys was imprecision in the definition of "formation." The glossary in Charles Lyell's *Principles* said it meant "a group, whether of alluvial deposits, sedimentary strata, or igneous rocks, referred to a common origin or period." In 1840 Charles T. Jackson told New Hampshire readers that formations were "groups of rocks, formed under certain conditions of the globe, or their relative ages" and in 1844 he said that they were "groups of rocks . . . formed under similar circumstances . . . and determined [in] relative age [by] organic beings, [which] denoted the circumstances under which the deposit was made." In his 1836 Pennsylvania report, Henry D. Rogers wrote that a formation was a species of rock with its associated minerals, and in his 1840 New Jersey volume he told readers that formations were distinct when they were "the result of a wholly different train of physical causes."[14] The term "formation" in the minds of early geologists carried an interpretation of time and origin, as well as descriptive content. Use of the words "group," "series," and so on was just as imprecise.

The first surveys varied in their devotion to paleontology. Since the legislators refused to pay for expensive plates, Henry D. Rogers published virtually nothing on fossils in his New Jersey volume and his Pennsylvania annual reports, although he noted in passing the existence of a fossil cactus in the coal formation, which may have startled contemporary paleobotanists. He and William B. Rogers of the Virginia survey did describe some of their fossil finds (not the cactus) in an American Philosophical Society paper.[15] William Mather said nothing about fossils in his Kentucky reconnaissance, and depended on his colleagues to identify specimens he found in his New York work. Charles T. Jackson shipped his Maine specimens to Augustus A. Gould of Boston, and Elisha Mitchell asked Timothy Abbot Conrad to name the North Carolina Tertiary fossils. David Dale Owen used fossils to date formations in his Indiana reconnaissance, as did Gerard Troost in

Tennessee. For overall devotion to biostratigraphy in their annual and district reports, the New York scientists exceeded the others in zeal, even if their accomplishments in the first phase were abridged by Conrad's failure to produce a final report.[16]

Given the heavy burden on the term "formation" and the erratic use of fossils among surveys, it is no surprise that the early state geologists showed no consensus on appropriate names to employ for strata or on other matters stratigraphic. In New Hampshire and Rhode Island, Charles T. Jackson said he preferred a numerical system for formations by analogy to the widely accepted European terms Primary, Secondary, and Tertiary.[17] Starting with Ebenezer Emmons and Conrad in 1836, the New York scientists favored geographical names, and Douglass Houghton in Michigan agreed with them. Conrad complained that the descriptive names used by other workers carried too much interpretation of origin, did not distinguish between strata of similar appearance in different parts of the column, and were based on lithological rather than paleontological evidence.[18] The New York corps, however, did not use place names consistently for rocks; a few fossil and descriptive terms crept into their nomenclature. Most of the other surveyors stuck to the older descriptive names hallowed in geological literature. Gerard Troost of Tennessee criticized the place-name practice:

> It is very desirable that a geological nomenclature should be introduced, but it must be a scientific nomenclature; no names of little spots or historical facts nearly forgotten. We have now an English, a French, a German, and an Eatonian Nomenclature. I expect we will soon have a nomenclature for each of our States also.[19]

Despite his objections and those of John Locke of Ohio, the New York practice prevailed, because it was based on fossil evidence and because the New York scientists, particularly James Hall and Conrad, were aggressive in extending their nomenclature into other states.[20]

The state geologists also interpreted geological structures. The Rogers brothers' theory of mountain building developed in part from their fieldwork in Virginia, New Jersey, and Pennsylvania. As Patsy Gerstner has remarked, rather little on their ideas appeared in their survey reports, although faint traces appear in the Pennsylvania reports of Henry D. Rogers' gradual shift from an emphasis on vertical forces to horizontal ones.[21] Mather also fashioned an original theory, which he detailed minutely in his final report on the first district of

New York, but it gained no adherents. Emmons complained that the Rogers' theory could not explain the Adirondacks, which must have been pushed up by an igneous body. Edward Hitchcock favored Elie de Beaumont's theory for the mountains of Massachusetts.[22] As for geosynclines, the state geologists had much of the evidence in their reports—all of them commented on patterns of varying thickness in Appalachian strata, and Henry D. Rogers' coal basins were a virtual microcosm of the trough—but they did not assemble it into a theory.

The structural theories of the early surveyors assumed that geological forces were the same kinds as those operating in the present, but that the rates had varied across time. None of the surveyors adhered to a radical uniformitarian theory of constant rates of forces. There was disagreement on the variance of the rates: some detected a gradual dampening down of forces across the ages for man's benefit, others believed the forces intensified in one era and decreased in others, and Mather thought the pattern was cyclical. Henry D. Rogers and David Dale Owen hypothesized that the forces could operate at the same time with different intensity in different districts.[23]

What are now termed Quaternary deposits fascinated many of the early state scientists. Elisha Mitchell said that diluvial features did not occur in North Carolina and that the Bible could not be contradicted by natural examples anyway. Floods played an important role in Henry D. Rogers' Pennsylvania reports, as spectacular erosional processes set in motion by sudden mountain uplifts. In his reports on New England from 1837 through 1841, Jackson equated the Biblical deluge with the flood that created diluvial formations; in 1842 he mentioned the glacial thesis, only to object that he could not trace diluvial materials as radiating outward from mountains. In 1844 he decided that icebergs transported boulders, and that the Biblical deluge could not have produced diluvium because no remains of man were found in that deposit. In Michigan, Houghton attributed drift features to a huge fresh water lake, with moraines as its boundaries. As for the New York scientists, Mather in 1836 adopted the theory that icebergs moved boulders, Conrad in 1838 embraced a modified version of Agassiz' continental glaciers, and Hall in 1839 accepted the theory that an enormous lake had burst its dam and spread the boulders. None of their final reports agreed on the cause of diluvial features. Edward Hitchcock's paper (1843) on diluvium was useful for reviewing the European literature and for assuring the geologists that even a Congregational minister did not expect Biblical literalism.[24]

To the reader faced with the reports of just one survey, the scientific output of the early state geologists seemed an unsorted jumble of information. Henry David Thoreau said of those for Massachusetts:

> These volumes deal much in measurements and minute descriptions, not interesting to the general reader, with only here and there a colored sentence to allure him, like those plants growing in dark forests, which bear only leaves without blossoms. But the ground was unbroken, and we will not complain of the pioneer, if he raises no flowers with his first crop. Let us not underrate the value of a fact; it will one day flower in a truth.[25]

When read collectively, however, the reports show that the surveyors were working on a set of hard problems, ranging from practical questions regarding agriculturally useful deposits of minerals to glacial theories. The dominant scientific concern was agreeing upon a uniform stratigraphy; until one was accepted, syntheses continued to be difficult. In handling this problem, American geologists adopted the systems-building approach used by Charles Lyell for Tertiary rocks and Roderick Murchison for Silurian formations. Indeed, these two geologists were more admired and imitated for their chronostratigraphic constructs than for their philosophies of geological change; American geologists tended to stick with the older, variable rate uniformitarianism of James Hutton.

In conclusion, then, it is noteworthy that the early American surveys were sensitive to more than economic and practical developments in farming, transportation systems, and mining industries. Early state geologists also contributed to the intellectual history of their era through their work in stratigraphy, paleontology, glaciology, topography, and even tectonics. Biographers of the great geologists of the nineteenth century frequently discover that their protagonists spent parts of their careers on one or more of these state ventures. It is not enough to study federally sponsored geology, development of the science in colleges and universities, and work on the discipline within scientific societies. To write the history of early American geology, one must add the state surveys.

Notes

This paper benefited from suggestions for revision by Clifford Nelson and Mark Aldrich. The original draft for the New Hampshire Conference had carried the surveys up to 1879.

1. Walter Hendrickson, "Nineteenth-Century State Geological Surveys: Early Government Support for Science," *Isis* (1961), reprinted in Nathan Reingold, ed., *Science in America Since 1820* (New York, Science History Publications, 1976), 131–145.
2. This paper is based upon the published annual and final reports of the surveys and upon secondary sources for each state when available. The author used copies of most of the volumes in the Robert Frost Library of Amherst College and consulted the remainder in the Library of the United States Geological Survey. The details on early survey publications come from Max Meisel, *A Bibliography of American Natural History: The Pioneer Century, 1769–1865* (Brooklyn, N.Y., Premier Publishing Company, 1926), II, 425–741.
3. Charles T. Jackson, *Third Annual Report on the Geology of the State of Maine* (Augusta, Smith and Robinson, 1839), viii. Emma Rogers, comp., *A Reprint of the Annual Reports and Other Papers on the Geology of the Virginias* (New York, Appleton, 1884), 546 pp. William W. Mather, *Report on the Geological Reconnoissance of Kentucky, Made in 1838* (n.p., 1839), 40 pp. James C. Booth, *Memoir of the Geological Survey of the State of Delaware: Including the Application of the Geological Observations to Agriculture* (Dover, S. Kimmey, 1841), 188 pp.
4. Evidence on politics and the Massachusetts survey is available throughout the professional correspondence of Edward Hitchcock in his papers in the Special Collections of the Robert Frost Library, Amherst College. For the situation in New York, see Michele L. Aldrich, "New York Natural History Survey, 1836–1845," Ph.D. dissertation, University of Texas at Austin (1974), 72–78, 255–258, 311–321.
5. Glossaries appeared in the following reports: William Barton Rogers, *Report of the Progress of the Geological Survey of the State of Virginia for the Year 1840* (Richmond, Samuel Shepherd, 1841), 125–132; George Fuller, ed., *Geological Reports of Douglass Houghton First State Geologist of Michigan 1837–1845* (Lansing, Michigan Historical Commission, 1928), "Report for 1838 [1839]," 327–338; Charles T. Jackson, *Final Report on the Geology and Mineralogy of the State of New Hampshire; With Contributions Toward the Improvement of Agriculture and Metallurgy* (Concord, Carroll and Baker, 1844), 365–370; Henry D. Rogers, *Second Annual Report on the Geological Exploration of the State of Pennsylvania* (Harrisburg, Packer, Barrett and Parke, 1838), 87–91; and Charles T. Jackson, *Report on the Geological and Agricul-*

tural Survey of the State of Rhode-Island, Made under a Resolve of the Legislature in the Year 1839 (Providence, B. Cranston, 1840), 303–310. Most of these glossaries were copied or abridged from the one in Charles Lyell's *Principles of Geology*. The surveys which used local residents extensively were Rhode Island, Ohio, Michigan, and New Hampshire; Mather in New York was also skilled at this time-saving technique.

6. The geologists' work on salt centered on brine springs, although they were hoping that beds of the mineral would eventually be detected. William B. Rogers summarized his tests of hydraulic limestone in *Report of the Progress of the Geological Survey of the State of Virginia for the Year 1838* (Richmond, n.p., 1839), 32 pp.
7. Walter Hendrickson's article in Reingold is excellent on the connection between state surveys and the internal improvements movement.
8. Aldrich, "New York Survey," 255–258. Emma Rogers, ed., *Life and Letters of William Barton Rogers* (Boston and New York, Houghton Mifflin, 1896), I, 179–184.
9. George P. Merrill, "Contributions to a History of American State Geological and Natural History Surveys," United States National Museum *Bulletin* (1920), 109:55, 154, 162, 365, 460.
10. Margaret Rossiter, *The Emergence of Agricultural Science: Justus Liebig and the Americans, 1840–1880* (New Haven, Yale University Press, 1975), 10–19.
11. James C. Booth, *First and Second Annual Reports of the Progress of the Geological and Mineralogical Survey of the State of Delaware* (Dover, S. Kimmey, 1839). Henry D. Rogers, *Description of the Geology of the State of New Jersey, Being a Final Report* (Philadelphia, C. Sherman, 1840), 206.
12. Rossiter, *Agricultural Science*, 20–46.
13. Merrill, "State Surveys," 142. At the New Hampshire Conference, Martha Bray asked why the contour method did not catch on from this early example. The copies of the Maryland reports in the U.S.G.S. library hint at the answer: the *geology* was not entered in a quantitative fashion on the Maryland contour maps; the geologist indicated rocks only by putting words ("sandstone") here and there over the topography, rather than by drawing in formation boundaries.
14. Charles T. Jackson, *First Annual Report on the Geology of the State of New Hampshire* (Concord, Cyrus Barton, 1841), 14. Jackson, *Final Report on . . . New Hampshire*, 11–12. Henry D.

Rogers, *Second Annual Report . . . of Pennsylvania*, 10. H. D. Rogers, *New Jersey, Being a Final Report*, 135.

15. H. D. Rogers, *Fourth Annual Report on the Geological Survey of the State of Pennsylvania* (Harrisburg, Holbrook, Henlock, and Bratton, 1840), 170. W. B. and H. D. Rogers, "Contributions to the Tertiary Formations of Virginia: Second Series, Being a Description of Several Species of Meiocene and Eocene Shells," American Philosophical Society *Transactions* (1834), n.s. 6:347–377.
16. C. T. Jackson, *First Report on the Geology of the State of Maine* (Augusta, Smith and Robinson, 1837), 119. Elisha Mitchell, *Elements of Geology, with an Outline of the Geology of North Carolina: For the Use of Students of the University* (n.p., 1842), 123–139. David Dale Owen, *Report of a Geological Reconnoisance of the State of Indiana: Made in the Year 1837, in Conformity to an Order of the Legislature* (Indianapolis, J. W. Osborn and J. S. Willets, 1838), 5. Gerard Troost, *Fifth Geological Report, to the Twenty Third General Assembly of Tennessee, Made November, 1839* (Nashville, J. George Hares, 1840), 44–75, and *Sixth Geological Report to the Twenty Fourth General Assembly of the State of Tennessee, Made October, 1841* (Nashville, B. R. McKennie, 1841), 9–25. Aldrich, "New York Survey," 145–196, 221–253, 287–289.
17. Note also the Rogers brothers' numerical system, detailed in Patsy Gerstner's paper in this volume. C. T. Jackson, *Final Report on . . . New Hampshire*, 12–13, and *Report on . . . Rhode Island*, 11.
18. Aldrich, "New York Survey," 147, 154, 156, 195–196, 299.
19. Gerard Troost, *Fourth Geological Report to the Twenty Second General Assembly of the State of Tennessee, Made October, 1837* (Nashville, S. Nye, 1837), 13.
20. Aldrich, "New York Survey," 298–300.
21. Patsy Gerstner, "A Dynamic Theory of Mountain Building: Henry Darwin Rogers, 1842," *Isis* (1975), reprinted in Reingold, 109.
22. Aldrich, "New York Survey," 303–304, 307. "Edward Hitchcock," in Charles C. Gillispie, ed., *Dictionary of Scientific Biography* (New York, Charles Scribner's Sons, 1972), 6:437–438.
23. Aldrich, "New York Survey," 303–304. H. D. Rogers, *Third Annual Report on the Geological Survey of the State of Pennsylvania* (Harrisburg, E. Guyer, 1839), 62. David Dale Owen, *Report of . . . Indiana.*
24. Elisha Mitchell, *North Carolina*, 112. H. D. Rogers, *New Jersey, Being a Final Report*, 113. C. T. Jackson, *First Report on the Geology of the State of Maine* (Augusta, Smith and Robinson,

1837), 110; *First Annual Report on the Geology of the State of New Hampshire* (Concord, Cyrus Barton, 1841), 55; "Third Year's Survey" (i.e. 1842), in *Final Report on . . . New Hampshire*, 127; and Introduction, ibid., 23–26. Fuller, *Michigan*, 464–473. Aldrich, "New York Survey," 145, 182–183, 235, 302–303.

25. Aldrich, "New York Survey," 67.

The Study of Ocean Currents in America Before 1930

Harold L. Burstyn and
Susan B. Schlee

When the White Star Line's newest ship, the *Titanic*, sank in the North Atlantic in April 1912 with great loss of life, the modern industrial world awoke to its lack of knowledge about ocean currents and the icebergs they carry. In response, the maritime nations of Europe and America, whose newspapers reported the tragic drowning of some of their most prominent citizens, established in 1913 the International Ice Patrol. Soon operated by the United States Coast Guard, the Patrol's mission was to find and keep track of the hundreds of icebergs that swept unpredictably into the sea lanes of the North Atlantic in the spring of each year.

Since direct observations of icebergs were hampered by fog, the officers of the Coast Guard turned to scientific techniques of calculating the currents. In the spring of 1913, before the Patrol was formally organized, the sightings of icebergs were augmented by measurements of winds, air and sea temperatures, salinity, and plankton (the microscopic plants and animals that float at the surface). Within a year modern dynamic oceanography—namely, calculating the velocity of currents from measurements of temperature and salinity—was introduced. So that the method of dynamic computation might be used to its fullest, a young Coast Guard officer, Edward H. Smith, was sent from the United States to Norway in 1924.

Smith went to Norway because Scandinavia had been the center for physical oceanography for more than a generation. Working in the Baltic and Norwegian Seas, scientists from Sweden, Norway, and Denmark had developed techniques for measuring temperatures and salinities in the open ocean and for computing the direction and strength of currents from these measurements. After a year at the Geophysical Institute at Bergen with Bjørn Helland-Hansen, the developer of the formula for computing currents, Smith brought back to the United States a full understanding of this latest method, the first paradigm in physical oceanography.

It probably occasions no surprise that a young American should go to Europe to learn the latest methods used in his science. After all, we think readily enough of prominent Americans who left for Europe to pursue their scientific careers—T. W. Richards in the 1880's, J. R. Oppenheimer in the 1920's, to name just two. But there are two things about Smith's sojourn in Scandinavia that are unique. First, it was Scandinavia to which he traveled rather than the scientific centers of Germany, England, or France; of this unlikely choice we shall have nothing further to say here. Second, to study physical oceanography, Edward Smith left the country that had been the leader in that subject not long before and in which the methods of the Scandinavian physical oceanographers were already being applied at the University of California. What was there about the work of Smith's predecessors in American physical oceanography and his contemporaries in California that made it necessary for him to travel across the Atlantic Ocean to a foreign country in order to come abreast of the work in his field? How had the United States lost its dominance of the study of ocean currents, and why did the efforts under way in La Jolla, California, seem to Smith less significant than those in Scandinavia?

America dominated the science of physical oceanography in the first half of the nineteenth century because of the way the American economy developed. A backward land whose chief resource lay in the vast area peopled only by nomadic Indians, the United States before the Civil War moved its goods almost entirely on the water. Steamboats on the rivers and sailing ships on the ocean brought to America the superior cultural and material resources of Europe and returned the raw materials with which to pay for them. From 1815 to 1860 the American merchant marine quadrupled in size. Trade expanded even faster, and the shipping route across the North Atlantic became the world's busiest. (It remained the world's busiest even after the decline of the American merchant marine that followed the Civil War.)

The importance of seaborne commerce to antebellum America meant that the techniques for bringing ships across the ocean and safely into harbor became serious subjects of study. Both those who were interested in science for its own sake and those who saw in science the means to improve the technology of seafaring could draw their sustenance from the shipping community. Before 1850 three agencies of the federal government were founded to furnish what we now call research and development services to maritime commerce, and the employees of these agencies brought American physical oceanography to its dominant position in world science.

The first of these agencies was the United States Coast Survey, founded in 1807 by President Thomas Jefferson. Under its first director, the Swiss immigrant F. R. Hassler, the Survey set high standards, but its work fell victim first to the War of 1812 and then, because of the surplus of naval officers created by the war, to a military takeover. In 1818 the civilian Survey was suspended. For fifteen years Hassler fought with Congress for its reinstatement. After his death in 1842, Alexander Dallas Bache, great-grandson of Benjamin Franklin (Franklin was the first American student of ocean currents) became superintendent. He rapidly expanded the Survey, sending vessels to sea to chart currents as well as surveying the shoreline, and providing employment for a number of American scientists. As Bache developed alliances with professors in the leading colleges and with scientists in Europe, the Coast Survey became the center of the American scientific community.

The second of the federal agencies concerned with oceanography was the Depot of Charts and Instruments. Founded in 1830 to provide a shore billet for the officer son of the Navy's Chief Clerk, the Depot soon gave Lieutenant Charles Wilkes the position from which he became leader of the U.S. Exploring Expedition, America's principal contribution to the geographical reconnaissance of the world's oceans. Unfortunately, Wilkes did little in physical oceanography during his four-year cruise. His successor at the Depot, the naval officer and astronomer James M. Gilliss, turned the agency's goals toward astronomy. However, just as he was building for the Navy an astronomical observatory second to none in the world, politics took the Depot from his hands and placed it in those of the ocean-minded Matthew Fontaine Maury. For almost two decades thereafter, Maury and Bache were rivals within the scientific community in the United States and abroad.

Their rivalry arose from their different perceptions of the respective roles of fundamental and applied science. Maury, first and foremost a naval officer, perceived no difference between them. Self-educated but not in the least self-effacing, his aims were to improve the fighting fitness of the Navy and to hasten the passage of vessels along the major shipping routes. He was also interested in preserving and expanding slavery, a cause unlikely to endear him to liberal-minded men of science. When to these intensely practical motives one adds to his character an incapacity for either self-correction or the acceptance of criticism, one can see why Maury was opposed by Bache. The latter, who graduated from West Point at the head of his class when only 19, stood for science as a profession in its own right. Given the penchant for governments to foster first, the military, and second, practical applications rather than fundamental results, it was difficult for Bache and his allies to develop sound science in a growing country like the United States. Science required standards; it was elitist rather than egalitarian. To outflank ignorance and parsimony required an enlightened leader who hired for the Coast Survey's practical work only those with training and interest in basic science. If provided with the means to earn their living, these individuals could (and did) make fundamental contributions in their spare time.

This reasoning was behind the successful effort of Bache and his friends to found the third of the federal agencies that served shipping through scientific research, the Nautical Almanac Office. Ostensibly a part of the Navy, like Maury's Depot, the Almanac Office was placed by its founder, Lieutenant Charles Henry Davis, in Cambridge, Massachusetts, so that it could use the Harvard College Observatory rather than Maury's National Observatory. More secure than the Coast Survey, since its job could never be finished, the Almanac also had more room for skilled scientists and mathematicians on its staff, though the work offered them was tedious.

At the Coast Survey Bache inaugurated a sophisticated investigation of the Gulf Stream, the most intense of the North Atlantic currents. From 1844 to 1860 Coast Survey ships ran fourteen lines of temperature measurements across the Gulf Stream between New Jersey and Florida, setting the pattern of research on the world's most carefully studied current. Maury at first confined himself to correlating other people's data. By the painstaking analysis of ships' logs, he compiled charts that showed the prevailing winds and currents over the world's oceans. But by 1849, he, too, wanted to direct research at sea. In the same appropriations bill that brought the Almanac into existence, two naval vessels for seagoing oceanography were given to Maury.

Such new ventures, especially in science, generally began in the deficiency appropriations bill whose passage was the last act of a lame-duck Congress. For the 1849 bill to contain both the Almanac Office and Maury's ships suggests that the Congressional supporters of Maury and his rivals may have struck a bargain.

Not content to restrict himself to calculating climatological averages for the world's oceans, Maury began writing down his ideas about their currents. First published as part of the *Sailing Directions* that accompanied his charts, Maury's theories were later collected into one of the all-time best-sellers of science, his *Physical Geography of the Sea* (1855). Unanimously rejected by those who understood best the problems he considered, Maury's book nonetheless had some influence on the history of marine science. So implausible were his theories of ocean currents that they stimulated a schoolteacher, William Ferrel of Nashville, Tennessee, to put forward his own. Thus began the career in dynamical meteorology and physical oceanography of the leading American practitioner of both.

Ferrel's first paper, his pioneering "Essay on the Winds and Currents of the Ocean," was published in Nashville in 1856.[1] In it he presented for the first time anywhere the principle that the earth's rotation accounts for the general circulation of the atmosphere and ocean. When this and other papers came to the attention of Bache and his circle, Ferrel was invited to Cambridge to join the staff of the Nautical Almanac Office. There he remained until 1867, when Bache's close friend and successor, the Harvard mathematician Benjamin Peirce, took Ferrel to Washington and the Coast Survey. Fifteen years later Ferrel left the Coast Survey for the Signal Service, predecessor of the Weather Bureau, where he completed his distinguished service in 1886. During this long and productive career Ferrel worked on theories of winds, ocean currents, and tides, and he built one of the earliest tide-predicting machines. In his prime he was probably the world's foremost physical oceanographer.

Yet William Ferrel did not leave behind him in Washington—or elsewhere in the United States—any successors. The period of American dominance over physical oceanography, of which his theoretical work made him the leader, came to an end. Until the establishment of the Ice Patrol in 1913, American efforts to understand the motions of the sea were fitful and short-lived. What happened to the structure erected by Bache, Maury, and their contemporaries?

The single most significant event that led to the decline of American marine science was the Civil War. When it began, Maury resigned from the Federal Navy and from all further study of the ocean. Bache launched himself into war work so strenuously that he suffered a stroke and died soon after Appomattox. In the years that followed the war, America turned inward, to the vast interior opened to exploitation by the spread of the railroads. Seaborne commerce gave way to mining as the theme of the American quest for riches, and the American merchant marine and Navy almost vanished from the oceans. In consequence, the three agencies that had been established to aid American shipping dwindled. The cuts in their budgets fell first on the marine science that had always been ancillary to the principal mission of each of the agencies.

Another reason for the decline of American physical oceanography was the technical difficulty of measuring currents at sea. The best of navigators considered himself lucky to come within ten miles of a calculated position when out of sight of land. Though the swiftly flowing Gulf Stream might be located with sufficient precision to test Ferrel's theory that the earth's rotation accounted for the pattern of circulation, the more slowly moving currents over the rest of the world's oceans could not be measured accurately enough. What interest there was in Ferrel's theories and those of his contemporaries in Europe was directed to the atmosphere, to which they applied equally well. Unlike a ship at sea, the earth itself formed a stable platform, from which the winds could be measured precisely by means of ascending balloons. Hence the small number of scientists attracted to the problems posed by the currents in the earth's fluid envelope studied the atmosphere rather than the ocean.

There was one exception. In 1884 Lieutenant John E. Pillsbury, commander of the Coast Survey steamer *Blake*, began a series of measurements of the Gulf Stream. His interest in the problems of direct measurement of currents led him to devise first a current meter and then a method of mooring his ship so that it could be held stationary in the strong Florida Current, the swiftest part of the Gulf Stream. From 1884 to 1890 the *Blake* spent part of each year moored along six sections that crossed the Gulf Stream between Cape Hatteras and the southern tip of Florida; the measurements that Pillsbury published in 1890 were the best ever made, but no one used them to prove or disprove theoretical ideas. Left in that limbo reserved for measurements unrelated to theory, they made little impact until the twentieth century.[2]

Meanwhile, an interest in physical oceanography developed elsewhere. In Scandinavia the apparent decline of the North Sea fisheries led to intensive study. The scientists who did the research were quick to see that biological productivity had to depend on the physical conditions of the marine environment. These they began to measure, and their work received a considerable boost from the immensely popular activities of the Arctic explorer Fridtjof Nansen. Nansen soon interested the most distinguished mathematical physicist in Scandinavia, Vilhelm Bjerknes, in the problem of ocean currents. Bjerknes' students J. W. Sändström and Bjørn Helland-Hansen worked out the techniques for calculating currents from the temperature and salinity data which were now routinely collected. By 1905 there existed a method to determine currents indirectly.

Another of Bjerknes's students, V. W. Ekman, proved to be the most talented theoretician of them all. Working from observations made by Nansen during the historic voyage of the *Fram*, Ekman developed elegant solutions to the problems of dead water and of the surface currents produced by the wind. Stimulated by these developments in Scandinavia, the maritime nations of Europe established the International Council for the Exploration of the Sea (ICES), with Nansen as its scientific director.

The renewed interest in ocean currents engendered by the Scandinavian techniques flowed back to the United States, but not through government channels. Rather, the intermediary was Alexander Agassiz at Harvard University. The wealthy son of the famous Swiss naturalist Louis Agassiz, Alexander had made a fortune in copper mining. His interests in marine biology and geology made him by the 1880's the leading American oceanographer, dispensing information and supporting private cruises on public vessels. With his extensive network of European colleagues and his numerous American students, Agassiz became the link between the new Scandinavian developments and the reactivation of physical oceanography in the United States.

One of Agassiz' students was William E. Ritter. After graduate study at Harvard, Ritter returned in 1891 to the University of California to teach zoology. Believing that marine zoology was at once the most neglected and the most promising part of the subject, Ritter set up a peripatetic seaside laboratory that moved along the California coast each summer. He and his students studied the marine organisms they could collect near the shore. Their efforts were haphazard until financial support from E. W. Scripps and his sister Ellen allowed Ritter to organize a marine laboratory on a permanent basis. In 1907 the present site of the Scripps Institution was purchased; several years later a building was erected for year-round research.

At this time one of Ritter's colleagues at Berkeley and another former student of Agassiz, Charles Kofoid, studied on a sabbatical in Europe the work of the marine laboratories there. He was particularly impressed by the modern marine stations in Scandinavia, supported "in the main from the profits of the state liquor business"; he found them to be no longer merely biological stations but environmental ones, equipped to carry out chemical, physical, and meteorological research. Kofoid brought back from Europe about $800 worth of modern oceanographic equipment and an appreciation of "the profounder problems of oceanography."[3]

When Ritter moved in 1910 into the new building of what was still called the Marine Biological Association of San Diego, he brought with him a graduate student in physics and mathematics, George F. McEwen. McEwen had joined Ritter's staff in 1908. From the start he planned to use the methods and instruments of the International Council, although he was severely handicapped by crude equipment that would not have been out of place on the Challenger Expedition of the 1870's. McEwen soon began to interpret the upwelling that occurs along the California coast in terms of V. W. Ekman's 1904 theory of wind-driven currents, the famous "Ekman spiral." By the time the Ice Patrol was organized in 1913, McEwen had collected thousands of density measurements, from which he was calculating currents.

Why, then, if McEwen was doing physical oceanography in the Scandinavian style and publishing his results in the newly-founded *University of California Publications in Zoology*, did Edward Smith of the Ice Patrol go to Europe rather than to California? Though our answers to this question cannot be definitive, we have some suggestions. McEwen, like his predecessor William Ferrel, was a shy and solitary man, as indeed are many mathematical physicists, who suffer from the "sense of frustration that is the lot of all theoretical physicists most of the time."[4] The efforts he made could be understood by few, and those few were likelier to remain at Berkeley working on problems whose applications were more readily grasped than they were to venture 500 miles south to La Jolla. McEwen further ensured his anonymity by selecting an obscure problem. The most successful theoreticians, our hindsight tells us, are those who pursue problems for which there are clear experimental tests. From those problems they choose the ones that seem most important to their times. In a United States largely indifferent to the oceans, the problem of ocean currents was not very high on the list of the mathematical physicists. Further, any particular problem demands a face-to-face group of active investigators if those who work on it are to receive sufficient psychic reward to continue in the face of inevitable frustrations. One needs a dynamic leader, a Bjerknes, a Sverdrup, a Stommel, a Munk, to attract students and colleagues. Such a group is harder to get started than to continue once begun, and neither McEwen nor Ferrel before him was able to establish a community of American physical oceanographers.

Yet Edward Smith's decision to seek information for the Ice Patrol in Norway rather than in California was influenced by more than McEwen's lack of celebrity and the isolation of La Jolla. Smith's adviser, Henry Bigelow, had close ties to European oceanographers. From 1914 when the Ice Patrol began its oceanographic program, it had relied on Bigelow, a curator at the Museum of Comparative Zoology at Harvard University. The Museum had been founded by Louis Agassiz and endowed by Alexander, whose student—like Ritter and Kofoid—Bigelow became. A member of a prominent Boston family, Bigelow succeeded to Alexander Agassiz' oceanographic researches at Harvard on the latter's death in 1910. From Agassiz, Bigelow also inherited close ties with scientists in Europe.

When Bigelow became special adviser to the Commandant of the Coast Guard, he suggested that an oceanographic observer be placed on the cutters of the Ice Patrol. In the spring of 1914 the first of these observers went to sea together with "a short round physicist [and a] long thin one," from the Bureau of Standards and the Weather Bureau respectively. Observers were to make hydrographic sections across the currents west and south of Greenland "to give us a line on some of the deeper eddy currents."[5] In addition, they were to sample the surface plankton to gain more information about the origin of the water masses. From both kinds of information a chart was drawn up and used to predict the probable drift of the icebergs.

Bigelow's observers copied the Scandinavian methods as best they could. They were coached by Bigelow himself, who had read almost everything published in Scandinavia and had used the dynamic method himself to compute the southern portion of the Gulf Stream. Bigelow and his co-workers soon realized that their mathematical comprehension was inferior to their ability to collect temperature and salinity measurements. When Smith went to Norway in 1924, then, he took with him several years' data. These he was to work up under the eye of Helland-Hansen, who could help him improve his understanding of the theory and the calculations necessary to apply it.

Thus both the derivative nature of the work in California and the close ties with Europe of his mentor led Edward Smith to Norway. From his return to the United States in 1925, American physical oceanography has had a continuous history. In this history the Rockefeller Foundation played a crucial role when it established the Woods Hole Oceanographic Institution in 1930. Bigelow became its first director, and at the same time the Foundation erected a new building for the Scripps Institution. At the behest of Ritter's successor as director, T. Wayland Vaughan (like Ritter, Kofoid, and Bigelow a student of Alexander Agassiz), the Scripps Institution "of Biological Research" became the Scripps Institution of Oceanography.[6]

So in spite of America's turning inward after the Civil War, in spite of the decline in the scientific stature of the Coast Survey, in spite of the loss of American dominance in physical oceanography by the end of the nineteenth century, continuity prevailed. In Alexander Agassiz and his students, none of them a physical oceanographer nor possessed of that mathematical talent that leads to the highest understanding of ocean currents, we find the link between the two periods of American preeminence in physical oceanography—that of the mid-nineteenth century and that of today, the period since 1945 when Americans have again become the leaders in advancing our understanding of the motions of the ocean.[7]

Notes

1. In the October issue of the *Nashville Journal of Medicine and Surgery*, republished in *Professional Papers of the U.S. Signal Service* (Washington, 1882), No. 12.
2. George Wüst's comparison of Pillsbury's direct measurements with calculations based on temperature and salinity data convinced oceanographers of the reliability of the dynamic method: "Florida- und Antillenstrom," Berlin University, Institut für Meereskunde, *Veröffentlichungen* n.f. *12 A* (1924). Henry Bigelow had attempted the same comparison a decade earlier. Unlike Wüst's calculations, Bigelow's did not agree with Pillsbury's measurements: "Explorations of the U.S.C. & G.S.S. *Bache* in the Western Atlantic," *Report of the U.S. Commissioner of Fisheries for 1915*, Appendix 5 (Washington, 1916).
3. C. A. Kofoid, *The Biological Stations of Europe*, U.S. Bureau of Education, Bulletin (Washington, 1910), 4:279, 287.
4. Lincoln Wolfenstein, "The tragedy of J. R. Oppenheimer," *Dissent* (1968), 15:81–85.
5. Bradley Patten, "Journal of the Seneca," Woods Hole Oceanographic Institution Archives, manuscript (1914), 2b, 5b.

6. H. L. Burstyn, "American Oceanography in the Twentieth Century," paper delivered at the Annual Meeting of the A.A.A.S., Boston, February 1976.
7. For fuller references see two previously published works: Susan B. Schlee, *The Edge of an Unfamiliar World: A History of Oceanography* (New York, 1973), and Harold L. Burstyn, "Seafaring and the Emergence of American Science," in *The Atlantic World of Robert G. Albion*, ed. B. W. Labaree (Middletown, 1975), 76–109, 223–228.

14

George Otis Smith as Fourth Director of the U.S. Geological Survey

Thomas G. Manning

At a Conference with a pronounced historical orientation it is fitting to remember George Otis Smith, fourth director of the U.S. Geological Survey, who presided over the bureau for twenty-three and one-half years, from May 1, 1907, to December 22, 1930; no other director comes within ten years of Smith's record. His long tenure takes in one quarter of the Survey's total experience and spans the Progressive Era, World War I, and the 1920's. This paper will consider salient characteristics of Smith the man, his conception of the directorship, and his purposes and behavior as he moved with the Geological Survey through the events and situations of a quarter century of American history and politics.[1]

It may seem paradoxical to begin a paper on the director of a twentieth-century scientific agency with a discussion of religion, but the fact that George Otis Smith was a good Baptist explains much about his public life and character. He was a faithful worshiper and trustee of the Calvary Baptist Church in Washington, and a close friend of the minister there, who often dined at the Smith home in Bancroft Place; he was also very prominent in the affairs of the Washington YMCA. The University of Chicago, which originated as a Baptist college, made him a board member to represent the Baptist denomination. And he even spoke once from the pulpit of a church in Bethany, Maine, near his residence there.[2]

Smith's religion made him a moral man and a very high-minded one for a government bureaucrat. He did not smoke, drink, or play cards. He knew that his colleagues spoke of his high ethics and called him an altruist, and he agreed with this judgment by saying that he believed in uplift and in the golden rule. Of course he was hard-working; he felt guilty one morning in Washington when seen taking a walk on the street after nine o'clock first by Mrs. Grace Coolidge and then by the president's private secretary. Hard work was part of the proper scheme of things, a truth which Smith never felt surer about than on the day before Thanksgiving in 1930. That Wednesday afternoon, November 26, he was busy in his office, although half a holiday had been declared. Thus he was prepared for the sudden turn in his life, which occurred at 4 P.M. when a summons came from the White House. Smith hurried over, and President Hoover offered him the chairmanship of the Federal Power Commission.

Smith's ethical preconceptions informed his major administrative decisions. He always liked the conception of the Survey as a fact-finding bureau, which told the truth about America's resources and refused to trade that truth for political advantage. In spending public money he liked to be sure that he was acting within the laws authorizing the expenditure. And during World War I he insisted that the Geological Survey was no profiteer; the bureau was not going to use the crisis to enlarge its functions or enhance its prestige; public service and high standards of work remained the proper goals of the Survey.[3]

Two years after beginning, George Otis Smith faced the event that was the supreme test of his own character and the philosophy of his directorship. This was the Ballinger-Pinchot affair, a controversy involving private claims on public coal lands in Alaska, which shattered the Taft administration and the Republican Party and might easily have ruined Smith and irreparably damaged the Survey.

Brutal infighting made Smith aware how critical the situation was. Several times in 1909 and 1910 Gifford Pinchot managed a personal confrontation with him. Pinchot said that Smith was going with the wrong crowd. Taking particular offense at the press notices of the Geological Survey, he accused the director of disloyalty. Pinchot said he knew of past illegal acts of the Survey, and the bureau would suffer in the eyes of the public if this knowledge was given to the press; Richard Ballinger, the forester said, was a yellow dog and an enemy of conservation. Smith had to react to this calculated treatment, if only to preserve his personal self-respect; so he complained to one of Pinchot's friends, former Secretary of the Interior H. R. Garfield, that he was being threatened, and before the joint investigatory committee of Congress he testified that Pinchot waved a club at him.[4]

More than transparent bullying separated Smith from Pinchot. Years of Survey experience with the conservation movement and Smith's own style of administration explain why he sided with Secretary Ballinger of the Interior Department against Pinchot, the chief of the Forest Service. For one thing Smith had found Ballinger easy to get along with, and enjoyed his confidence. The Secretary made no demands that Survey policy be changed; he fully supported the classifying and valuing of the coal lands which Smith considered so important. Since becoming director in 1907, Smith had observed the methods of the ardent conservationists, and he did not like what he saw. Once he had fallen into disagreement with Pinchot about the expenditure of Survey appropriations for maps of national forests; Smith wanted to do field work and make the maps through topographical surveys, which he felt the law required, while Pinchot wanted the maps compiled in Washington from knowledge at hand. H. R. Garfield, Secretary of the Interior under Theodore Roosevelt, had overruled Smith. As a strict constructionist of the law, Smith did not approve of withdrawals by Secretary Garfield of water-power sites under the reclamation law. When the Geological Survey was asked to examine a forest in Puerto Rico and Smith was waiting to hear from the Comptroller that the mission was legal, he did not take kindly to the statement of a high official in the Forest Service that the Survey should go ahead without waiting for the ruling of legality. Smith told the official that he was not taking orders from the Forest Service.[5]

Ultimately Smith wanted Survey geologists to be impartial, disinterested, and scientific about their inventory of the nation's resources. He did not want them to behave as if they were members of a regulatory and punitive organization; they should not consider themselves special agents probing the nature of entries on the public lands; they should not appear on the witness stand in court trials. Smith contrasted his geologists with the rangers of the Forest Service; the geologist classified the coal lands fairly and understood that the mineral entryman was after coal and willing to pay the price for it. The ranger, faced with a report unfavorable to the mineral entryman, accepted that report; but when a favorable report for the entryman came to hand, this same ranger asked for another study to reach a conclusion unfavorable to the entryman. Some of the members of the Forest Service, Smith asserted, sought to oust the mineral entryman by proving in any way possible the nonmineral character of the land. In congressional committee Smith spoke contemptuously of the investigatory zeal of employees of the Forest Service; this was "gum-shoe" work, he testified. He particularly meant L. R. Glavis, whose earnest search of Alaskan land claims precipitated the Ballinger-Pinchot dispute.[6]

World War I was only an interval in the history of the Survey, but its course must have reassured Smith that he had been right in stressing practical mineral geology within the bureau, for it became imperative in 1917 and 1918 to find strategic minerals in a hurry. We are often told that wars undermine morals; certainly Smith's ethics were not as high during this struggle; in February 1917, two months before the war began, he made a personal attack on a geologist named Emil Böse at the University of Texas. Director Smith told the chief of the Division of Investigation in the Department of Justice that Böse was a German who would do "anything in the way of intrigue, perjury and theft for the sake of his father land;" Böse was a dangerous inhabitant to have in the United States, Smith went on, "especially in a border state."[7] Furthermore, during the war Smith used his connections with the anthracite coal operators to steer coal to Skowhegan, Maine, his hometown. Not long afterward the secretary of the coal operators' association was soliciting Smith's support for a person whom the secretary wanted to place in the federal government.[8]

In the 1920's Smith and the Geological Survey reversed their historical position of the early 1900's. Whereas earlier they had been looked upon as spoilers of the conservation movement, unwilling to take the steps that would assure its final victory, now the director and his bureau were the public leaders in defending and advancing the cause. The Mineral Leasing Act of 1920 had brought the Survey into the business of administering the prospecting and leasing of the public lands in coal, oil, gas, and potash. Of course the bureau sought to avoid waste and improper conduct. Smith now felt comfortable with the word "conservation," which earlier in the century had radical connotations for public opinion. So in the mid-twenties the director introduced the word into his organizational arrangements for the Survey, where it has remained ever since.[9] The famous national event in which he participated was the Teapot Dome scandal, which, apart from its story of corruption, clearly demonstrated that Smith's superior, Secretary Albert B. Fall of the Department of the Interior, was an out and out anticonservationist in theory and practice. Director Smith professed no knowledge of the initial stage in the scandal—the transfer of naval oil reserves from the Navy Department to the Interior Department; but he criticized the secrecy of this maneuver and revealed that the Bureau of Mines supported Secretary Fall's transfer plan. For a time, however, Smith was willing to defend E. L. Doheny as an honest, naive person who might see no impropriety in making a loan to his friend, Secretary Fall.[10] When the scandal broke, however, and the government perforce sought to retrieve the situation and restore public confidence in good government, President Coolidge appointed Smith chairman of the commission to recommend a policy for the Teapot Dome and Elk Hill reserves. Smith gave this ad hoc group its name—President's Commission on Naval Oil Reserves, and President Coolidge told him that his presence on the Commission was reassurance for the public that the interest of conservation would be attended to.[11]

Smith always thought the most difficult task of his regime came in the spring of 1929, eighteen months before his directorship ceased, and it was about conservation. At that time his important intergovernmental job was chairmanship of the technical advisory committee to the Federal Oil Conservation Board, and President Hoover asked him to go to Kettleman Hills, California, to obtain a voluntary agreement to restrict oil drilling there. This was a one-man job which Smith must execute without the many efficient helpers who went with the directorship in Washington. He was reminded of the burden of sole responsibility in constructing the 30-minute quadrangles of the Survey geological folios. At Kettleman Hills he described himself as

> teaching simple kindergarten economics to a self-centered industry—putting across the general welfare idea to a short-sighted, present-profit seeking business—getting concerted action from individualistic captains of industry—seeing the other fellows side and showing him the public's side—measuring wits with some very bright men and accepting everything they have of value and letting them grab off some of my ideas without my knowing it—being firm for the essentials without being hard set for petty details—being a good fellow without being an easy mark—but above all showing everyone that Government at Washington is their government and is seeking nothing but what will benefit all concerned in the long run.[12]

He had five interviews and spent twelve hours with John Hays Hammond, Jr., before he could persuade him to come in with the other oil operators. After a month of talking, the independents and the big companies agreed to limit drastically for a year and a half their drilling in the north dome of the hills.

This paper ends, as it began, with a unique feature of George Otis Smith. More than any other director he was in demand as a popular speaker and writer. It was almost as if he were a member of the cabinet. In one three-week period in 1928 he traveled through twenty-four states to give eleven speeches in six of them. He also wrote steadily for general magazines; his skill there was to include hard facts from one of the Survey geologists, but sugarcoat these facts in the first part of the article; which part, he thought, was more likely to be read. George Otis Smith became a popular, public spokesman because he gave his audiences what they wanted to hear from a director of a scientific bureau. He articulated pervasive views and goals of the early twentieth century; he communicated the practical purposes, the optimistic mood, and the mild idealism of science and scientists of that period.

Smith came to his audience not as a geologist, but as an engineer. This role was more than a pose on his part; the two men whom he addressed for years as chief were engineers: Herbert Hoover and the senior John Hays Hammond. Referring to a favorite topic of his own, he said that conservation and engineering were "almost synonomous terms"; going further, he defined conservation as utilization with maximum efficiency. Smith wrote and talked so much about coal as a mineral resource that Chief Justice William Howard Taft called him the old coal digger. The subject pleased Smith and his audiences because it allowed him to celebrate the greatness of America. "Better the Coal Age of America," he wrote in the *Atlantic Monthly*, "than the Golden Age of Greece."[13] Without this fuel, America's great industrial cities would be cold, dark, and silent. After the eulogies he pleaded for the ever increasing use of mechanical power and improved machines, so that American standards of wages and living could continue.

In large part the spirit and appeal of the director's message to the American public derived from his evocation of group solidarity. Whether in Maine or in the District of Columbia, he was active in community affairs, and no doubt his national preaching came easily after his local experience with group effort and his sense of its importance. The director interpolated liberally the statements or symbols of American unity through such words as general welfare, service to humanity, and the common good. Of course he stressed the contribution of science to good citizenship; science made men better and raised business ideals, he said. He also introduced the forces of disruption, which if allowed to flourish would destroy the fabric of American society. He pointed to privilege, to monopoly and speculation, and to the politics of government and labor unions. The stable, unified society for Smith was a partnership of the individual and the community. The individual exercised his initiative, government through regulation represented the interests of the community, and enlightened public opinion motivated and controlled the relation of the two polar forces.

George Otis Smith was the first thoroughgoing conservative to be appointed director of the Geological Survey. His outlook accorded well with his great wealth, but no doubt his own temperament and his quick and steady success as an administrator also contributed to the final shaping of the man. He was useful to a number of presidents, and he fitted well into the growth of the interdepartmental committees and commissions which were increasingly a feature of the federal bureaucracy in the twentieth century. It is no surprise, then, to find that the Geological Survey did not change much during Smith's regime; the federal appropriation in 1924 was the same as the appropriation in 1906. Fifteen years after leaving the Survey, C. D. Walcott studied carefully its Forty-third Annual Report in 1922; the great difference that Walcott saw from his directorship was in the engraving, printing, and photography of the publication branch.[14] Thus to a manifest and rare virtue of the present-day Survey—its continuity—Smith made a large contribution; how many other scientific bureaus have today the same name and function with which they began in the nineteenth century? During Smith's regime the Survey put down deep roots in the consciousness of its members. The dying Clarence E. Dutton remembered all his friends in the Geological Survey; W. C. Mendenhall, Smith's successor, once said that the Survey was his religion;[15] and Smith himself predicted confidently in 1925 that the Survey would have its accepted place in the government fifty years later. Perhaps the most serious criticism that can be made of Director Smith related to his outspoken advocacy of popular thinking. Americans needed to hear a message in the 1920's, but not the flattering and hopeful one that George Otis Smith brought. More somber and critical statements were necessary to expose the serious defects of the American economic system, which for lack of understanding them or doing anything about them led to the catastrophe of the Great Depression.[16]

Notes

1. This paper is based upon the George Otis Smith Collection, Conservation History and Research Center, University of Wyoming; hereafter referred to as Smith Papers. I am indebted to Director Gene M. Gressley and Research Historian David Crosson for their courtesy and assistance at the Center. For travel expenses I am

obligated to the Department of History and to the Office of Research Services, Texas Tech University.

2. Biographical details about Smith in this paper come mostly from his daily journals of the 1920's.
3. Smith to F. W. Clarke, C. W. Cross, and G. F. Becker, May 27, 1918, Smith Papers.
4. "Investigation of the Department of the Interior and of the Bureau of Forestry. Hearings before Committee, April, 1910," *Sen. Doc. No. 719*, 61 Cong., 1st Sess., p. 3504; see also pp. 3311, 3321–3323, 3333.
5. Ibid., p. 3317.
6. Ibid., pp. 3306, 3499, 3500.
7. Smith to A. Bruce Bielaski, Chief, Division of Investigation, Department of Justice, February 21, 1917, Smith Papers.
8. E. W. Parker, Director, Anthracite Bureau of Information, to Smith, July 1, 1918; Smith to E. W. Parker, July 13, 1918; Smith Papers.
9. Journal of Smith, January, 1924, ibid.
10. Ibid., April, 1924.
11. Ibid.
12. Ibid., June 1929.
13. *Atlantic Monthly* (October 1922), 130:534.
14. C. D. Walcott to Smith, December 21, 1922, Smith Papers.
15. Personal Interview, November 1, 1950.
16. For a parallel interpretation of Smith's directorship, with somewhat different emphasis, see Thomas G. Manning, *Government in Science: The U.S. Geological Survey, 1867–1894* (Lexington, University Press of Kentucky, 1967), 222–226.

From the State Surveys to a Continental Science

The Men

James C. Booth and the First Delaware Geological Survey, 1837–1841

Thomas E. Pickett

James Curtis Booth (1810–1888) was an early American analytical chemist-mineralogist whose significant contributions to regional geology are not widely recognized. His work in Delaware is important primarily because of his accurate descriptions and analyses of contemporary outcrops, rather than a synthesis of regional geologic history. His descriptive and analytical work has practical value to modern geologists and soil scientists. Studying Booth's careful descriptions of outcrops, forgotten since 1841, often results in reinterpretations in the light of modern geologic thought and inclusion on geologic maps prepared by the present Delaware Geological Survey. In addition, Booth's ideas on the uses of greensand, the sedimentary origin of some of Delaware's Piedmont rocks, and the ecologic value of coastal wetlands were ahead of his time. This paper examines Booth's scientific activities and applies some of his work to modern geologic investigations.

Education

Booth was born on July 28, 1810. He attended public schools in Philadelphia and graduated from Hartsville Seminary in Bucks County, Pennsylvania. In 1825 he entered the University of Pennsylvania and studied under William Keating, a chemist and mineralogist who had prepared at the School of Mines in Paris, and Robert Hare, of the Medical School,[1] who was known as the best lecturer on chemistry of his time.[2]

In 1829, after graduating, Booth went to Rensselaer Institute, Troy, New York, to study under the famous geologist, Amos Eaton. Eaton's emphasis on learning by doing, a change from a more prevalent fashion of learning from listening to lectures and watching demonstrations, appealed to Booth, whose student notebooks indicate that his lifetime penchant for careful, exacting work was being developed by his geological field work and analyses under Eaton. Clearly, Eaton was a major influence on his life.

Booth taught chemistry in the winter of 1831–32 in Flushing, New York.[3] In December 1832, he went to HesseCassel, Germany, to study at Friedrich Wöhler's Laboratory. Booth was probably the first American to study analytical chemistry in Germany.[4] Wöhler assessed Booth in these terms:

> He has fair attainments in chemistry, physics, and mineralogy . . . He has a very good head, but I do not think he will be productive or originate new or broad ideas. He is very industrious, scrupulously conscientious, and accurate . . . He is discrete, quiet, and cheerful; withal a young fellow of culture.[5]

In 1834 Booth went to Berlin to study under Gustav Magnus for nine months. The following year he traveled to Vienna, Austria, to visit chemical industries, and then to England before returning to Philadelphia in either late 1835 or early 1836.[6]

Booth's great interest in applied chemistry provided him with the tools necessary for his careful, analytical examination of Delaware's geology. He probably had the best education in applied chemistry of any American of his time; in addition to the practical geologic training from Amos Eaton.

First Professional Activities

One of Booth's first activities after returning home was to establish his own chemistry laboratory, which was probably the first private lab offering practical training in analytical chemistry in the United States.[7] This lab still exists as Booth, Garrett, and Blair Incorporated, Ambler, Pennsylvania, who advertise themselves as the oldest commercial laboratory in the United States. In an advertisement for the laboratory in the Philadelphia papers of March 3, 1836, Booth stated:

> It is not intended to deliver a course of lectures nor is it contemplated to communicate recipes for processes employed by the manufacturing chemist, but the instruction to be given will consist in manipulations made by the student himself, either for researches directed to the extension of the science, or to the discovery and improvements of the processes of chemistry applied to the Arts.[8]

In May 1836 Booth was offered a teaching position at the Franklin Institute, Philadelphia. His letter of acceptance gives a further clue to the philosophy he followed in his Delaware geological survey and other work:

> The advantage generally held out for the delivery of lectures on "Chemistry applied to the Arts" is that the artisan, by becoming acquainted with the theory of his profession, may be made capable of improving it. In my opinion another equally important consideration is that during the progress of such a course he is initiated into the processes of many other arts from which he may draw valuable hints tending to the advancement of analogous points in his own profession.[9]

Booth was a well-rounded scientist who could apply his knowledge and abilities to almost any scientific task.

In May 1836 Booth was hired as an assistant to Henry D. Rogers, who made the first geological survey of Pennsylvania. Booth did field research in the Paleozoic section of Pennsylvania in the Susquehanna and Juniata River valleys.[10] He resigned in the fall of 1836, to teach at the Franklin Institute.

The Delaware Survey

The Delaware Legislature on February 18, 1837, passed an act "to procure to make a geologic and mineralogic survey of the State." Three commissioners, one from each county, were named to hire a geologist: Thomas Stockton, Jonathan Jenkins, and Henry F. Hall. On June 1, 1837, they hired Booth as the Delaware State Geologist at a salary of $1200 a year. He was paid this amount for two years. In addition to the requirements of the act of the Legislature, Booth agreed to give information on any useful discovery to the owner of the land and to collect and deposit with the commissioners specimens of all the minerals he found.[11] The location of these samples is unknown today. The commissioners were to spend some time in the field with Booth. Stockton spent forty days, Hall twenty-eight, and Jenkins eleven.[12] They were paid three dollars a day.

An amendment to the act to establish the survey required that an equal amount of money and, presumably, time be spent in each county.[13] Although this appears politically sound, it was not geologically sound, and Booth was forced to ignore it, rationalizing his actions thus:

> A detailed and careful examination of the mineral contents of each tract of land or farm in the State would occupy the time and labor of many individuals for years and therefore, I laid down this principle of action—to institute such examination in each county, according to the time allotted to me, as would prove of the greatest benefit to the greatest number of individuals.[14]

He further declared:

> The geologist has it generally in his power to state the nature of the whole by an examination of a part—nay, further he may frequently predict with some degree of certainty what formations will be found in a locality not visited and what probability there is of discovering valuable deposits.[15]

With this plan for the survey, Booth started his field work in the summer of 1837. In the winter of 1837–38, he analyzed samples he had taken. He returned in the field season of 1838 with an auger capable of drilling up to 20 feet, which he used for the evaluation of greensand.[16] He was assisted by John F. Frazer, his former co-investigator in the Pennsylvania survey.

The field work was apparently concluded by the fall of 1838. Booth analyzed more samples, becoming more oriented toward geology for agricultural purposes as he realized that most of Delaware's mineral wealth was in the soil and could be used to benefit crops.

In 1839–40 he wrote his "Memoir of the Geological Survey of Delaware: Including the Application of the Geological Observations to Agriculture," which he submitted to the Legislature on May 4, 1841. In a letter with his report, Booth concluded that there was no formation of special value except some "excellent clay," which modern tests have confirmed; there was a "moderate amount of iron ore" (mostly bog iron); and "the 2000 square miles of Delaware should be devoted to agriculture, because the whole state is peculiarly well adapted to it, and because there is no other general object to which it is as well adapted." Booth wrote that his main purpose during his field investigations was to serve as "a travelling instructor in agriculture, without exhibiting the formality of teacher among the people to be taught."[17]

Geological formations delineated by Booth in his report were divided into: (1) Primary (modern Piedmont crystalline rocks); (2) Upper Secondary (modern Cretaceous formations and lower Tertiary greensands); (3) Tertiary (modern Miocene); and (4) Recent (modern Quaternary). This was the first attempt to describe Delaware's stratigraphy. Copies of the published report were rare as early as 1891 and are exceedingly so now. As a bicentennial project the Delaware Geological Survey has reprinted a facsimile.[18]

Booth's colleague John Frazer constructed a geologic map of Delaware to go with the survey, based on "an old but excellent map of the state by Mr. Varley." Booth declined to publish the map, however, asserting that the minute descriptions of the survey alleviated the need somewhat, and recommended that Maryland, Delaware, and Virginia cooperate in publishing a map of the Delmarva Peninsula, since the entire area contained similar geologic formations.[19] The Commissioner's report on the survey to Charles Marim, Secretary of State, said that the "map which was to accompany the *Memoir* is still in the hands of Mr. Booth, but will be forwarded to you in a few days."[20] A search by the author and others revealed no trace of this map, which could have been useful in locating some of Booth's outcrops. Perhaps it was never completed.

Later Activities

After publication of the *Memoir* in 1841, Booth returned to Philadelphia, where he continued his teaching of chemistry, first at the Franklin Institute (1836–1845) and later as professor of Chemistry at the University of Pennsylvania (1851–1855). In 1855 he was a co-author on a geologic map of a mining district in Michigan.[21] He presumably consulted on other mining projects.[22] There is no record of other geologic activities. He wrote many papers in Chemistry, however, and in 1849 President Taylor appointed him Melter and Refiner at the U.S. Mint in Philadelphia.[23] At this time large amounts of gold were beginning to arrive from the California gold fields. He remained at the Mint until 1887.

Among Booth's other concerns were the conduct of his private laboratory for instruction; membership in the American Philosophical Society, which he joined in 1839; and membership in the Academy of Natural Sciences, which he joined in 1852. He was president of the American Chemical Society 1883–84. Booth received an LL.D. degree from the University of Lewisburg (Bucknell) in 1867 and a Ph.D. from Rensselaer Polytechnic Institute in 1884. He died in Haverford, Pennsylvania, on March 21, 1888.[24]

Conclusions

James C. Booth has been overlooked as a contributor to American regional geology, largely because of his identification as a chemist. His practical application of geologic data to agriculture was of invaluable service to Delawareans.

Of great value to the present Delaware Geological Survey, which was founded in 1951, are his outcrop descriptions and analyses. Because natural outcrops are rare in Delaware's Coastal Plain, Booth's report is a useful guide to the rediscovery of long-lost outcrops. This has had direct application to the Geologic Quadrangle Mapping Program. Other state surveys may find it useful to study carefully the original surveys of their states.

In his *Memoir* Booth was interested in promoting the use of greensand (glauconite) as a natural source of potash for fertilizer. He devoted many pages to this subject and to chemical analyses of Delaware greensands. Today the Delaware Geological Survey has a grant from the U.S. Bureau of Mines to investigate the interesting ability of greensand to remove objectionable metallic ions from wastewater. Delaware has some of the purest glauconite deposits in the United States.

The origin of the Delaware Piedmont crystalline rocks is today enigmatic. Booth reasoned that they are, at least in part, sedimentary, because their weathering revealed relict sedimentary structures.[25] That theory coincides more with the views of modern geologists than with those of geologists between 1841 and the present.

Booth was modern in his concern for ecology. In 1841 he extolled the value of

> the splendid Bay and Ocean Front, directly contributing food to man, of the best quality, in ample supply, and through all time, only needing proper legislative influence to regulate it, for the welfare of all the citizens of the State.[26]

Thus Booth was ahead of his time in calling for legislative regulation of Delaware's important coastal wetlands. Delaware was one of the first states to vote such legislation, though not until 1971 was the Delaware Coastal Zone Act finally passed.

Notes

This research on Booth was greatly aided by Stewart Rafert, graduate student in History at the University of Delaware. Mr. Rafert checked documents at the Edgar F. Smith Collection of the library at the University of Pennsylvania, Franklin Institute Archives, and Booth, Garrett, and Blair Incorporated, Consulting Chemists, Ambler, Pennsylvania. Mr. Frank Stewart of that company has Booth's notebooks. Michele Aldrich was helpful and encouraging by sending copies of Booth's letters in the Joseph Henry Papers at the Smithsonian Institution and by suggesting other sources of Booth material. Kenneth Woodruff, of the Delaware Geological Survey, and my wife, Suzanne Steinmetz, reviewed the paper.

1. Edgar F. Smith, "James Curtis Booth, Chemist, 1810–1888," *Journal of Chemical Education*, 20 (1943) 315.
2. C. A. Brown, "The History of Chemical Education in America Between the Years 1820 and 1870," *Journal of Chemical Education*, 9 (1932) 706.
3. Smith, 315.
4. Ibid.
5. Ibid.
6. Wyndham D. Miles, "With J. C. Booth in Europe—1834," *Chymia* 11 (1966) 139–149.
7. Brown, 713.
8. In the possession of Booth, Garrett, and Blair, Ambler, Pa.
9. James C. Booth to the Committee of Instruction, Franklin Institute, May 23, 1836, Franklin Institute Archives.
10. Patterson Dubois, "Sketch of James Curtis Booth," *Popular Science Monthly* (November 1891), 118.
11. Delaware Senate Journal (1841), 38.
12. Ibid., 39, 40.
13. Delaware House Journal (1837), 231.
14. James C. Booth, *First and Second Annual Reports of the Progress of the Geological and Mineralogical Survey of the State of Delaware* (Dover, Del., S. Kimmey, printer, 1839), 3.

15. Ibid., 7.
16. Autograph letter with the geological survey of Delaware, February 3, 1841, University of Pennsylvania, Edgar F. Smith Memorial Collection, p. 9.
17. Ibid., p. 8.
18. Dubois, 118. James C. Booth, *Memoir of the Geological Survey of Delaware* (Dover, Del., S. Kimmey, printer, 1841). Pickett, T. E., *James C. Booth and the First Delaware Geological Survey—A Facsimile of the 1841 Memoir by James Booth* (Newark, Del., University of Delaware Bicentennial Committee and Delaware Geological Survey, 1976), 188 pp.
19. *Memoir*, p. x.
20. Delaware Senate Journal, p. 37.
21. James C. Booth, and E. J. Hulbert, *Geologic and Topographic Map of the Mining District of Lake Superior, Michigan*, 1855.
22. Smith, 316.
23. Ibid., 317.
24. Ibid., 357.
25. Autograph letter (above, n. 16), p. 4.
26. Ibid., p. 10.

16

Henry Darwin Rogers and William Barton Rogers on the Nomenclature of the American Paleozoic Rocks

Patsy A. Gerstner

The Rogers brothers are well known for their work in geology in the United States in the 1830's and 1840's. Henry as director of the first Pennsylvania and first New Jersey state surveys and William as director of the first Virginia state survey were in position to lay down guidelines for the naming of the Paleozoic rocks of the Appalachian area. Their ideas, though not widely accepted, represent an interesting stage in naming the geologic column and illustrate the major problems involved. Their development of a nomenclature advanced over a period of about fifteen years from a numerical format to a complex series of names designating time.

The first suggestion of a numerical system of nomenclature is found in the first annual report of the Pennsylvania State Geological Survey covering the activities of 1836, the year in which the survey began. Henry numbered twelve distinct sets of rocks, the first nine of which he designated the Appalachian system and the last three the Carboniferous system.[1] In the second annual report he systematized his numerical system of identification by discussing formations by number.[2] In the 1837 annual report of the Virginia State Geological Survey, William, too, used a formalized numerical system of nomenclature. Which of the brothers suggested this system first is impossible to ascertain. Henry expressed pleasure at "the excellent progress you [William] have made in developing your intricate Appalachian geology,"[3] but the brothers worked so closely together and so often developed their ideas during personal meetings that it is at best uncertain in many cases who originated a particular idea or approach. Whoever suggested it first, however, the numerical system was used uniformly by both for years.

Several things troubled the Rogers regarding nomenclature and classification, and their concern for these is behind the system they developed. At the time they were placing the Appalachian rocks in thirteen defined formations, it was the custom among most American geologists to define American formations in accordance with usages in vogue in Europe, and to give individual names of local significance to the various formations. The tendency to identify with European classification and nomenclature was, in fact, historically the tradition in America. William Maclure, in the early nineteenth century, applied the classification to American rocks that had developed from the late eighteenth-century work of Abraham Gottlob Werner. Maclure arranged the rocks in four classes: Primitive, Transition, Secondary, and Alluvial.[4] Each group contained certain kinds of rocks identified according to lithological bases. Thus the Transition was composed of Transition limestone, graywacke, Transition gypsum, etc. Such a generalized system was modified in America and Europe in the 1820's to make it more precise, with widespread use of fossils as indices. Work in England, particularly, paved the way for a more mature and intricate classification as the mass of materials called the Transition came under close study, and individual formations within the whole, as well as the broad time divisions of Cambrian, Silurian, and Devonian, were recognized. When Henry and William began to study the Appalachian region, most American geologists thought in terms of correlating not only broad time divisions in America and Europe, but individual formations as well. This to the Rogers was a grave mistake. As early as 1835 Henry commented to J. Vanderkemp that geologists in America were all too often guided by a liking for a certain rock classification and tended to force rocks into that scheme. This could only hinder the advance of geology in the United States.[5] It was futile, Henry thought, to assign age or European equivalency to rocks that had scarcely been studied.[6]

The Rogers were equally concerned about an overuse of fossils as opposed to lithological characters. By no means was either brother opposed to the use of fossils, but both felt strongly that lithological characters should be weighed as heavily. Still another point, and a most significant one in their eyes, was the tendency to name formations according to locality. The Rogers felt that this was a serious mistake because such local names obscured natural relationships in and among the formations, and they recognized that the practice could lead to confusion as names for the same formations proliferated on a geographical basis.

The Rogers had a definite philosophy of classification and nomenclature in mind as they rejected these common tendencies of American geologists. We find a clue to that philosophy in Henry's suggestion to William that he read William Whewell on the *History of the Inductive Sciences* because it contains "many sound and broad views . . . in agreement with our own."[7] To Whewell the whole process of giving things names to describe them was inductive, requiring the discovery of fixed characters that relate things on a permanent and real basis:

> Our classification of objects must be made consistent and systematic, in order to be scientific; we must discover marks and characters, properties and conditions which are constant in their occurrence and relations; we must form classes, we must impose our names, according to such marks. We can thus, and thus alone, arrive at that precise, certain and systematic knowledge, which we seek; that is, at science. The object, then, of classificatory sciences is to obtain FIXED CHARACTERS of the kinds of things, and the criterion of the fitness of names is, that THEY MAKE GENERAL PROPOSITIONS POSSIBLE.[8]

Whewell found particular fault with geological nomenclature because it clearly lacked the fixed characters that he felt were so essential. Names based on local terms or local places were not good descriptive names because the character on which the name was based might not be essential and might obscure natural marks of connection. The natural mark of connection must come first, before the classification and nomenclature. This discovery of the natural relationship makes all systems of classification truly natural, although they may necessarily use an artificially chosen criterion of classification once the real relativity is established.[9]

Although Whewell believed that in botany the species were clearly fixed and recognized, to him this did not seem to be true in geology.

The identification of a formation or layer of rock was likely to be doubtful and should be approached with great caution, lest the assumption that there were universal formations such as those suggested by Werner lead one to construct an incorrect history of the earth in a country other than the one where the rocks were originally named. Strata had to be studied and compared within their own country or area so that their natural relationship with one another could be understood before they were submitted to a broader, perhaps more universal, scheme of artificial nomenclature which might reduce them ultimately to a nomenclature in keeping with that used elsewhere.[10]

The Rogers' use of a numerical system, at least at the outset of their studies, allowed them to avoid the pitfalls of classification and nomenclature which Whewell discussed at length. The Appalachian rocks could be studied and named and understood in their own context under this numerical designation. After this was done, it would be possible to find their natural relationships with other areas, especially if workers in other areas would not yield to the temptation of using local names and searching prematurely for relationships with other areas.

By the late 1830's, however, this approach to classification seemed nearly impossible to achieve, and the survey of the state of New York which began in 1836 eventually made it completely impossible. From the early years of the New York survey the geologists of that state had used a local nomenclature. It was in these early years, however—as apparent to some of them as to the Rogers—that such usage would lead to confusion. Therefore, in 1838 Lardner Vanuxem, geologist of the Third District of New York, suggested a meeting of all state geologists to discuss the matter of uniform nomenclature. Henry Rogers' use of numbers was well known and of special interest; Vanuxem wanted him at the meeting.[11] The meeting did not take place until 1840. Although it signaled the beginning of the Association of American Geologists and Naturalists, which in 1848 became the American Association for the Advancement of Science, and although Henry was there, the problem of a uniform nomenclature was not resolved. The geologists of each survey went their own way, gradually adopting the New York nomenclature as a base for all Paleozoic nomenclature. In addition, the New York geologists were intent on correlation with Europe. From the early years of the New York survey, Timothy Conrad, who was first assigned to the Third District as geologist but transferred to the Paleontology Department when it was organized in 1837, was determined to make it "a leading object to compare and identify as far as possible the formations of the state of New York with those of

Europe, at least so far as to ascertain their geological equivalents."[12] To do this he emphasized primarily the fossil content far and above the lithological character.

The Rogers' position, then, was the opposite from that of New York and, in general, other areas. The difference between the Rogers' approach and that of the New York geologists brought the basic issues into sharp focus when Conrad attacked Henry in the pages of the *American Journal of Science*. Commenting on the importance of the Silurian system and the necessity of the fossil assemblage for its proper identification, and noting that "without such knowledge, every step will be embarrassed, and years of labor may be unprofitably devoted to the subject," Conrad said:

> An instance of error on a large scale may be observed in the second annual report of the geological exploration of Pennsylvania, where the graywacke of the Hudson river is confounded with a rock somewhat similar, it is true, in mineral character, which abounds in Oswego County and forms the banks of the Salmon River. Not a single species of shells or plants is common to both. The former is highly inclined, and on its edges rest unconformably the calciferous sandrock of Eaton, then follows the sparry limerock of the same author, with some fossils peculiar to it; above these the limestones and shales of the Trenton series, several hundred feet thick, and then the Salmon river sandstone follows in the ascending order. This shows the great danger of error in endeavoring to identify strata over large areas, if we neglect to appeal to the evidence afforded by paleontology and rely too exclusively upon the ever varying mineral composition of rocks . . ."[13]

In "Remarks by the Editors" appended to Conrad's article it was pointed out that Rogers felt his reasoning was as valid as Conrad's and for the better understanding of all geologists Rogers' ideas and reasoning were appended. Rogers, too, the editor notes, had carefully examined the inclined strata and "he considers the argument based on the want of identity in the fossils as inconclusive, until it shall appear that a large number of species from each formation have been compared, and this because he places more confidence in conclusions drawn from following the rocks themselves over wide areas . . ."[14] Rogers thought that the seeming unconformity was in reality only the result of the dip of the rocks in that area and that the calciferous sandrock of Eaton lay beneath the graywacke rather than above, and following the formations led him to conclude that the formation beneath the Trenton and forming the banks of the Salmon was indeed the same as the calciferous sandrock.

In spite of criticism concerning their use of fossils, their apparent inability to convince the New York geologists or others that correlation attempts were basically unsound at such an early time in the development of American stratigraphy, and the failure to adopt any uniform nomenclature, the brothers began to think of grouping their formations in larger categories. In 1839 William suggested that their formations be divided into three great groups encompassing the numbered formations of the survey reports.[15] By 1841 the larger groupings were further defined by broad groups headed marine and terrestrial deposits.[16] Although the numbers were still used by the brothers, they began to find problems with a numerical designation as early as 1841. In that year Henry proposed that all formations below Formation XII, which was the dividing point between marine and terrestrial formations, be called the Appalachian System and divided into three groups, each designated by a Greek word meaning Appalachian Morning, Appalachian Afternoon, and Appalachian Evening. By giving the various groups names designating time they would be "exempt from the difficulties of either numerical or geographical reasoning."[17] Within a few days Henry proposed still another division, this time a five-fold division: Eoan (Formations I and II); Ante Meridian (Formations III–VII); Meridian (Formations VIII–IX); Post Meridian (Formations X–XI); and Hesperion (Formations XII–XIII).[18] With these names signifying divisions of time, the numbers fell to the background, and gradually the fivefold division gave way to even more elaborate systems. Most of the evolution of thought on the subject was known only to the brothers and probably to a few close associates, until it was published in 1844. We gather some ideas of the progress of the nomenclature before 1844 from a manuscript of 120 consecutively numbered pages found in the William Barton Rogers papers at the Massachusetts Institute of Technology. Examination of the manuscript reveals that it is not all part of one treatise, but parts of not less than four outlines for works or lectures on nomenclature.[19] Three of the parts appear to be in Henry's writing; one part is probably by William. One part by Henry is titled "A System of Classification and Nomenclature of the Paleozoic Rocks of the United States, With an Account of Their Distribution More Particularly in the Appalachian Mountain Chain." The "System" is a detailed version very similar to that suggested above, although it extends in detail only through the Ante Meridian, with a broad division of the Meridian given and the Post Meridian and Hesperion only mentioned. Three criteria for any system of nomenclature are given as an introduction to the "System."

> First it should comprise in a symmetrical form of terms all the wider as well as the more restricted groups of strata, presenting in due subordination the great systems of rocks and the several subdivisions into which their individual members group themselves.
>
> Secondly. It should possess such pliancy as to admit of expressing by some simple adjunct all the modifications of type exhibited by the minor subdivisions in different or distant regions.
>
> Thirdly. The primary idea suggested by the names of the great divisions should be that of their order in time. This in the subordinate divisions is connected with characteristic mineral or paleontological features.

In keeping with the above, each division name, Eoan, etc., was the basic part of a stratigraphic name. Thus the Eoan was composed of the Eoan Conglomerate, the Eoan Sandstone Slate, the Eoan Vitreous Sandstone, and the Eoan Ferriferous Slate. Together these were members of the Eoan Sandstone Group. Following this group came the Eoan Limestone Group, composed of the Eoan Magnesian Limestone and the Eoan Fossiliferous Limestone. Next came the Eoan Slate Group. The manuscript contains occasional references to the New York equivalents, indicating both the thought that the Appalachian geology was well enough understood now to make such correlations, and the recognition that the New York nomenclature was becoming a standard basis of reference. Thus the Magnesian Limestone of the Eoan Limestone Group is described as "blue or bluish grey subcrystalline limestone" in general but with lower beds "generally aranaceous and in some districts also talcose." These lower beds of the Magnesian Limestone are identical with the New York Calciferous Sandrock. The Fossiliferous Limestone is made the equivalent of the Trenton Limestone, and so forth.

The names used in the "System" were first suggested late in 1841, and the manuscript probably dates from soon after that, because by late 1842 William had proposed a new system.[20] Although this new system was not outlined in 1842, only mentioned, it may be the one that Henry presented at a meeting of the Association of American Geologists and Naturalists in 1844. The nomenclature of the New York geologists had become even more firmly established by 1844; nevertheless, Henry took the opportunity presented to him as president of the

AAGN in delivering the annual address to summarize developments in American geology and to give this newest version of the nomenclature as an alternative to the New York system.[21] After outlining the work of the New York geologists, Henry pointed out the need for a general, rather than a localized nomenclature, one based on time and using nine divisions representing periods of the day:

Primal	dawn
Matinal	morning
Levant	sunrise
Premedidial	forenoon
Medidial	afternoon
Postmedidial	sunset
Ponent	evening
Vespertine	twilight
Seral	dark

> Subdividing each *series* in obedience to natural and obvious relations of the organic remains and mineral boundaries, we have named each ultimate subdivision or *formation*, calling the time during which each formation was produced an epoch, and between the series and formation, we have constructed *groups* in all cases where the natural affinities of the formations require that two or more of these latter shall be united into associations subordinate to the series. . . . The title given to any formation is composed first, of the name of the period to which it appertains, and secondly, of a word or words descriptive of the *ruling* mineral *character* of the rock; and to these is appended, when we wish to specify the type under which the formation is referred to, the name of the district or place where it is so developed . . . The well characterized formation called in the New York survey the Marcellus shales, is named by us the *Postmedidial* older black slate . . . and a member of the Clinton group of New York . . . we propose to call the *Levant iron sandstone*.[22]

Like the earlier "System," this nomenclature suggests relative ages and defines "the fundamental relationship of *succession in time*."

The nomenclature was easily adaptable to all situations. Although rocks of a given age might vary from state to state, the application of this system would preclude confusion caused by calling a rock by one name in New York and by another in Tennessee. Thus the Trenton limestone of New York would be called the Matinal newer limestone and the Utica slate the Matinal older slate, and in Virginia where these are blended, the rock would be called the Matinal argillaceous limestone group, and in Tennessee it would be called the Matinal encrinal limestone.[23]

In 1852 Henry adopted a still further altered nomenclature and classification, a version that was also used in Henry's final report of the Pennsylvania survey published in 1858:[24]

Primal	dawn
Auroral	daybreak
Matinal	morning
Levant	sunrise
Surgent	ascending day
Scalent	high morning
Pre-Meridian	forenoon
Meridian	highnoon
Post-Meridian	afternoon
Cadent	waning day
Vergent	descending day
Ponent	sunset
Vespertine	evening
Umbral	dusk
Seral	nightfall

In the changes and revisions that are evident in the development of the nomenclature, we see the Rogers' continuing struggle with a basic problem in the development of American geology. The years during which they worked on the problem were years during which all American geologists were concerned with nomenclature. James Hall of the New York survey shared some of the Rogers' basic notions and recognized that the many independent surveys would lead "to the adoption of terms, which, however applicable, cannot all be well retained without overburdening the science with synonyms."[25] Hall thought a systematic arrangement of classes, orders, genera, and species was the right way to proceed in geological classification (thereby voicing an opinion not too unlike Whewell's) and that the Rogers were in agreement with this view since the early years.[26] Neither Hall nor any of the other geologists believed their system to be perfect, but as Hall noted, it was "easier to make objections than to propose a more acceptable substitute."[27] It is possible that the Rogers' system might have been considered more thoroughly had it been better known. The use of numbers for formations was well publicized through the annual reports of the surveys with which the brothers were associated, but with the exception of close associates, little was known to other geologists about the greater development of the system of nomenclature before the 1844 report to the AAGN. By that time the New York system was well developed and used as a system of reference. Hall felt, in spite of his reservations, that "the territory of New York, from possessing the most complete series and abundance of fossils, together with the undisturbed position of the strata, offers the most interesting field of investigation and reference, and will be found the best point of departure for the geologist who is making more extended researches."[28] Few disagreed with him, and the Rogers' concepts of nomenclature had little effect.

Notes

This paper is abstracted from a biography of Henry Darwin Rogers now in progress. The subject of Rogers' work on classification and nomenclature will be explored in greater detail in the completed biography, and relationships between the Rogers' work and that of others in this area will be examined.

1. H. D. Rogers, *First Annual Report of the State Geologist* (Harrisburg, printed by Samuel D. Patterson, 1836), 12. In that same year Henry published his *Report of the Geological Survey of New Jersey* (Philadelphia, DeSilver, Thomas, 1836), in which there is no mention of a numerical arrangement of strata. The 1836 New Jersey report was printed in Freehold, N.J., by B. Connolly. The final report of the New Jersey Survey (1840) did have the information: *Description of the Geology of the State of New Jersey, Being a Final Report* (Philadelphia, C. Sherman, 1840).
2. *Second Annual Report of the Geological Exploration of the State of Pennsylvania* (Harrisburg, Thompson and Clark, printers, 1838). In this report the number of formations was enlarged to thirteen by adding an additional one in the Carboniferous system. Whereas in 1836 there were three sets belonging to the Carboniferous, there are now four, an additional one created by separating a thick layer of conglomerates from the Coal Measures, or Formation No. 3, of the original list.
3. August 29, 1838. *Life and Letters of William Barton Rogers*, ed. Emma Rogers (Boston, Houghton Mifflin, 1896), 1:160. This volume will be referred to in other notes as LL-1.
4. William Maclure, *Observations on the Geology of the United States*

of America (Munich, Werner Fritsch, 1966), reprint of the 1817 Philadelphia edition, 28–30.

5. Letter, December 27, 1835, American Philosophical Society Collections.
6. *Report of the Geological Survey of the State of New Jersey* (1836), 94. This disenchantment with correlation was not unique. Henry was quick to point out to William that he had spoken with Charles G. B. Daubeny, an English chemist and geologist who was then in America, concerning the use of European names, and that Daubeny supported him and had told him that George Greenough would certainly agree. Dec. 11, 1837, LL-1, 150.
7. Ibid.
8. *The History of the Inductive Sciences from the Earliest to the Present Time* (London, John W. Parker, 1837), 3:188–189.
9. Ibid., 528–529.
10. Ibid., 532–533.
11. "[A brief history of the Association of American Geologists and Naturalists]" and "A History of the [New York] Geological Survey and the Conditions Precedent Which Led to It." Typescripts in the James Hall Papers at the New York State Library, presumably written by Hall. Edward Hitchcock whose pioneering survey of Massachusetts had set a tone for surveys and whose studies covered areas likely to be of great importance in establishing the chronology of the American Paleozoic rocks was also to be contacted about the meeting.
12. "Citations from, and Abstract of, the Geological Reports of the State of New York, for 1837–8, Being State Document No. 200," *American Journal of Science* (1839), 36:12.
13. Timothy A. Conrad, "Observations on Characteristic Fossils, and upon a Fall of Temperature in Different Geological Epochs," ibid. (1839), 35:243–244.
14. Ibid., 250–251.
15. April 22, 1839, LL-1, 165.
16. October 16, 1841, LL-1, 196–197.
17. Ibid. What the difficulties were is not stated.
18. October 29, 1841, LL-1, 199. Henry noted that these names were suggested by McIlvaine and were better than those suggested by Parke. What Parke suggested is not known.
19. In some cases parts of the manuscript that should go together are not in consecutive order.
20. In the fall of 1842 a note by Robert Rogers, youngest of the broth-

ers, appended to a letter to William from another brother, James, said that Henry was working to settle every point of William's proposed new system. September 2, 1842, William Barton Rogers Papers, Massachusetts Institute of Technology.

21. "Address Delivered at the Meeting of the Association of American Geologists and Naturalists, Held in Washington, May 1844," *American Journal of Science* (1844), 47:137–160. This address was also printed separately. Henry had urged that they prepare their ideas for publication, and apparently there were plans to send the nomenclature to Europe, but Henry cautioned that they must see its fate in the United States first. December 24, 1842, LL-1, 220. He was hopeful at last of presenting the nomenclature at the 1843 meeting of the Association in Albany, but not until 1844 was this done. It may not have been presented at Albany because of a poor attendance or because William was not there—or because they were not ready.
22. "Address Delivered," 157.
23. Ibid., 160.
24. December 11, 1852, LL-1, 328–329. *The Geology of Pennsylvania. A Government Survey* (2 vols., Philadelphia, J. B. Lippincott, 1858).
25. James Hall, *Geology of New York. Part IV Comprising the Survey of the Fourth Geological District* (Albany, Carroll and Cook, printers, 1843), 23.
26. Ibid., 3.
27. Ibid., 2.
28. Ibid., 24.

17

The Ante-Bellum Collaboration of Meek and Hayden in Stratigraphy

Clifford M. Nelson and Fritiof M. Fryxell

American stratigraphic geology in the nineteenth century was an unsettled science, stirred by controversy over the interpretation of the geologic record as it was slowly being unveiled in the American West. Many scientific reputations were staked on this important issue. E. D. Cope and O. C. Marsh clashed over the description and significance of vertebrate faunas; F. H. Knowlton and T. W. Stanton quarreled over the age of various strata; other celebrated collaborations ended with equal acrimony. In describing such vigorous debate, historians of American geology have tended to overlook the quiet partnership of Fielding Bradford Meek and Ferdinand Vandeveer Hayden, and to underestimate the impressive results of this collaboration for reconnaissance stratigraphy. An exception is Karl M. Waage's recent perceptive analysis of Meek and Hayden's investigation of the Cretaceous strata in the western interior. During their long and harmonious collaboration, which extended from 1853 until Meek's death in 1876, Hayden evolved as a resourceful field geologist and naturalist and Meek matured as a gifted stratigrapher and systematist. The present paper examines the nature and assesses the results of the antebellum portion of this fruitful collaboration.

Their association began in May 1853 as field workers sent to St. Louis by James Hall, the State Geologist of New York, then at work on his map of the trans-Mississippi West. They were to geologize in the upper Paleozoic, Cretaceous, and Tertiary strata along the Missouri River to Fort Pierre and thence westward to the White River Mauvaises Terres in present southwestern South Dakota. John Evans' newly published account of these badlands and Joseph Leidy's description of their fossil vertebrates had prompted Hall's mini-expedition.[1] Hall requested Meek and Hayden to be especially vigilant for fossil plants and to "gather facts to give some idea of the Geology of the region and its connection with the geology farther east."[2]

They were and remained an odd but effectively matched pair. Meek, the senior at age 36, was tall, slender, blue-eyed, and fair. Moderately vigorous and self-reliant, he was as yet only slightly troubled by the beginnings of tuberculosis and deafness and not as taciturn as in postbellum years. Modest and unassuming, Meek was also extremely sensitive, kind, and a thoroughly agreeable companion who made longlasting friendships. With only a common-school training in Madison, Indiana, he was largely self-educated. In Owensboro, Kentucky, during the 1840's he had geologized and botanized with George Scarborough, a teacher-naturalist who had been James Hall's classmate at Rensselaer in 1830. Recommended by Scarborough and the Louisville naturalists to David Dale Owen, Meek had been tutored at New Harmony for the survey of Wisconsin, Iowa, and Minnesota in 1848–49, and earned by his ability the assignment to examine Wisconsin's Rush River area. Comparisons of Meek's journals reveal how significantly this experience contributed to his professional growth. Under Owen's tutelage and subsequently that of others, Meek developed an extraordinary ability to delineate delicately and accurately fossil and living invertebrates and to sketch landscapes. Late in 1849, Owen and Scarborough recommended Meek to Hall, who was searching for an assistant. After three years of correspondence, during which Meek refused Owen's offer of a place on the Indiana Survey, he went to Albany as Hall's resident assistant and illustrator, principally to aid in the preparation of the third volume of the *Palaeontology of New York*. The nature of the 50-odd scientific volumes that Meek left in storage with Scarborough also indicates that he was joining Hall as much more than a trainee, although something less than a junior colleague. By May 1853 Meek had acquired a year's matchless experience with Hall and Ezekiel Jewett, later curator of the State Museum of Natural History, in the stratigraphic techniques by which Hall was deciphering the New York Paleozoic.

Hayden, then nearly 24, short and slight in build but hardy, was extraordinarily energetic, contagiously enthusiastic, excitable but persevering, and temperamentally outgoing. He was very much the extrovert, and as such far different from the boyish and diffident person of his undergraduate years. Born in western Massachusetts, Hayden had lived in New York and Ohio before attending Oberlin, from which he graduated in 1850 with a decided taste for natural history and languages. Removing to Cleveland, Hayden trained privately with the paleobotanist-physician John Strong Newberry, himself a protégé of Hall, and attended Jared Potter Kirtland's class at the Cleveland Medical College. After meeting Hall there, Hayden petitioned the Cleveland naturalists, and through them their colleagues, in seeking a field collector's position with Hall or, through Spencer Baird, the Smithsonian's Assistant Secretary and "exploration manager," with a government expedition. Hall, when his initial selectees for collector in the Badlands proved unavailable, turned to the eager if inexperienced Hayden, especially after receiving Newberry's positive, although cautious, recommendation. Before departing Cleveland for St. Louis, Hayden spent a week during April in Albany. There he met Meek and added Hall's specific instructions to the general training he had received from Newberry. In May, Meek was sent to St. Louis to reinforce Hayden, so that "the expedition may be strong enough to make good work."[3]

On their geologic reconnaissance in 1853, Meek's only field season in the upper Missouri country, they used Nicollet and Frémont's excellent topographic map (1843) of the upper Mississippi hydrographic basin. In traveling upriver to Fort Pierre between May 22 and June 19, Meek and Hayden's examination of upper Paleozoic and Cretaceous strata was restricted essentially to observations from the burly sidewheeler *Robert Campbell* and to brief shore excursions at wood stops and overnight tie-ups; they did manage a two-day hike across the Great Bend. Between June 21 and July 18, when their field work was terminated prematurely by the activities of the Sioux, Meek measured geologic sections and sketched landforms, while both men collected specimens in the Upper Cretaceous and Tertiary sequences in the Bad (Teton), White, and Cheyenne basins. They used John Evans' sketch map of 1849 as a reference and conversed with Evans himself before he and his party left the Badlands on July 7. The trip downriver by fur company mackinaw during July 27 to August 7 enabled Meek and Hayden to examine in much greater detail the riverine exposures from

Fort Pierre to the mouth of the Platte. Meek recorded the natural history, soils, climate, and agricultural prospects of the valley, and the genesis of its landforms. Owen had emphasized this sort of broadly based survey, and in later years Hayden always stressed the economic implications of his field investigations. Meek and Hayden reached St. Louis in the *Robert Campbell* on August 21 and were back in Albany by September 3.

During the fall and winter of 1853–54 Meek and Hayden (the latter only intermittently) worked up their fossil invertebrates. Meek drew figures of them, wrote up his field notes, and constructed geologic profiles of their route. These would illustrate the descriptions of the Cretaceous invertebrates and the stratigraphy of the routes traversed, which Hall and Meek were preparing for publication. Meek's field data and sketches were sought eagerly by Leidy, who had received the fossil vertebrates collected by Meek and Hayden, and by Evans.

Hall, in lieu of two month's work due him, allowed Hayden to enroll at the Albany Medical College in a 16-week course of lectures toward an M.D. awarded him in January 1854. Meek, and to a much lesser extent Hall, subsequently trained Hayden further in geologic and paleontologic techniques. Unwilling to become an Army surgeon, as Baird suggested, and unable to obtain a position with an exploring expedition, Hayden worked briefly with Chester Dewey on the University of Rochester's collections. Then, severing his connection with Hall, and under the auspices of American Fur Company factors and officials of the Upper Missouri Indian Agency—and with the advice of Baird, Meek, Leidy, and botanist-physician George Engelmann—Hayden conducted under very adverse conditions a single-handed geological and natural history survey of the vast, newly organized Nebraska Territory. Hayden divided the voluminous and diverse collections made by this survey during 1854–55 in the Cheyenne, lower Yellowstone, and upper Missouri basins (as far west as Fort Benton) among Baird, Meek, Leidy, Engelmann, Newberry, and the St. Louis Academy of Sciences. Also, Hayden initiated an unsuccessful campaign to promote a territorially or federally funded geological survey of Nebraska.

Before departing for the Badlands with Hayden, Meek missed an opportunity to serve as geologist-botanist on Robert Williamson's survey of southern California, a position accepted by William Phipps Blake. Meek used Baird's offer of this western post in renegotiating his contract with Hall before leaving for St. Louis in June 1854 to join George Clinton Swallow's Missouri Survey as assistant geologist. During 1854–56 Meek spent six to eight months of each year at Columbia, Missouri, investigating the geology of four nearby counties and illustrating the paleontology of the survey. He resigned this position to concentrate on his post with Hall in Albany when the press of work for his collaboration with Hayden and for Hall's numerous projects became too great. While at Columbia, Meek had to refuse an offer to join William Emory's Mexican Boundary Survey. During his six years at Albany, Meek formed friendships with William More Gabb, Robert Parr Whitfield, Alexander Winchell, and others, thereby significantly expanding his paleontological contacts.

Hoping to function somewhat independently of Hall, Meek had suggested to Hayden early in 1854 that:

> If you go to the U[pper]. M[issouri]. country again do not fail to collect all the new Cretaceous fossils you can, and if you can find no other person that would suit you better, I would like to join you in investigating them.[4]

Hayden readily agreed. Hall, while emphasizing to Meek the value of continuing their own joint studies of the upper Missouri country, then in press or nearing completion, approved his collaboration with Hayden. In the new research team Meek was mentor and chief analyst. He assumed the major responsibility for interpreting the stratigraphic and systematic significance of the collections and geologic data, illustrating the specimens, and co-authoring and guiding the papers through to publication. Hayden, with his energetic and restless spirit, applied his truly remarkable capacity for making the essential field observations despite working alone or with a single, youthful assistant (from 1856 with James Stevenson of Kentucky), minimal facilities and poor food, difficult travel over long distances and adverse terrain, and the continuing threat of hostile Indians.

> Hayden's quick mind, drive and ego fitted him to the role of field man and explorer, whereas the self-effacing Meek, with his greater discipline and keener, intuitive mind, fitted naturally the role of savant, advisor, and laboratory man: the combination of qualities made for a fruitful and untroubled partnership.[5]

Meek and Hayden's tacit recognition of these complementary qualities allowed them to accomplish together what otherwise would have been impossible.

Meek's initial directions to Hayden were comprehensive. He briefed Hayden again on the superpositional relations, thicknesses, lithic characters, orientation, and distribution of the Cretaceous and Tertiary strata they had seen. He suggested where and how critical observations and significant collections might resolve key questions of age, correlation (the Bijou Hills section, south of the mouth of the White River), distribution and economic value (the "Great Lignite beds," later their Fort Union Group, above Fort Pierre), and environments of deposition (the Badlands Tertiary sequence for which Meek favored a lacustrine origin on the basis of contained mollusks). In subsequent letters Meek forwarded the preliminary results of his analyses and asked for additional data and specimens. He advised on their collection and preservation, requested a systematization in field labeling and shipping of collections, and suggested the use of certain pocket-sized instruments. Meek also gently restrained several of Hayden's overly enthusiastic field identifications and age determinations, such as those relating to the Black Hills beds discovered in 1857, which Hayden believed might be equivalent to the New Red Sandstone. Meek's careful identification of the contained ammonoids demonstrated their Jurassic age.

However, it was one thing for Meek to propose plans and methods of operation and quite another for Hayden to apply them in the field, where the vagaries of travel and weather, difficulties in safely packaging and shipping collections, and frustrating delays in exchanging letters greatly hampered close coordination of their efforts. They were almost totally dependent on correspondence, for during the first four years of their collaboration they worked together, either in Albany or in Washington, D.C., for less than two months. Later, the opportunities provided Hayden as surgeon-naturalist in the explorations of the upper Missouri country by Gouverneur Kemble Warren (1856–57) and William Franklin Raynolds (1859–60) of the Corps of Topographical Engineers favored his determination to map "every stream and important locality from [the] mouth of [the] Platte to [the] sources of [the] Yellowstone & Mo.,"[6] but they tied him closely as well to the route, pace, and requirements of these military wagon road expeditions.

Early in 1856, Meek emphasized that:

> It should be understood that it is not merely the introduction to the scientific world of about one hundred species, we propose, but that we possess the means, and the right, to give the classification and nomenclature to the formations of the great Cretaceous & Tertiary Systems so grandly developed over an area of country more extensive than all that of Great Britain.[7]

This accentuation of the stratigraphic significance of their work reflects a recognition of Hayden's growing capabilities. Meek sent him progressively fewer directions in 1856–57 for making his field operations more effective. Meek's next letter contains a more sophisticated brief and suggestions for testing age determinations and correlations, especially of the supposed "Wealden" Cretaceous beds of the Judith River basin and the Tertiary outliers along the Missouri, the latter as bearing on the time of uplift of the Cretaceous strata.

> I hope you will not think I am giving too many instructions as though you were acting under my direction, for my only object is to give you the benefit of any little advantage I may possess in the way of experience.[8]

The first three papers co-authored by Meek and Hayden were prepared together in Albany during three weeks in March 1856 and published in the *Proceedings of the Academy of Natural Sciences of Philadelphia*. The initial article included their first superpositional stratigraphic classification of the western interior Cretaceous and Tertiary strata, providing a fivefold, numbered division of the former. Since preprints of all three papers were available as a 16-page pamphlet prior to June 8, it predates what is essentially the same section published by Hall and Meek between June 12 and July 17.

In November, Hayden returned to Washington with notes for a complete profile from Fort Benton and the Big Horn River southward to Fort Leavenworth, to accompany a geologic map of the upper Missouri country. Hereafter, Hayden based his operations at the Smithsonian, and by December had interested its Secretary, Joseph Henry, in the possibility of publishing a memoir illustrating the fossils he and Meek were describing and for which Meek and Frank Swinton were preparing figures. Earlier, Meek had studied major collections in Philadelphia and Washington and corresponded with Timothy Abbott Conrad, George Hammell Cook, James Merrill Safford, Michael Tuomey, and Alexander Winchell while establishing intra- and intercontinental correlations. Now, at Hayden's urging, they used the Smithsonian mailings to reach a wider audience abroad, one that initially included Joachim Barrande, John Jeremiah Bigsby, Heinrich Georg Bronn, Thomas Davidson, Eduard Desor, Hanns Bruno Geinitz, Laurent Guillaume de Koninck, Charles Lyell, Roderick Murchison, Alcide d'Orbigny, Carl Ferdinand Roemer, John Salter, and Philippe Edouard de Verneuil.

Notices of their papers and subsequent publications in *The American Journal of Science and the Arts*, and their occasional republication with revisions in European journals, assured Meek and Hayden of an increased audience and a prospective network for the exchange of data and specimens. The first (1857) and second (1858) editions of Hayden's geologic map on Warren's topographic base, printed by Julius Bien of New York, presented their interpretations in their most striking form.

Meek's name had been removed for fear of offending Hall. The second edition limned the newly determined geology of the Black Hills and noted the first "Potsdam-age" (Cambrian) strata discovered in the interior. For the large region treated in common with the contemporary maps of Hall, Edward Hitchcock, and Jules Marcou, Hayden's map was superior for the western interior, as it was based on his direct field observations and he refused to extrapolate beyond its boundaries.

In April 1858 controversy over priority in the discovery of the Permian in North America culminated for Meek in a verbal exchange with Hall and Swallow at the Baltimore meeting of the American Association for the Advancement of Science. The conflict was partially symptomatic of Meek's increasing difficulties with the tyrannical Hall, who, Meek believed, doubted his integrity and had deprived him of due credit in publication. Meek abruptly left Albany for Washington in May, as Hayden had strongly urged him to do while also offering to share living quarters and suggesting that Henry (or so said Baird) would appoint him curator to prepare collections as they were received at the Smithsonian.[9] Its collections were expanded significantly in July by those of recent federal expeditions and surveys in natural history transferred from the Patent Office. Henry viewed the collections as research tools. Principal type specimens were retained after publication; other type and "duplicate" specimens were distributed as aids to museum curators and in education.

Meek lived in Washington for the remainder of his life and after 1861 resided in the russet Smithsonian castle as the Institution's unsalaried Collaborator in Paleontology. Henry, with geologic interests dating from his early training with Amos Eaton, could well appreciate the value of Meek's expertise in determining the relative ages of strata by their contained fossils. Also, Henry had encouraged and supported minimally the Meek-Hayden work since 1856; by 1858 the pair had both an established program and significant results. Henry offered to consider their paleontological memoir for publication if Warren's final report on the Upper Missouri was unduly delayed or never issued.[10]

That summer, with Warren's explorations postponed, Meek and Hayden decided to visit northeastern and central Kansas from August to October to determine definitely the age of the disputed Permian strata and alleged Triassic or Jurassic red sandstones. The latter they had tentatively identified as equivalent to their Cretaceous "Formation No. 1" ("Dakota Group," 1861) of the Nebraska section based on its containing dicot angiosperm leaves previously received for dating. Meek and Hayden collected these plants, several types of which Hayden's former mentor, J. S. Newberry, thought diagnostic of Cretaceous floras, at several localities in the upper, lignite-bearing, red clastics of the Kansas River basin. The Cretaceous age of these beds was confirmed when they found them directly underlying strata equivalent to the Nebraska Cretaceous "Formations No. 2 and 3." They were unable to find fossils in the gypsum-bearing Jurassic(?) clays of the lower portion of the disputed sequence between the Cretaceous and Permian units. As for the Permian strata, it was important not only to position them accurately within the relative time scale and establish their areal distribution, but also to correlate them with the Permian of Europe.

In Europe the validity of the Permian System-Period was under vigorous attack by Jules Marcou as poorly defined stratotypically, undeserving of separate rank, and subordinate to the Carboniferous. Nevertheless, Lyell welcomed Meek and Hayden's earlier announcement of the existence of the Permian System in Kansas as completing the recognition in North America of all the major European chronostratigraphic units. Meek and Hayden referred most of Swallow's "lower" Permian of Kansas, which contained fossils from both system-periods, to an age intermediate between the Permian and upper Carboniferous. Swallow's "upper" Permian strata, which bore only Permian fossils and into which the lower beds graded, were considered equivalent to the Permian of Europe.

The last of Hayden's antebellum explorations, with the Corps of Engineers' Raynolds expedition, extended knowledge of the Permian to age-equivalent outcrops near the headwaters of the Powder River in the eastern Big Horn Mountains and in the Wind River range. Also, Hayden discovered Cambrian and Jurassic strata near these exposures. Beds assigned to the Cambrian System were noted in the Laramie Mountains and Jurassic strata in the Red Buttes area on the North Platte. Hayden, although correct, was unable to verify his suggested Triassic age for the widespread, nonfossiliferous, red arenaceous beds and gypsum-bearing marls lying between the Permian and Jurassic sequences in the Black Hills, Laramie, Big Horn, and Wind River Mountains.

As early as 1856 Meek suggested to Hayden that the Smithsonian's fossil collection, in which Hayden's specimens were being deposited, should be arranged stratigraphically rather than zoologically. Meek's stratigraphic methodology, like the practices of Hall, Gabb, and Charles Abiathar White, required detailed knowledge of both teilzones and biozones of species-group taxa, constantly revised by investigation and data exchange. Meek and Hayden's stratigraphic boundaries, based on serial extinctions and creations, were rather sharply defined:

> So far as my observation goes there is not a single [Nebraska] species that passes any of the lower formations. If this is really the case, there must have been a total destruction of life at the close of each of these epochs.[11]

Doubtless influenced by James Dwight Dana's "Thoughts on Species,"[12] they maintained this view throughout their ante-bellum investigations. The complete replacement of upper Missouri faunas across the supposed lithically gradational Cretaceous-Tertiary boundary, Meek and Hayden ascribed to the elevation of the area above sea level at the close of the Cretaceous, which "caused the total destruction of the whole Cretaceous fauna."[13]

To summarize: in the decade after 1853, Meek and Hayden mapped and described in reconnaissance the geology of much of the northern Great Plains and northeastern Rocky Mountains. They confirmed the existence of Cambrian, Permian, and Jurassic rocks in the western interior, established a classification for its Cretaceous strata that persists today as the framework of our standard section, demonstrated the synchroneity of these units with the European standard, and made perceptive interpretations of the paleoenvironmental and paleobiological significance of the Cretaceous marine faunas. Also, they deciphered the structure and geologic history of the Black Hills. Many of these interpretations and explanations were adopted by Dana in his celebrated *Manual of Geology* (1863), in which he acknowledged his debt to Meek's artistic skill and paleontological expertise during the preparation of the first edition. Subsequent editions still bore "prominent evidence of his knowledge, judgement, and scrupulous exactness."[14] Meek's excellent woodcuts in this textbook exerted great influence in teaching and in field identification.

Waage evaluated the results of this "first purposeful study on the interior Cretaceous,"[15] emphasizing:

> What endures is the synthesis of Cretaceous and Tertiary stratigraphy and the perceptive generalizations about the interior Cretaceous sea and its faunas. Their work, outstanding exploration reconnaissance for its time, obviously cannot be judged in the context of detailed stratigraphic work.[16]

For the group-rank lithostratigraphic units they established, their investigations are "the original source of information, furnishing name, gross lithology, approximate stratigraphic position, and general area of outcrop."[17]

Their investigations disclosed the eastern, fine clastic, transgressive facies of the interior Cretaceous. To the west, the great coarse clastic wedges of the partially age-equivalent regressive facies remained to be discovered, save for the last of these deposits represented by the time-transgressive, inshore and onshore clastics of the Fox Hills Formation. John Bell Hatcher's work later in the century would begin to decipher the interior Cretaceous through the application of concepts of depositional facies. Although Meek and Hayden's emphasis on superpositional relations tended to stress vertical relations at the expense of horizontal continuity, their classification, with its strong biostratigraphic as well as lithostratigraphic base, would be applicable throughout most of the interior, except where the clastic wedges coalesced.[18] Waage believed their investigations "reached a high point in 1861 with their most detailed paper on Upper Missouri stratigraphy in which names replaced numbers on the five units of their original classification of the Cretaceous sequence."[18]

Their article also completed the substitution of names for their original, letter-based classification of the known Tertiary strata. In descending order, they were the Pliocene "Loup River beds" (a mixture of Miocene and Pliocene strata), the Miocene White River Group (now Oligocene, a series-epoch introduced in 1854 but not then widely used), the Wind River deposits of uncertain age (now Eocene), and the "Eocene?" Fort Union or "Great Lignite" Group. This last unit Meek and Hayden divided into upper "true lignite" beds and the lower estuarine strata of the Judith River and Green River basins, all of which represent strata now assigned to the Eocene, Paleocene, or uppermost Cretaceous. Hayden's assumption of two or three widespread, age-equivalent, major units of lacustrine origin made less effective the correlation of these vertebrate-bearing Tertiary strata, which represented different ages and depositional environments, as did the less age-diagnostic, non-marine mollusks they contained.

William Goetzmann termed Meek and Hayden the first geologists trained in the native empiricism that characterized American earth science in most of the nineteenth century to apply that methodology to investigations of the upper Missouri country.[20] Waage demonstrated that Meek, especially, was more than an empiricist. The paleontological pages he contributed to his own publications, to those by Meek and Hayden, and to Dana's *Manual* exhibit an inductive appreciation of the paleoecological value of their fossils and accompanying geologic data in determining depositional environments and in discerning environmentally based, sequential changes in the interior Cretaceous marine faunas.[21]

Although perhaps less successful at generalization, Hayden in the antebellum articles published under his name, but with original input from Meek, outlined the geologic history of the Black Hills and northeastern Rocky Mountains and the initial suggestions for a general lacustrine origin for the Tertiary continental strata of the western interior. The former work correctly dated the interval of major uplift of the Cretaceous strata. These antebellum years influenced all of Hayden's subsequent investigations and account for the often criticized "descriptive geology of the routes" presentation which characterized the narrative portions of his later, larger reports. Hayden's 1862 summary of his investigations of the upper Missouri country prior to 1859, to which Baird, William Greene Binney, Cope, Dewey, Engelmann, Theodore Nicholas Gill, Isaac Lea, Meek, and Newberry all contributed, is representative of the broad coverage of the antebellum, federally funded, military-civilian surveys in which Hayden was partially trained.[22] The 1862 paper is also a microcosm of his publications and those of others in the postbellum U.S. Geological and Geographical Survey of the Territories (First Division), of which he was director, exploration planner, research coordinator, and lobbyist. Hayden's postbellum survey similarly provided extensive field and laboratory opportunities for well established collaborators, as well as splendid educational experiences for many promising graduate trainees.[23]

Notes

This article blends investigations toward separate biographies of Meek and Hayden, begun under the auspices of postdoctoral grants and fellowships from Augustana College, the National Science Foundation, and the John Simon Guggenheim Foundation (to Fryxell), and the Smithsonian Institution (to Nelson). Gene M. Gressley, Director of the University of Wyoming's Western History Research Center, generously made available the resources of the extensive Hayden Collection assembled by Fryxell and J. V. Howell and now at Laramie. We thank Michele L. Aldrich and Donald Zochert for their critical and constructive reading of the manuscript.

1. John Evans, "Incidental Observations on the Missouri River, and on the Mauvaises Terres (Bad Lands)," in David Dale Owen, *Report of A Geological Survey of Wisconsin, Iowa, and Minnesota;*

and Incidentally of A Portion of Nebraska Territory. Made Under Instructions From the United States Treasury Department (Philadelphia, Lippincott, Grambo, 1852), 194–206, 2 text figs., 1 map [Atlas]. Joseph Leidy, "Description of the Remains of Extinct Mammalia and Chelonia from Nebraska Territory, Collected During the Geological Survey Under the Direction of Dr. D. D. Owen," ibid., [533] 539–572, pls. IX–XII, XIIA, XIIB, XIII–XV [Atlas].

2. Hall to Meek, Albany, August 12, 1853, Fielding B. Meek Papers, Record Unit 7062, Smithsonian Institution Archives, Washington, D.C. (hereafter SIA).
3. Hall to Joseph Henry, Albany, April 24, 1853, Joseph Henry Collection, Record Unit 7001, SIA.
4. Meek to Hayden, Albany, March 29, 1854, Hayden Letters Received, Records of the Geological and Geographical Survey of the Territories, Record Group 57, National Archives and Records Service, Washington, D.C. (hereafter NA).
5. Karl Mensch Waage, "Deciphering the Basic Sedimentary Structure of the Cretaceous System in the Western Interior," in William E. G. Caldwell, ed., *The Cretaceous System in the Western Interior of North America*, Geological Association of Canada, Special Paper No. 13 (1975), 59.
6. Hayden to Baird, Ft. Pierre [Nebr. Terr.], April 6, 1855, Spencer F. Baird Papers, Record Unit 7002, SIA.
7. Meek to Hayden, Albany, February 26, 1856, Hayden Letters Received, NA.
8. Ibid., March 30, 1856.
9. Hayden to Meek, Washington, D.C., March 25, 1858, Meek Papers, SIA.
10. Henry, Desk Diary, January 25, 1858, Henry Collection, SIA, and Hayden to Meek, Washington, D.C., December 11, 1856, Meek Papers, SIA. In 1865 the Smithsonian published Meek and Hayden's memoir on the Cambrian through Jurassic invertebrate fossils of the Upper Missouri country. See also Henry, Desk Diary, May 3, 1865, Henry Collection, SIA.
11. Diary, July 17, 1856.
12. James Dwight Dana, "Thoughts on Species," *The American Journal of Science and Arts* (1857), 2nd ser., 24:307–316.
13. Meek and Hayden, "Descriptions of New Lower Silurian, (Primordial), Jurassic, Cretaceous, and Tertiary Fossils, Collected in Nebraska, by the Exploring Expedition Under the Command of Capt. Wm. F. Raynolds, U.S. Top. Engrs.; With Some Remarks

on the Rocks From Which They Were Obtained," *Proceedings of the Academy of Natural Sciences of Philadelphia for 1861* (1862), 13:432.

14. James Dwight Dana, *Manual of Geology. Treating of the Principles of the Science with Special Reference to American Geological History* (4th ed., New York, American Book Co., 1895), 4.
15. Waage, 59.
16. K. M. Waage, "The Type Fox Hills Formation, Cretaceous (Maestrichtian), South Dakota. Part 1. Stratigraphy and Paleoenvironments," Peabody Museum of Natural History, Yale University, *Bulletin* 27 (1968), 20.
17. Ibid., 36.
18. Waage, "Deciphering . . . the Cretaceous System," 72–77, and New Haven, Conn., June 18, 1975, (personal communication).
19. Waage, "The Type Fox Hills Formation," 22.
20. William Harry Goetzmann, *Army Exploration in the American West 1803–1863* (New Haven, Yale University Press, 1959), 422–424, and *Exploration and Empire: The Explorer and the Scientist in the Winning of the American West* (New York, Knopf, 1966), 494, where Goetzmann emphasizes that "the science of geology in Western America" had become "firmly established in local induction" with the end of Hayden's field work with Raynolds in 1860.
21. Waage, "Deciphering . . . the Cretaceous System," 63.
22. Hayden, "On the Geology and Natural History of the Upper Missouri," *Transactions of the American Philosophical Society*, new ser., 12 (1862), 218 pp., 1 map.
23. Goetzmann, *Exploration and Empire*, 527–529; see also 494–495. Goetzmann's evaluation of Hayden's accomplishments and those of his postbellum survey is far more accurate and fair than most earlier appraisals. See also Jesse Victor Howell, "Geology Plus Adventure: The Story of the Hayden Survey," *Journal of the Washington Academy of Sciences* (1959), 49:220–224.

18

Raymond Thomassy and the Practical Geology of Louisiana

Hubert C. Skinner

About 1850 Marie-Joseph Raymond Thomassy (1810–1863), a French hydraulic engineer, came to the United States. During his sojourn he resided for some years in New Orleans, where he became fascinated with the dynamic power of the Mississippi River and intrigued by the physical characteristics of the coastal plain of Louisiana and how it must have formed. His intense interest and scientific curiosity resulted in his *Géologie pratique de la Louisiane*, published in New Orleans in 1860.[1] This pioneer work on Louisiana geology, significant and important, is based on Thomassy's personal observations during four or five field excursions along the shores of the Mississippi River and throughout the lower Mississippi basin. It is the first work to describe the emergent domes of the "Five Islands" as salt intrusions and the first to attempt to explain their origin. Thomassy diagramed and described the "mud lumps" near the mouths of the Mississippi River and attempted to explain them. He presented a series of maps illustrating changes in the Louisiana coastline between 1684 and 1859, especially in the area of the present Balize delta. Another map illustrates the lower basin of the Red River in considerable detail, with an inset diagram of the log jam on the river near Fort St. Jean Baptiste.

Thomassy was proud, arrogant, ambitious, and an ardent French nationalist—factors that combined to produce an unfortunate bias in his work which must be reconciled before his ideas can be reviewed and evaluated. His volatile nature is clearly revealed by an incident during his stay in New Orleans. Thomassy had published a lengthy series of letters on the hydraulics of the Mississippi River, in which he proposed plans for its control or the channeling of its waters. One day while he was expounding to a group of citizens the perfection of his system for river control, a Creole gentleman ventured to remark that the Mississippi was a very headstrong stream and that calculations based on the smaller rivers of Europe might not be applicable to so mighty a river.[2] Thomassy's intemperate reply, accompanied by a gesture of contempt, provoked the Creole to respond with, "Sir, I will never allow the Mississippi to be insulted or disparaged in my presence by an arrogant pretender to knowledge," and forthwith to challenge him to a duel. Thomassy lost the duel but survived the encounter with only a painful and disfiguring facial injury to remind him of the incident. His own account of the result of the duel follows: "I should have killed my adversary if it were not for the miserable character of your American steel. My sword, sir, doubled like lead. Had it been a genuine *colichemarde*, he would have been properly punished for having brutally outraged the sensibilities of a French gentleman." Thomassy continued with a lecture on the carbonization of iron, which could nowhere be effected properly except with wood cut from a certain forest in France.[2]

The Practical Geology of Louisiana

The prologue to the *Practical Geology* presents a lengthy justification for the work, its basis in extensive observation, the broad geographic framework for the area described, and Thomassy's conviction that the entire area is best studied at the mouths of the Mississippi River, where it makes ceaseless encroachments upon the Gulf of Mexico (Figure 1). He suggests that study of the alluvial lands alone will provide the key to the great problems of Louisiana geology, and throw light upon "analogous terrains of anterior periods." In reviewing the content he proposes to cover in subsequent chapters, Thomassy makes a clear uniformitarian statement and alludes to the active agents which formed the mineral springs, artesian wells, mud springs, and mud

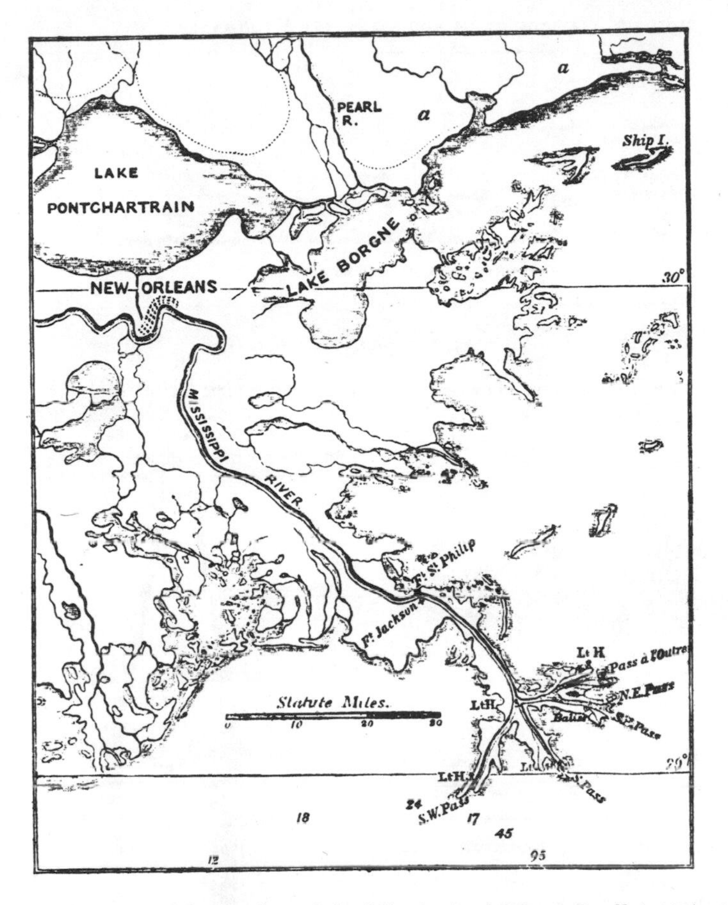

1. The lower delta region of the Mississippi River. Lyell (1872), 1:444.

lumps; he discusses hydrothermal and volcanic forces and the evidence for subterranean convulsions in forming the "Five Islands" of southern Louisiana. He asserts the need for a map of the country depicting soils and agriculture and the distribution of the anterior formations, marshlands, and ancient marine alluviums. The foreword of the prologue ends with a list of native raw materials—plaster, kaolin, plastic clays, limes, and hydraulic mortars—the "treasures" that should be exploited for the good of the people. He dedicates the work to that great majority of people who have a lively interest in science.

The lands of southern Louisiana were formed by the alluvial deposits of the Mississippi River alternating with salt water marine sands and muds, volcanic forces which held back alluviums within the original

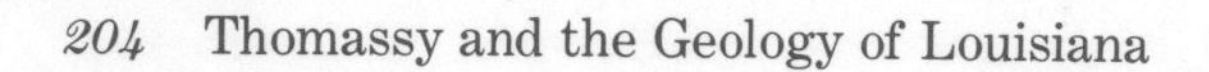

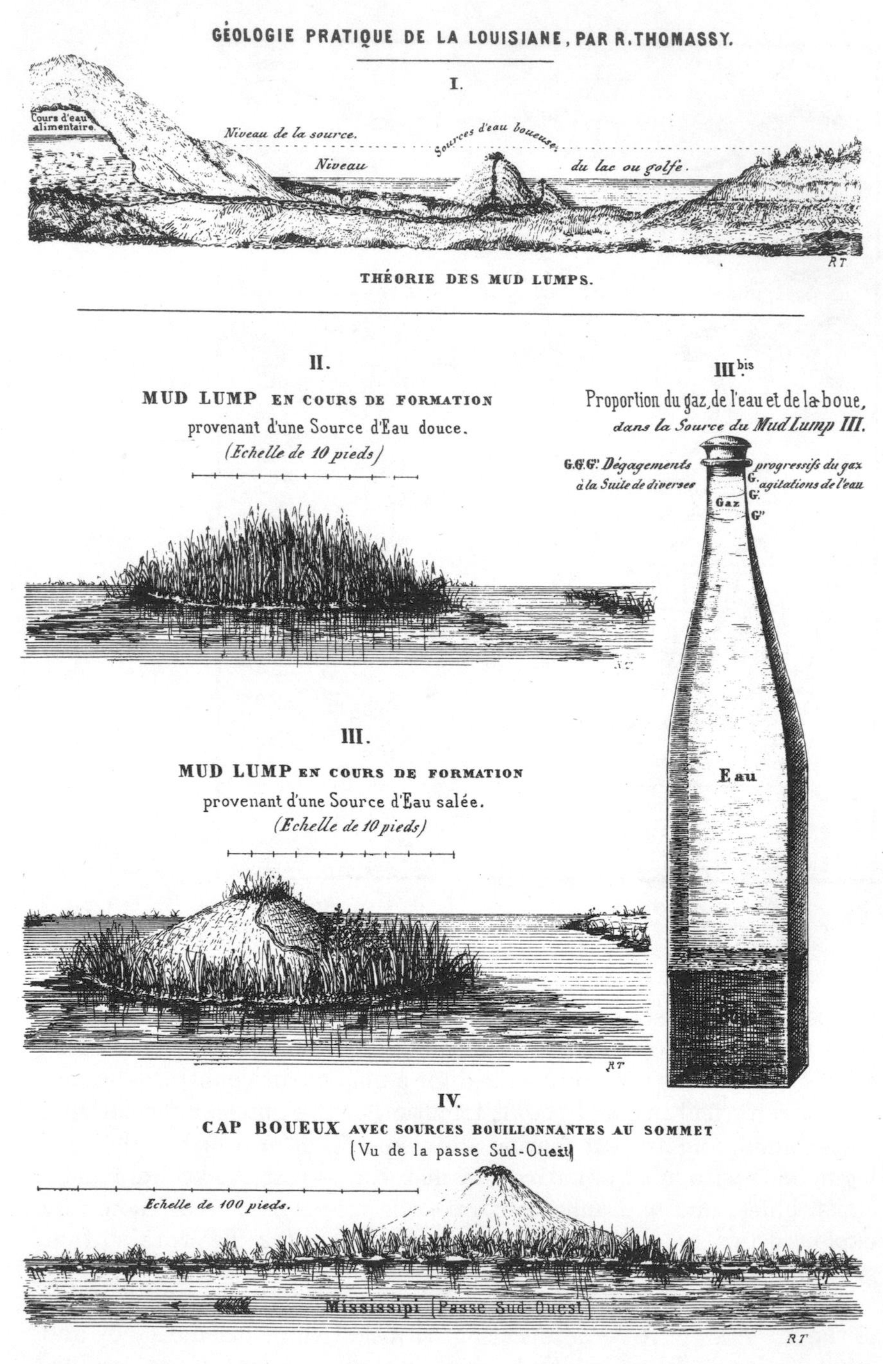

GÉOLOGIE PRATIQUE DE LA LOUISIANE, PAR R.THOMASSY.
I.
Cours d'eau alimentaire.
Niveau de la source.
Sources d'eau boueuse.
Niveau
du lac ou golfe.
RT
THÉORIE DES MUD LUMPS.
II.
MUD LUMP EN COURS DE FORMATION
provenant d'une Source d'Eau douce.
(Echelle de 10 pieds)
III.
MUD LUMP EN COURS DE FORMATION
provenant d'une Source d'Eau salée.
(Echelle de 10 pieds)
RT
III.bis
Proportion du gaz, de l'eau et de la boue,
dans la Source du Mud Lump III.
G.G'.G". Dégagements progressifs du gaz
à la Suite de diverses agitations de l'eau
G
G'
Gaz
G"
Eau
IV.
CAP BOUEUX AVEC SOURCES BOUILLONNANTES AU SOMMET
(Vu de la passe Sud-Ouest)
Echelle de 100 pieds.
Mississipi (Passe Sud-Ouest)
RT

2a.

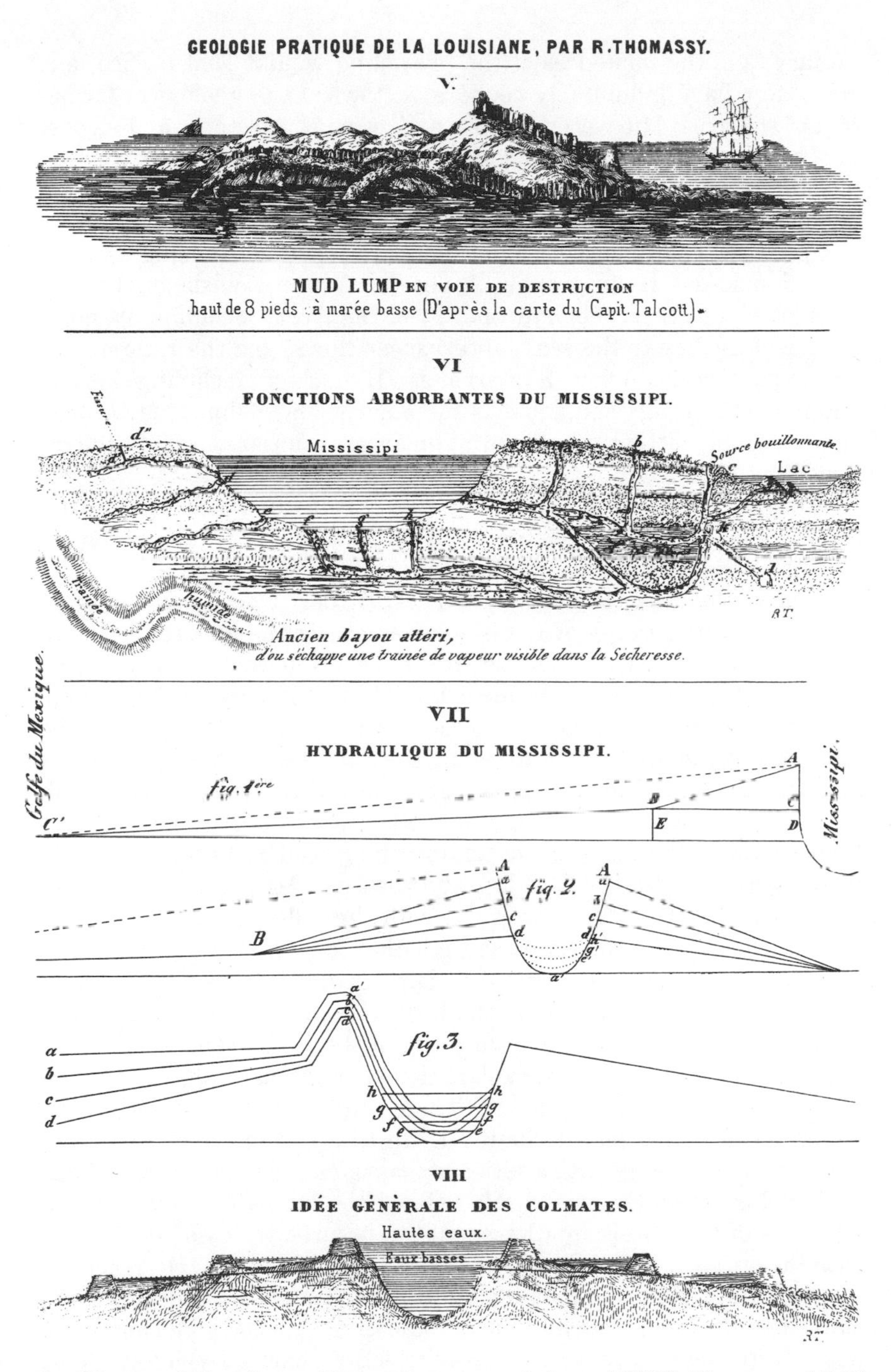

2b. *Theory of Mud Lump Formation, stages I through VIII. Thomassy (1860), plate IV.*

estuary, and the influences of the prevailing winds. That is, fire, air, and water havé indubitably cooperated toward the geological formation of the area, but water is most active, most constant, and easiest to observe, and will have to be given account above all others. He emphasizes that the lower delta is the "key to Louisiana geology" and delineates five separate source areas for the sediments of the delta. The prologue continues with eight numbered sections, dealing with the river-dominated Hydrographic Aspect of Lower Louisiana; the Hydrology of the Area; the Hydrometry of the Area, including evaporation, surface flow to the sea, subterranean flows, and the ratio of surface to subterranean flow; Subterranean Hydrology, including thermal and mineral springs and aqueous metamorphism; Submarine Hydrology, including littoral lines and submarine eruptions; and Problems Solved by Hydrology.

The actual text of the *Practical Geology* is divided into three parts, the first consisting of nine numbered sections on geography and geology; the second, of eight, on practical geology and engineering problems; and, the third, of nine appendices. In Part One, Thomassy considers the Cartography of Louisiana, reviewing the geographic maps left from earlier days as documents illustrating the changes in the coastline of Louisiana and the measure of time taken by the Mississippi to create its gigantic delta. He recounts in detail the history of the discovery of the Mississippi River by the Sieur de la Salle, including a reproduction of the map of 1684 (plate I), and continues with subsequent explorations by Iberville and others. The fifth section of this part considers the functions of an absorbing well and its relationship to subterranean flow, sink holes, and the New Madrid earthquake of 1811, and compares the Mississippi with the Nile and the Euphrates rivers, whose main point of measure is their deltas.

The sixth section deals with Mud Springs and Islands and the influence of subterranean waters in the formation of the Mississippi Delta. This is one of the most significant parts of the work, representing the first attempt at a geological explanation of the "mud lumps" near the mouths of the Mississippi River. These mud and mineral springs he described in detail and attributed them to a sort of volcanic action, illustrating his theory with a series of diagrams showing the role of gas in their formation (Figure 2). He related the lumps to the absorbing functions of the Mississippi River and to fissures which he correlated with the springs, and their muddy monticules or *tumuli*. He recorded the most important fact that the muds of the lumps are different from alluvial muds, being "stickier" and possessing "special adhering qualities." Here and elsewhere Thomassy took Sir Charles Lyell to task for

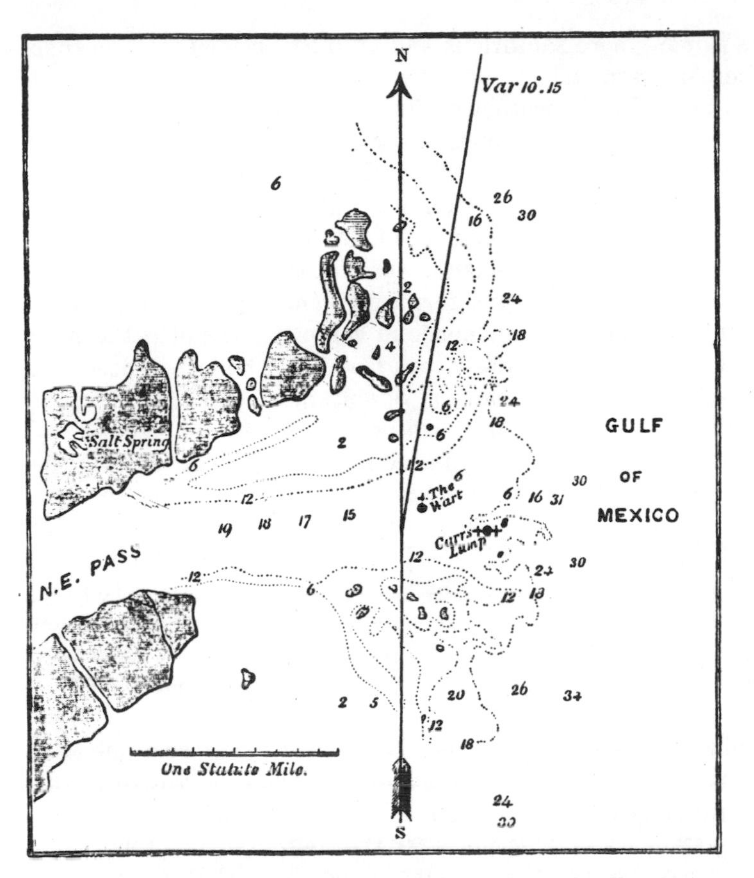

3. Location of Mud Lumps at the Northeast Pass. Lyell (1872), 1:445.

his "silence" about the mud lumps; "Through a mistake, or an oversight equally regrettable, that celebrated geologist did not bring out the part played by *mud springs* and *mud lumps*, springs and mounds of mud, which emerge from the waves of the Gulf of Mexico, sometime even in the middle of the passes of the river, revealing the presence of a new active agent in the progressive formation of the delta." At least four times, Thomassy criticizes grave errors of omission by Lyell, "all the more regrettable on the part of so eminent a mind." It should be recorded that the subject of mud lumps in the passes of the Mississippi River is treated fully in later editions of Lyell's *Principles of Geology* (Figure 3), though these had not appeared by 1860.[3] The seventh section of this part describes the "refrigerating function" of the Mississippi River in affecting the temperatures in southern Louisiana and making its climate more salubrious and with less risk of malaria than that of other southern states.

The eighth section, Hydrothermal and Volcanic Forces in the Formation of Lower-Louisiana, is another significant part. Here Thomassy introduces his ideas of littoral lines: upraised cones or domes separating the alluvial lands from the open sea and serving to stabilize the coastline. He uses the "Five Islands" of southern Louisiana, a series of emergent salt domes, as his example of such features.[4] He describes them as raised domes, of distinct formation, and aligned, considering them another volcanic phenomenon and compares them to the mud lumps at the mouths of the Mississippi River. He believed these salt intrusions to be formed by some sort of volcanic action, concentrated from the saline waters through hydrothermal forces and elevated by aqueous eruption. He described exposed inclined gravel beds atop the domes, salt springs, calcareous, ferrous oxide, and sulphurous springs, and suggested that the salt masses could be exploited as a source of salt. In 1861–62 the conflict between the states resulted in the sinking of pits or shafts at Avery Island (then called Île Petite Anse) to mine salt for use in the local area. Upon receiving word of the confirmation of his salt intrusion theory in the origin of these domes, Thomassy returned to Louisiana from France to view the salt pits and to write a supplement to his earlier work.[5]

Section nine, Of the Formation and the Present Progress of the Mississippi, contains Thomassy's estimates of the rate of growth of the Mississippi River delta. He notes a most significant salient characteristic, the deviation of its lower course to the southeast, and reports on marine shells recovered from sediments under New Orleans and subsequently identified by Deshayes, the famed French paleontologist. He describes the delta as due to a mixture of fluvial and marine origins, and gauges its advance at 100 meters per year or one mile in sixteen years. He presents a series of maps (plates II, III, and VI) illustrating changes in the lower delta. Again, he attacks the conclusions of Lyell, this time concentrating on Lyell's estimates of the age of the delta and his comments made about changes in the configuration of the delta.

Part Two of the *Practical Geology* consists of eight numbered sections on practical or applied geology. The first section, Geology of Louisiana in respect to her Hydrography, extolls the virtues of the area for future growth and development, its salubrious climate and temperatures compared to other southern areas, and comments on the economy of water transportation as the first agent of riches, and calls for a complete hydrographic report on Louisiana and for the reclamation of at least ten million acres from the marshlands. In the second section, he discusses the geology of Louisiana in relation to her drainages, marine-alluvial origin, littoral lines, and reclamation procedures. Twelve letters that were published in the local press follow. They are proposals for river control structures and dams and projects designed to improve the sanitary conditions and to reclaim lands through natural hydraulics, and levee and canal construction. There is much repetition and argumentative entreaty in these letters. Some deal with the costs of projects proposed by others as well as those by Thomassy. The next six sections of this part deal with proposals for building dams on Bayou Plaquemine and Bayou Lafourche, additional problems of river hydraulics, saline deposits, native salt production, and imported salt. Some include extensive case histories from other regions for comparison with engineering problems and proposals in Louisiana.

The nine appendices include lists of the manuscripts and excerpts from de la Salle, lists of formerly published maps of the Louisiana area, a description of the lower basin of the Red River, and additional engineering data and case histories.

Conclusion

This pioneer work on Louisiana geology is significant and important to American geology. It is the first work to describe the emergent domes of the "Five Islands" as salt intrusions and the first to attempt to explain their origin. Thomassy diagramed and described the "mud lumps" near the mouths of the Mississippi River and attempted to explain them. He presented a series of maps illustrating changes in the Louisiana coastline between 1684 and 1859, especially in the area of the present Balize delta. He computed the rate of advance of the delta at 100 meters per year. Another map illustrates the lower basin of the Red River in detail with an inset diagram of the log jam on the river near Fort St. Jean Baptiste (Figure 4). In general, citations of Thomassy's work in the subsequent geological literature have treated it with less respect than it deserves.[6–11] It comprises the first comprehensive effort to describe the geology of southern Louisiana, a monumental project at this early date.

Notes

1. Raymond Thomassy, *Géologie pratique de la Louisiane* (New Orleans, Louisiana, the author, 1860), lxvii + 264 pp., 6 folding plates.

This work, a quarto, is preceded by a long prologue (of 68 pages) and is accompanied by six folding plates (hand-colored in a very few copies). A second edition [?] is reported to exist, issued in Paris, 1861. Some copies show a paste-up change in the publisher's name and may comprise a second issue of the work.

2. Stuart O. Landry, *Duelling in Old New Orleans* (New Orleans, Louisiana, Harmanson, 1950), 36–37.
3. Hubert C. Skinner, "Charles Lyell in Louisiana," *Tulane Studies in Geology and Paleontology* (1976), 12:243–248.
4. Hubert C. Skinner, "A Comparison of the Mississippi Submarine Trench with the Iberian Trough," *Gulf Coast Association of Geological Societies, Transactions* (1960), 10:1–6.
5. Raymond Thomassy, "Supplément à la géologie pratique de la Louisiane," *Société Géologique de France, Bulletin* (1863), 2nd ser., 8:542–544.
6. Everette L. DeGolyer, "Origin of North American Salt Domes," in *Geology of Salt Dome Oil Fields*, American Association of Petroleum Geologists, Symposium Volume (Tulsa, Oklahoma, 1926), 1–44.
7. Gilbert D. Harris and A. C. Veatch, *A Preliminary Report on the Geology of Louisiana*, Louisiana Geological Survey (Baton Rouge, Louisiana, Louisiana State University, 1899), 21 ff.
8. George P. Merrill, *Contributions to the History of American Geology*, United States National Museum, Report for 1904 (Washington, 1906), 504.
9. Eugene Wesley Shaw, *The Mud Lumps at the Mouths of the Mississippi*, United States Geological Survey, Professional Paper 85-B (Washington, 1913), 11–27.
10. Francis E. Vaughan, "The Five Islands, Louisiana," in *Geology of Salt Dome Oil Fields* (above), 356–397, 12 figs.
11. A. C. Veatch, "The Five Islands," in *A Preliminary Report on the Geology of Louisiana* (above), 214–215.

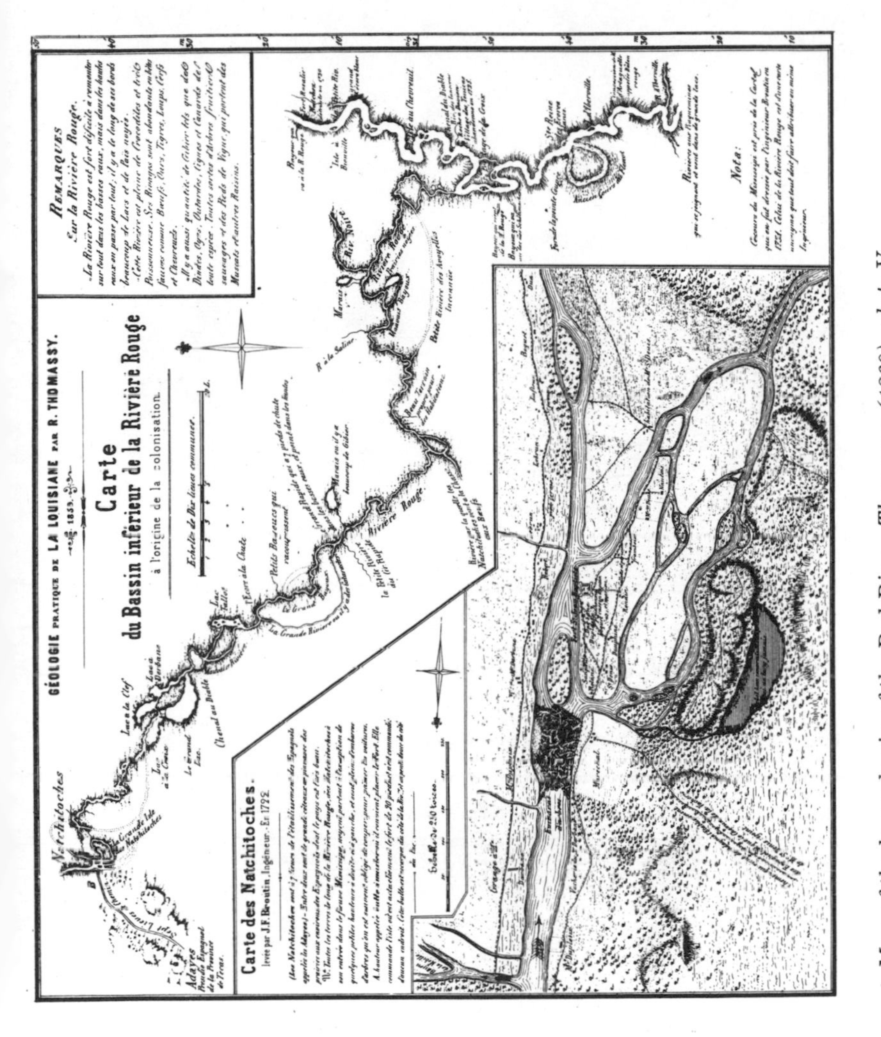

4. Map of the lower basin of the Red River. Thomassy (1860), plate V.

The Work of Edwin Theodore Dumble on the East Texas Lignite Deposits (1888–1892)

Nancy Alexander

Within East Texas lies an enormous storehouse of energy in the form of lignite. Though reported by explorers and geologists and mined locally as early as 1819,[1] not until Edwin Theodore Dumble's concentrated studies between 1888 and 1892 was the potential value of Texas lignite disclosed.

Prior to these studies Dumble was an amateur geologist whose investigations were conducted during his spare time at his own expense. He depended for support upon his talents at bookkeeping, accounting, and management. For ten years he was a "lamp drummer"—that is, a buyer of china and glassware—and manager of his father's business in Houston.[2]

The year 1888 marked a turning point in Dumble's life as well as in the history of geological investigations in Texas. The third Texas Geological Survey was established and Dumble was appointed State Geologist.

While administering his duties, Dumble researched the lignite deposits in East Texas. The work was a continuation of studies he had begun during the 1870's when he had drawn up preliminary maps of the deposits and had conducted laboratory experiments on them.[3] In his judgment these coal beds constituted a virtually inexhaustible supply of fuel. Crossed by major railroads, they were readily accessible for exploitation. In 1890 he affirmed that while European lignites were inferior to those in Texas, they were nevertheless used "for every purpose for which bituminous coal is available . . ."[4] Dumble concluded that "there can remain no shadow of doubt of the adaptability of the great lignite fields of Texas, and other parts of America as well, to meet the wants of the people for cheap fuel." In 1890 he forecast that the

> ease and cheapness of mining, the small cost of preparation, and its value when prepared, will enable it to compete with wood in the best wooded portions of the State, with coal in close proximity to the coal mine, and it will prove of inestimable value in those localities in which it is the only fuel.[5]

Dumble's field observations and laboratory experiments were not enough; it was essential that he gather first-hand information about an active lignite industry. "As this character of fuel has not formed the subject of such detailed investigation in this country as it has in Europe," he wrote, "any description of the coal or its utilization must draw largely from foreign sources, as it is only there that machinery and appliances can be found especially designed for its use, the adaptability of which have been proved by actual trial."[6]

In 1891 Dumble acquired a legislative appropriation of $3000 to visit Germany and Austria.[7] This expenditure provided one of several weapons later used to attack him during one of the more vituperative epochs in the history of Texas geology. Around September 1890 a methodical plot was being hatched to force Dumble's removal from office. Formal charges, dated April 2, 1891, were submitted to the authorities by J. H. Herndon, Survey Chemist and the official accuser. Five specific charges and additional subcharges boiled down to incompetency, plagiarism, and maladministration in office.[8]

The State Geologist had produced nothing of *economic* importance, or so it was claimed by scientists on his staff. Not the least of the charges involved his work on brown coal: "His articles on lignite coal as a fuel, which caused an appropriation of $3,000 . . . are misleading and reckless"; he was merely promoting investments and extravagant appropriations.[9] One of the assistant geologists on the survey was "thoroughly convinced that he has mistaken the shadow for the substance in making all his fuss about east Texas lignites."[10] The ironic twist here was that pressures for instant economic results of geological work and accusations for failure to so produce originated not from without by the politicians, as had been the usual case, but from within the Survey by scientists under Dumble's own direction.

Rumors of the trouble reached Dumble on April 27, and he realized its gravity on May 7 upon receipt of the court order. He was informed by the Commissioner of Agriculture, Insurance, Statistics, and History, Hon. J. E. Hollingsworth, that an official investigation and hearing would be held at 4 P.M. on May 8. Dumble had less than twenty-four hours to prepare his defense.[11] He sensed that his chemist was not the sole accuser. The charges, he wrote, "should have had still another signature—that of Robert T. Hill . . ."[12]

Robert Hill was nationally recognized as the leading specialist in Texas geology. He was also known to be personally sensitive, irascible, and hot tempered. Formerly he had been associated with the University of Texas as a professor but had resigned at the end of three semesters because of a disagreement with the Acting Chairman of the faculty. He joined the Texas Survey in February 1890, and resigned in August because of a disagreement with Dumble. No more than a month after his resignation, the plot against Dumble was seeded. Hill, at that time with the United States Geological Survey, would serve as producer of the drama.

The leading villain on stage, J. H. Herndon, was another seemingly excitable, sensitive, confident, and determined young man. Not coincidentally, he was an ardent admirer and ex-student of Robert Hill at the University of Texas.

Dumble's reaction bespoke his self-assurance. In a written statement to Commissioner Hollingsworth on May 8, he declared:

> Mr. Herndon wishes a full investigation. This is my only point of agreement with him. I not only desire a full investigation but request that it be made in every part strictly under oath, and that a full and complete record be kept of the proceedings.
>
> In conclusion I desire to state, that the whole charges are the outgrowth of malice, and taken together with the proof submitted, which consists principally of letters written to Mr. Hill, show that gentleman . . . is in this case "reluctantly" using Mr. Herndon, as a cat's paw.
>
> Let justice be done both.[13]

The first witness called for testimony was the Survey's prestigious mining and civil engineer, Professor W. H. von Streeruwitz. Hollingsworth questioned him: "Did any one ask you to testify?" Streeruwitz answered:

> Mr. Herndon asked me some days ago if I would testify. I answered that I would answer any questions put by any one who was authorized to ask questions, but if they were asked by parties not authorized to ask them I would kick them all over the Capitol grounds.[14]

Hollingsworth discovered that not only Herndon but Hill had approached Streeruwitz. He then wished to know if either had offered him "any inducement" for testimony against Dumble. "I could read between the lines that if Mr. Dumble was removed that I would be removed. I said that I would not be bought by promises nor frightened by threats."[15] The hearing was quickly in full and surprising swing against Herndon, who had entered the courtroom confident of victory.

L. E. Magnenat, another chemist for the Survey whose associate in the organization was Herndon, was called to the stand. To Herndon's shock, Magnenat revealed that Herndon had worked on the charges for six months, and when asked if "Mr. Herndon read you the questions?" he answered:

> Yes. He showed them to me and told me how to answer. He told me the duty of the State Geologist. He read "Do you regard Prof. Dumble qualified to be State Geologist" and told me to answer "No."[16]

He had been subjected to other pressures. It seems that Herndon was toting a gun around his potential witnesses. "Mr. Herndon . . . you told me that you would shoot the stuff out of me if I did not testify as you wanted me to." Herndon asked, "Was that said in a joking, laughing tone?" to which his associate answered, ". . . there is laughing and laughing and as I seen you carrying a pistol and heard you prefer threats against other members of the survey, I did not take it exactly as a joke."[17]

William Kennedy, Geologist for East Texas, testified that after he had informed Herndon that he "would not be made the cat's paw of either John Henry Herndon or Robert T. Hill . . . he has been threatening to shoot me." Kennedy turned on Herndon and questioned his competence as chemist. Herndon objected to the "irrelevant" matters . . . I have turned over the witness and he has become the attorney."[18]

E. S. Ellsworth, Assistant Geologist, testified that he had been badgered with bribes: "He said . . . if I were a friend of his, if I would come here and swear to what he wanted me to, that if I could give my evidence as he wanted me to, that I could have a position on the re-organized Survey at $1200.00 a year."[19] Ellsworth related the story of his involvement with Herndon. At its completion, Herndon was excited:

> That was a pretty story you told, how long did it take you to learn it? Do you know that it took you an hour and five minutes to tell that pretty story? How long did it take you to write it out? Did you write it down?[20]

Things were not going well for Herndon.

Herndon's choice witness, Dr. Otto Lerch, distinguished geologist from Germany, who was estranged from Dumble for personal reasons, expressed his opinion that Dumble was a plagiarist and incompetent "to fill the office . . ." The removal of Dumble, he contended, "would be an advantage to Texas . . ." Yet when examined by Dumble on the question of the East Texas lignites, Lerch, who had himself published papers on the subject and had observed the industry in his homeland, could answer no less than professional truth. When asked if he evaluated Dumble's lignite studies in Germany as "misleading and reckless," and if he felt that the Texas deposits were of value, Lerch answered, "From the material which has been submitted to me, I think they are far superior to the German lignites." Furthermore, he stated that lignites had been in practical use in Germany for years.[21] Streeruwitz fully substantiated Lerch's contentions on the value of German versus Texas lignite deposits.

Justice reigned those days of May 8 and 9, 1891, in the Texas courtroom. Dumble was vindicated of all charges and retained his position as State Geologist. Herndon, in humiliation, anger, and wounded pride, protested that Dumble had "not been placed upon the stand as one whom charges had been brought against, but I seem to have been placed upon the stand and my ability brought into question."[22]

The culprits did now allow the matter to rest for some five months after the trial. The hearing, Herndon publicized, had been a "sham trial," a "mock trial," a "conspiracy to whitewash" the State Geologist.[23] The news media savored the lingering scandal. One newspaper offered "a standing reward for the man who will show that [Dumble] or his office ever benefited the State ten cents."[24] In Austin one newspaper accused Dumble of fighting facts "in the face of the testimony of the most experienced and best scientists in this country" concerning lignite.[25] Herndon continued to assert:

> TO MEN OF SCIENCE AND LEARNING the whole matter becomes ridiculous . . . there are able men in Austin today who have assisted in making experiments on lignite coal as fuel, and who are able to show how fallacious those claims of Dumble's are.[26]

Another newspaper reminded the public of Dumble's past business status when it wrote: "These opinions are strengthened when we remember that Mr. Dumble was a lamp drummer prior to his appointment as State Geologist and that he was utterly and entirely unknown in scientific circles of the smallest radius."[27]

One version of Dumble's administration was even depicted in verse (anonymously) and printed in the news:

Preparatory Course in Geology on Texas Survey

> First be an crockery drummer, a parson, or his son,
> Or a worn-out politician; get a banner or a gun.
> Find a big appropriation (don't forget your fishing line);
> Talk wise about creation, and find a lignite mine.
>
> Be only called Professor, be sure you don't forget.
> Even when you are transgressor, you must be very set.
> Never hear the last suggestion, even though you fall,
> That there be any question that you do not know at all.
>
> Push into the papers, if you find a little tin,
> Cut all parts of capers (that will make a din).
> Disparage all your rivals, call the other man a liar,
> Attend the church revivals, exhort about Hell-fire.
>
> Read the old "transactus," "the proceedings," too
> Make numerous extractions, pass 'em off for new.
> Get out a publication, never print a fact,
> Avoid all demonstrations, you need not be exact.
>
> Quarrel with your assistants, then withhold their pay,
> And when they show resistance, don't forget to pray.
> Never be veracious, but slowly, meek, and humble,
> And, "by gatlins, you may yet become a (Dumble)."
>
> Geologist[28]

The charges, recapitulation of the trial, newspaper critiques, and derogatory letters were reviewed by Herndon in a privately printed pamphlet which was widely distributed.[29] The continued provocation to the public taxpayer resulted in a general feeling that the Survey should be abolished. In time, however, the scandal subsided and the Survey under the direction of E. T. Dumble continued in efficient operation.

The following year, 1892, the climax of Dumble's findings on lignites was published by the Geological Survey of Texas, *Report on the Brown Coal and Lignite of Texas. Character, Formation, and Fuel Uses.* As complete a coverage of the subject as was then possible was presented. Included in the 243-page book were sections on the history of the fuel use in Texas, its origin, its physical and chemical properties, distribution stratigraphically and geographically by counties with associated maps, methods and economics of utilization, and descriptions and drawings of appropriate equipment of utilization of the substance; and comparisons between the Texas brown coals and those in Austria, Bohemia, Hungary, Italy, and Germany.

Although brown coal was not "the equal of the best raw bituminous coal as a fuel, weight for weight," Dumble was confident that it would serve in the same capacities as more superior coals in Texas for many purposes. He proved its value, but he was aware that "its progress into popular favor" would probably be slow inasmuch as people are innately conservative and normally "continue to use of that with which they are entirely familiar, unless decided advantages are to be obtained by making a change."[30]

European governmental reports and figures reflected progressive annual increases of up to 150 percent in their use of lignite. The fuel was not only being used in the vicinity of the mines, but the Bohemian coal was "freighted equal distances with the bituminous coal . . . and meets it in open competition on a basis of actual value as fuel."[31]

By now, 1892, Dumble had estimated that the great East Texas lignite belt stretched 650 miles in length, with a maximum width of 200 miles, extending through 84 counties and covering an area of some 60,000 square miles. Available fuel, he emphasized, "has a most important bearing on the development of a country":

> During the earlier stages of its history wood, when sufficiently abundant, answers for fuel, but with increased population and consequent demand for manufactures comes the necessity for a better combustible, such as is found among the fossil fuels . . .[32]

Hollingsworth wrote of Dumble's work:

> The result at last achieved has fully repaid him for all work done. Texas will reap the grand benefits, for the capitalists will come and establish industries, the wants of the people will be supplied at home, and prosperity will abound.

> That the brown coal is the cheap fuel so long needed is no longer a question, for which all Texas should rejoice.[33]

Lignite production gradually increased, reaching two peaks in activity prior to present-day industry, one between 1914 and 1920, the other just prior to 1930.[34] These peaks were interrupted by lags in production and the closing of mines, which may be attributed to competition resulting from the discovery of oil at Spindletop in Beaumont and the East Texas oil field respectively. Dumble's faith in the future of brown coal never diminished, and in 1918 he forecast that in time, lignite fuel would be "one of the principal source of supply for the entire Gulf Coast region."[35]

In 1974 W. R. Kaiser of the Bureau of Economic Geology, Texas, reported lignite production figures for East Texas alone to be 8 to 10 million short tons per year; reserves within 90 feet of the surface were estimated at 3.3 billion tons, at less than 200 feet at 10.4 billion, in addition to far more substantial deep basin deposits (200–5,000 feet)—a total of over 100 billion tons and equivalent to 277 billion barrels of oil. It was projected at this time that by 1980 over 25 million tons a year would probably be in production.[36] Today a lignite boom has struck the state comparable to the oil rush propelled by Spindletop in 1901. Production figures will surely exceed former expectations.

In 1924 Dumble turned to the past and remembered his early dreams:

> The courses of my last two years at College were directed by the interest awakened in me through a slight knowledge of our lignite and iron deposits. I had undoubting faith in the possibility of the use of geological knowledge and investigation for the development of those and other resources, and to me, as I considered and pondered over it, it seemed the opportunity of a lifetime to demonstrate this . . . for our great State of Texas and prove . . . the practical value of scientific geology. This was my dream.[37]

Acknowledgments

The writer acknowledges appreciation to Stephen F. Austin State University, for support of the writing of this paper through a Faculty Research Grant. Dr. Raymond D. Steinhoff, Chairman, Geology Department, Stephen F. Austin State University, and Mr. James G. Stephens, head librarian, Science & Engineering Library, Southern Methodist University, provided facilities for research at the respective universities. Mrs. Debby Petty, secretary to Dr. Raymond Steinhoff, typed the final manuscript. A special note of gratitude is extended to Dr. Claude C. Albritton, Jr., Hamilton Professor of Geology, Southern Methodist University, who not only edited the manuscript but was also adviser to the writer during its preparation.

Notes

1. L. F. L'Heritier, *Le Champ–D'Asile tableau topographique et historique du Texas, etc.*, 2nd ed. (Paris, 1819), map of establishments.
2. E. T. Dumble, *A Dream of Fifty Years Ago* (privately printed, 1924), in James R. Underwood, Jr., "Edwin Theodore Dumble," *Southwestern Historical Quarterly*, 68 (1965), 54, 55.
3. ETD, *Report on the Brown Coal and Lignite of Texas. Character, Formation, Occurence, and Fuel Uses* (Austin, Geological Survey of Texas, 1892), 21–22.
4. ETD, "Report of the State Geologist," *Second Annual Report of the Geological Survey of Texas, 1890* (Austin, 1891), xxxviii.
5. Ibid., xl.
6. ETD, "Letter of Transmittal to Hon. J. E. Hollingsworth, Commissioner of Agriculture, Insurance, Statistics, and History," in ETD, *Report on the Brown Coal and Lignite of Texas*, 7.
7. ETD, "Financial Statement, Appropriation for Geological Survey of Texas, Jan. 1, 1891 to Dec. 31, 1891," *Third Annual Report of the Geological Survey of Texas, 1891* (Austin, 1892), vi.
8. "Charges Preferred against Prof. E. T. Dumble, State Geologist of Texas," Records of Dumble Hearing, Austin, Texas, May 8–9, 1891, Dumble Papers, Texas State Library, Austin.
9. Ibid., charges 1, 5.
10. Wilson T. Davidson to J. H. Herndon, April 2, 1891, reprinted in J. H. Herndon, *Plea for the Life of the Geological and Mineralogical Survey of Texas and Review of the Charges Preferred against Prof. E. T. Dumble, State Geologist, for Incompetency, Plagiarism, and Maladministration in Office, and the Sham Trial Thereof* (Austin: privately printed, 1891), Hill Collection, Southern Methodist University, 35.
11. ETD, "Mr. Herndon's Charges, The Investigation of Them and the Result Adduced therefrom. To the People of Texas," (typescript, 1891), pp. 2–3, Dumble Papers.
12. ETD to Hon. Jno. E. Hollingsworth, May 8, 1891, Dumble Papers.

13. Ibid.
14. "Testimony Taken on Charges Preferred by J. H. Herndon against Prof. E. T. Dumble, State Geologist," Records of Dumble Hearing, Austin, May 8–9, 1891, p. 1, Dumble Papers.
15. Ibid., 5.
16. Ibid., 7b.
17. Ibid., 20, 21.
18. Ibid., 15.
19. Ibid., 8.
20. Ibid., 10.
21. Ibid., 20, 18.
22. Ibid., 34.
23. Herndon, *Plea for the Life*, 3, 4.
24. *Weatherford Republic*, reprinted in *Plea for the Life*, 10.
25. *Capitolian* (Austin), June 19, 1891, reprinted in *Plea for the Life*, 10.
26. *Plea for the Life*, 25.
27. *Jewett Messenger*, May 1891, reprinted in *Plea for the Life*, 10.
28. Reprinted in *Plea for the Life*, 13, 14. Most phases refer to Dumble. The references to the parson, creation, and church indicate W. F. Cummins, one of Dumble's loyal defenders. The worn-out politician is probably directed toward Gov. James S. Hogg, who refused to investigate the case personally, and over Herndon's objections referred the case to a newly appointed Commissioner, J. E. Hollingsworth, whom Herndon felt was prejudiced in favor of Dumble.
29. *Plea for the Life*.
30. ETD, Report on the Brown Coal and Lignite of Texas, 206.
31. Ibid., 209.
32. Ibid., 17.
33. Jno. E. Hollingsworth, "Letter of Transmittal to James S. Hogg, Governor of Texas," in ETD, *Report of the Brown Coal and Lignite of Texas*, 5.
34. H. B. Stenzel, "Review of Coal Production in Texas," *Texas Mineral Resources*, University of Texas Publication 4301 (Austin, 1943), 205.
35. ETD, *The Geology of East Texas*, University of Texas Bulletin 1869 (Austin, 1918), 276.
36. W. R. Kaiser, *Texas Lignite: Near Surface and Deep-Basin Resources*, Bureau of Economic Geology Report of Investigation, 79 (Austin, 1974), 1, 3, 29.
37. ETD, *A Dream of Fifty Years Ago*, 55.

VII

Geology Comes of Age: The American Themes, I

Certain Allied Problems in Mechanics: Grove Karl Gilbert at the Henry Mountains

Stephen J. Pyne

I

Grove Karl Gilbert was one of the grand figures from the heroic age of American geology. He is eminent as an explorer (especially as a member of the Powell and Wheeler surveys), as a scientific administrator with the U.S. Geological Survey, and as the author of such monographs as *Geology of the Henry Mountains*, *Lake Bonneville*, *Hydraulic Mining Debris in the Sierra Nevada*, and *The Transportation of Debris by Running Water*. In brief, Gilbert was a durable, all-purpose geological thinker—an experimentalist, theoretician, and methodological critic. Born in 1843, ten years after Charles Lyell published the final volume of *Principles of Geology*, Gilbert died in 1918, ten years before the published proceedings of the first symposium on continental drift. The seventy-five years of his life witnessed the institutional and intellectual consolidation of the classical period of American geology.[1]

That Gilbert was an outstanding participant in this era is incontestable. Less apparent is that in method and perception he was curiously at odds with the assumptions, metaphors, and techniques which guided the scientific inquiry of the earth as conducted by his contemporaries. Of all his works, the *Henry Mountains* best illuminates the traits which made Gilbert at once representative and unique. The *Henry Mountains* reveals a style of thinking as much as a logical system of concepts and evidence. The first of his major publications, it also established a stylistic and conceptual pattern for the rest.

Gilbert's science can be understood as much from his personality as from his philosophy. At heart Gilbert was a classicist, ever searching for proportion and balance, for old patterns among new landscapes. As a scientist, he was an engineer as much as a geophysicist. Reportedly his last major undertaking while on his deathbed was to struggle through his financial accounts, trying to make the books balance. He did. And so he did with nature. Like Gilbert nature would never act to excess; nature would always be expected to balance her books, especially her energy expenditures. Nature seemed to exist in an unending present: Gilbert was as uninterested in the evolutionary future of nature, or of society, as he was in its glamorous past. Unlike many romantic contemporaries, Gilbert did not see in earth history the melancholy wreckage of ancient geologic empires. Instead he dealt almost exclusively with contemporary events, rarely venturing farther back than the Pleistocene. On the shores of Lake Bonneville, for example, it was not the decaying legacy of a vanished lake that fascinated him, but the gleam of the pebbles on the shorebars, so fresh that one could still imagine the slap and dash of waves upon them.

Grove Karl Gilbert was a pure Newtonian, who found in nature an equilibrium of forces and resistances, whose meaning and relationships he sought to interpret by way of bold mechanical analogies. Nature for him was a machine, not an organism. The reasons for this preference were probably based as much in his psychology and education as in his philosophy of nature. His education at the University of Rochester had followed a traditional curriculum. It left Gilbert, on the one hand, with an indelible bias of classicism which saturated equally his temperament, prose, and science; and on the other, it meant the science he absorbed belonged more to natural philosophy than to natural history. He had only one course in geology, but many in mathematics, physics, and astronomy. In particular, he seems to have derived a great deal of his science from William Rankine's popular *Manual of Applied Mechanics*, a systematic attempt to merge physics and engineering. This impression is strengthened by a remark he made in later years: that, upon graduating from college, he had found engineering more interesting than geology. The great insight of his scientific career was to bring the techniques of mechanics to the subject matter of geology, rather as Rankine had done with physics and engineering. In examining geologic events he tended to analyze their processes as Rankine might have studied the composition of forces in a mechanical or structural system.

These preferences seem equally grounded in Gilbert's temperament. Contemporaries spoke often in praise of Gilbert's exceptional capacity for self-control, for an almost gyroscopic psychological balance, a talent for self-regulation not unlike the self-regulation that Gilbert the geologist perceived in nature. Again there is a touch of the classicist in his preference for an equilibrium model, as well as that of the bookkeeper—a man, who, in his own words, was a stickler for accountability. Such a man was unlikely to be swept away by the romantic spectacle of universal processes or events modifying everything in their path. The panorama of an Ice Age with its global changes in climate, for example, meant little to him in the abstract; instead he carefully showed how a universal event, such as a lowering of temperature, would produce different outcomes—some diametrically opposed—depending on the circumstances of the local environment. That is, one had to appreciate the resistances as well as the force.[2]

Throughout his career Gilbert showed, overtly and covertly, a passion for just such a Newtonian universe, one constructed from mathematical and mechanical blueprints. When the San Francisco earthquake struck, Gilbert was in bed in Berkeley, but immediately he began to study the swing of the light fixture above him, trying to deduce from it the direction of the earth waves. In San Francisco during the fire he instinctively began to measure events—in this case the burning time for wooden buildings. He referred to Niagara Falls as a physiographic engine, and sought to measure its efficiency. He compared the laccoliths of the Henry Mountains to pistons, and hydraulic presses. The floor of Lake Bonneville, swelling upward after the draining of the lake, he likened to the flexures of an elastic beam, and sought to calculate the amount of loading involved. The flumes he erected to study hydraulic mining debris differed in important ways from natural rivers, he admitted, but added that his results might nevertheless apply to streams that were geometrically similar to them. In like manner Gilbert analyzed the process of scientific thinking: all creative thought proceeded by way of analogies, and such "analogies of proportion" were constructed by a process of almost geometric reasoning.[3] Even in lighter moments such impulses were evident. Guiding an Australian geologist through the Sierras, Gilbert hurried him past the groves of giant Sequoias in an eager search for the web of a particular spider, a web that resembled a paraboloid.

In all of this Gilbert's mechanistic bias propelled him to the fore of nineteenth-century geology. But another aspect of his Newtonianism brought him into conflict with the science of his contemporaries. This was his conception of geologic time. William Goetzmann has described a Second Great Age of Discovery, the period between the voyages of Cook and the conquest of the poles, during which the continental interiors of the earth were exposed to Western science. Before the Second Age of Discovery, the geographic knowledge of the globe by Europe and its colonies was limited to Europe and a few coastlines, while the age of the earth was frequently thought, based on Biblical genealogies like Bishop Ussher's, to be less than 6000 years. By the time the era ended, the age of the earth was determined to be about 4.5 billion years old, and its size and shape were accurately mapped on a global scale. How to organize this million-fold increase in the landscape of time was the chief preoccupation of the earth sciences. On the one hand, evolution (through paleontology) furnished an organizing schema; on the other, geophysics (guided by the newly discovered second law of thermodynamics) provided another. The paleontological schema resulted from the temporalization of the Great Chain of Being.[4] Stratigraphy, the first specifically geologic subject, was organized around the succession of fossils. Much of the remaining development of geology involved a search for the equivalents of fossils in the earth that could be developed on a similar pattern. This steady assimilation of geologic topics into the realm of historical geology is probably the ruling theme to the history of the earth sciences of this period. Geophysics, meanwhile, sought out new forms of entropy clocks, like heat loss or tidal friction, by which to develop a similarly progressive history. For both, the object was history; the mechanics, some form of evolution. Thus, although geophysics and geology might quarrel, the contest was more over technique and metaphor than over purpose. Geophysics might work within the confines of a contractional hypothesis and geology within the context of evolution, but the ambition of both was ultimately historical. Both were staunchly committed to historicism, whether as something continually gained or continually lost —that is, to the belief that earth history revealed the linear progression of what Arthur Eddington termed "time's arrow."[5]

Gilbert's perception of geologic time, however, led him to solve the historical riddles posed by the earth in ways quite different from his contemporaries. Gilbert as classicist, accountant, engineer, and Newtonian philosopher created unique solutions to the landscapes discovered by Gilbert the explorer, solutions manifest in his metaphors, prose style, methodology, and theoretical constructions. Gilbert distrusted paleontology and never practiced stratigraphy. Even his geophysics harked back to an earlier era in which time was more or less reversible. There was no necessity for the world described by the older Newtonian mechanics to proceed in a particular direction; one could, for example, reverse the algebraic signs of the equations and have the planets move in opposite order. And it is this perspective which is implied in Gilbert's geologic monographs. Ignoring the debate about the age of the earth, he preferred states of equilibrium to stages of evolution. In place of thermodynamic or life cycles, earth history proceeded as a series of discontinuous systems, each bracketed by its own state of equilibrium. Time was not a process, but a quantity like mass. Causality was not historical. Instead of entropy clocks Gilbert sought diligently for "rhythmic" chronometers such as astronomical cycles of the frictional rhythms arising from the continued action of force and resistance along some boundary.

His preferred unit of historical analysis was not the evolutionary stage, but a period of equilibrium. Nearly all of Gilbert's monographs begin with the onset of some disequilibrating force—an intrusion of magma, the flooding of a basin, the overcharging of a stream with debris—and end when the system has reestablished equilibrium. Whereas most of his contemporaries considered the shape of a landform to be evidence of the stage of its life history, Gilbert saw it as merely the reflection of a temporary equilibrium between erosive or geophysical forces and structural resistances. This is what Gilbert meant when he once described landforms as a sort of accidental by-product of dynamic geology. Geologic history consisted of events having a "plexus of antecedents and consequences," underscored by a multiplicity of natural rhythms.

Gilbert actually proposed several such rhythmic timepieces. His favorite was the earth's precession. The precession could systematically alter climate, and the resulting climatic cycles might be locally manifest as cycles of deposition. He experimented with correlating repeating series of strata in Colorado to such a cycle. In 1900, during a presidential address to the AAAS, he criticized both geologists and geophysicists for their addiction to a linear time scale. As an example of an alternative chronometer, he again proposed the precessional wobble, calling it a "frictionless pendulum" pulsing through the ages. It was valuable precisely because it did not degenerate through time, and was, in fact, a perpetual motion machine.[6]

II

It was the Second Age of Discovery that transported Gilbert to the barren beauty of the Henry Mountains, and it was in the imagery of the machine that he fashioned an explanation for them.[7] Perhaps the most salient fact about his report is the speed with which Gilbert produced it. He examined the peaks in the field for a little more than one week in 1875, and little more than two months in 1876. Most of those weeks were spent in topographic mapping. He prepared the report in three month's time, and much of that went into constructing plaster models and stereograms. Although only a lean 150 pages, the monograph—clearly a *tour de force*—was highly concentrated in its ideas and crystalline in its composition. The conclusion is inescapable that Gilbert had a sharp conception of the process of mountain formation and its erosional dynamics at the Henrys from very early in his research, if not before. Indeed, his analysis opens with a comparison between a volcano and a laccolith, a contrast that harks back to his earlier report for the Wheeler Survey (1875) in which he paired the Zuni Plateau with Mount Taylor.

In structure Gilbert's monograph is simple. It consists of three nearly equal parts. The first part systematically describes the geography of the Henrys (in Utah). The mountains had been sighted and named previously, but Gilbert's account was the first methodical survey of the region, and the one which brought it into the realm of formal knowledge. The second part describes the formation of the Henrys, a range which represented a new form of mountain structure. Gilbert named this type of mountain a "laccolite," meaning "lake-rock"; to avoid confusion with minerals, Dana altered the Greek to "laccolith," though Gilbert persevered in his original spelling. The laccolith thus became the second of his newly proposed mountain structures, the other being that of the Basin and Range structure. The third part of the monograph describes the present processes of land sculpture acting on the Henrys. It systematically codified the mechanics of erosion and added a basic contribution to the American fluvial tradition by describing the dynamics of what later became known as the "graded stream." Thus Gilbert gave a dynamic explanation for the new mountains and landforms of the West when his Survey director, John Wesley Powell, had given a taxonomic order and his associate, Clarence Dutton, an historical context. When he finished, Gilbert's reputation was assured.

Gilbert's analysis of laccolith formation, part two of the monograph, developed from a key observation, a pair of assumptions in rheology, two mechanical analogies, and a fundamental comparison to volcanoes. He began with a contrast between the volcanism of the Henrys and that typical of most volcanic piles. The critical observation which made the *Henry Mountains* more than a mere recapitulation of his earlier contrast in the Wheeler Report between Mount Taylor and the Zuni Plateau was first that there existed a correlation between the size of a laccolith and its depth, and second, the corollary observation that there were no small laccoliths. The rising column of magma behaved according to the principle of hydrostatic equilibrium. It ascended through the rock until it reached the point at which its driving pressure equaled the weight of overburden. At this point it began to spread laterally as a sill, in an approximately circular pattern. Gilbert imagined the sill as a hydraulic press, trying to raise its overburden. This required that it have a certain area, and that this necessary area would increase with the depth at which the sill was buried. The deeper the sill, the larger the "limital" area. When the sill became sufficiently large, a laccolith resulted.

To calculate the forces involved, Gilbert converted the hemispheric shape of the laccolith into a cylinder. This he did through a clever pair of assumptions. He likened the laccolith dome to a monocline (a simple steplike fold) rotated about a point. The monocline, in turn, was structurally homologous to a fault (a crustal fracture with one side of the break moved relative to the other). Hence, one could replace the circular monocline with a circular fault. The laccolith became a mountainous piston. Gilbert concluded his analysis, however, by returning to the case of the dome. "Guided by certain allied problems in mechanics"—he cited Rankine's *Manual*—Gilbert likened the stresses in the flexed strata to those involved in the bending of a structural beam.[8] Considered in thermodynamic terms, the resulting laccolith was analogous to a volcano formed under high confining pressure. Precisely because this pressure was so intense, the strata behaved plastically rather than brittlely under the stress of flexure. Otherwise a series of fractures and dikes would have spilled the magma onto the surface, creating a more familiar volcanic landscape of cones, piles, and lava flows.

Gilbert summarized the shaping of the laccolith in this way:

> The laccolith in its formation is constantly solving a problem of "least force," and its form is the result. Below, above, and on all sides its expansion is resisted, and where resistance is greatest its contour is least convex. The floor of its chamber is unyielding, and the bottom of the laccolith is flat. The roof and walls alike yield

> reluctantly to the pressure, but the weight of the lava diminishes its pressure on the roof. Hence the top of the laccolite is broadly convex, and its edges acutely. Local accidents excepted, the walls oppose an equal resistance on every side; and the base of the laccolite is rendered circular.[9]

Gilbert's structural analysis brought a novel perspective to the rugged mountains of the West. Where others, such as his companions Powell and Dutton, had brought geologic views hybridized with esthetics, literature, and biology, Gilbert brought a perception derived from physics and engineering. On one hand, his physical model furnished a concise logical set of relationships among the data, and, on the other, it allowed for predictions that were not inherent or obvious in that data. In concluding his description of the laccolith, he offered exactly such a prediction. He tried to estimate the depth of overburden (all of it Tertiary strata) which had been removed by erosion since the formation of the laccoliths. He arrived at a figure of 7700 feet. The enormous scale of erosion whose discovery by earlier explorers to the Colorado Plateau, such as John Newberry, had startled the scientific community, the ruinous decay of landscape on which his associates on the Survey, Dutton and Powell, had lavished their purple prose, now had a quantitative meaning.

The final and third part of the monograph, perhaps more than any other work of the century, erected the foundations for modern geomorphology. Under the general rubric of land sculpture, Gilbert described the erosional dynamics operating on the modern Henrys. He divided this analysis into three parts, each bringing the subject to a more elaborate stage. First, Gilbert codified the erosive processes and the manner in which environmental conditions, such as slope, lithology, and climate, modified their rates. Second, he tried to axiomatize these processes, especially fluvial erosion, into a set of three laws which governed land sculpture. Third, he integrated land sculpture, again through rivers, on the regional scale of the drainage basin. In each case examples were drawn from, and conclusions applied to, the Henrys.

Gilbert modestly claimed that he was only restating certain "principles of erosion which have been derived or enforced by the study of the Colorado Plateau."[10] In one respect this was true. Newberry, Powell, and Dutton had all written about the evidence for massive erosion on the plateau. It was not in the novelty of particular ideas that made Gilbert's own report remarkable, but in its brilliant codification of known processes and in the axiomatic format in which they were presented. Actual observations were verified by derivation from these axioms as though they were lemmas to a geometric theorem. Of his three laws of land sculpture, the first was the law of declivities, which derived from the evident fact "that if steep slopes are worn more rapidly than gentle, the tendency is to abolish all differences of slope and produce uniformity."[11] The second law was the law of structure, derived from the differential erodability of rocks. The third law, the law of divides, held simply that the grade of a river steepened as the divide was approached.

In themselves the laws are simple. But Gilbert's special insight was to set them in mutual opposition. "The law of uniform slope opposes diversity of topography, and if not complemented by other laws, would reduce all drainage basins to plains,"[12] but this required a uniformity of conditions which nowhere exists, so the law is opposed by the law of structure, which generates variety of form. This action, or differentiation, "continues until an equilibrium is reached . . . When the ratio of erosive action as dependent on declivities becomes equal to the ratio of resistances and dependence on rock character, there is equality of action."[13] Similarly for the law of divides. Thus the laws of declivities and divides are responsible for the erosive forces; the law of structure for resisting forces. The actual landscape represented a dynamic equilibrium between their antagonistic tendencies.

It is obvious that Gilbert sought to organize the explanation of the Henry Mountains by analogy to mechanics. What is less obvious is his indifference to the second law of thermodynamics. Gilbert organized his description of the Henrys, both as they were formed and as they exist today, by presenting them as separate systems in equilibrium. To his contemporaries the more important facts would have been the sequences of rocks piled up or swept away, the degradation of energy involved in the orogenic events, and the subsequent systematic denudation of the landscape that would have followed—in short, the life history of the mountain—and, certainly, where the creation of the laccoliths belonged in the great evolutionary chronology of the earth. Gilbert not only ignored the degradation of the systems over time, but in his enthusiasm for his ideal model of laccolith formation nearly forgot to mention just when the event occurred.

The frontispiece to the *Henry Mountains* shows two historical systems, each complete in themselves, juxtaposed into the same block diagram (Figure 1). In his text Gilbert explains how the two sets of forces, one in each system, acted against an opposing resistance to create the form shown in the diagram. The magma column and the stream had each, in its time, reached a profile of equilibrium in acting against

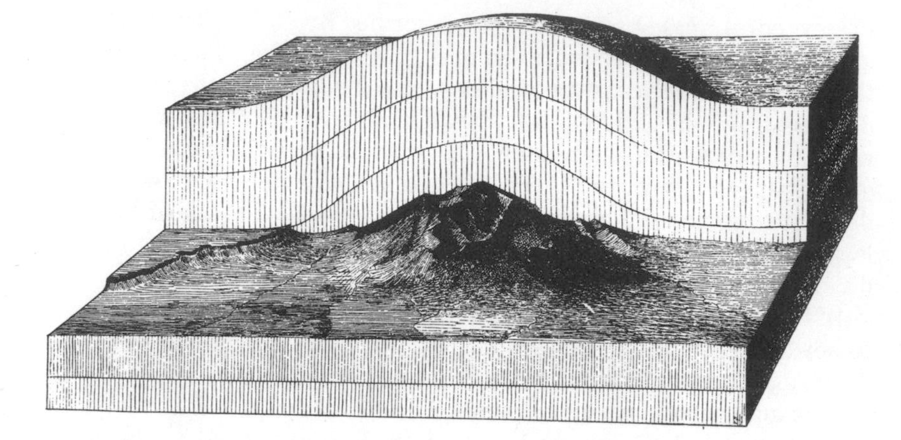

1. Frontispiece to the Henry Mountains. *"Half-stereogram of Mount Ellsworth, drawn to illustrate the form of the displacement and the progress of the erosion. The base of the figure represents the sea-level. The remote half shows the result of uplift alone; the near half, the result of uplift and erosion or the actual condition."*

the resistance offered by the rocks of the region. In both cases a path of least force was taken. In the case of the laccolith, the resulting equilibrium was static; in the case of streams and landforms, it was dynamic. So long as there existed an equilibrium of action, the ratio of force to resistance effectively shaping the landscape remained constant. So long as this ratio remained constant, there was a similar conservation of form, whether as the profile of a river, the dome of a laccolith, or the face of a cliff. When Gilbert depicted some of the streams draining the slopes of the Henrys as shifting their channels laterally, the river—like the laccolith—was only solving the problem of least force. Gilbert's description of lateral planation by streams was probably the first in geologic literature, but it followed almost as a corollary from the peculiar physical conditions of the Henrys in which the streams worked, just as the laccolith rather than a volcano followed from the circumstances of magmatic intrusion. Lateral planation was not a stage in the evolution of a river any more than the laccolith was a stage in the evolution of a volcano. The two systems are nowhere joined by an historical narrative describing what happened, era by era, between the time the laccoliths were formed and their present condition. Gilbert's philosophy of history thus shaped his monograph, allowing him to ignore precisely those questions typically asked most eagerly by geologists of his day—namely, the age of the events and their systematic change over time.

The things Gilbert did not do are equally instructive. There is almost total indifference to historical geology and to contemporary interpretations of volcanism. His insights are physical rather than chemical, dynamical rather than topographical, typical rather than historical. His neglect to describe the historical erosion of the thousands of feet of strata that, to the south, awed Dutton, Newberry, and Powell—the very phenomenon in fact, at the heart of their monographs on the Colorado Plateau—was especially revealing. Gilbert's impulse was to find the essence of a scene, not to celebrate its variety. Yet the danger in his work was never that he would generalize too broadly beyond the problem at hand, but that he would fail to piece the individual puzzle parts into a larger mosaic, that he would merely solve a number of particular riddles in the manner of an engineer. The *Henry Mountains*, like all his great works, was spared either extreme. Guided by its theme of equilibrium, it seemed itself to show between the driving force of its generalizations and the opposing resistance of its data a dynamic equilibrium of insight.

Gilbert's labors for the Powell Survey paralleled those of his two comrades: Capt. Clarence Dutton's *High Plateaus of Utah* (1880) and *Tertiary History of the Grand Canyon District* (1882), and Major Powell's *Exploration of the Colorado River* (1875) and *Geology of the Uinta Mountains* (1876). Both men were scientific soldiers of fortune, stamped by their Civil War experiences, who found in exploration geology a moral equivalent to war and an intellectual challenge that they relished. Collectively, they, along with the more scholarly Gilbert, formed one of the most successful collaborations in the history of geology. After 100 years the larger interpretation of the Colorado Plateau is still the one these three gave to it.

The Powell Survey, nonetheless, never founded a formal school of thought, although the impact of Gilbert, Dutton, and Powell on the institutional composition and theoretical predictions of the U.S. Geological Survey was considerable. The academic synthesis of their work was left largely to the Harvard schoolmaster, William Morris Davis. Instead it was their function to dramatize new geographic discoveries and to establish conceptual links between their new theories and existing intellectual traditions—politics and social thought for Powell; esthetics and chemistry for Dutton; engineering and physics for Gilbert.

What particularly differentiated Dutton and Powell from Gilbert was their historical sense, the evolutionary vista. Powell saw in the slow flexing of the Uinta Mountains the majestic patience of geologic processes, and in the mountain's stratigraphic profile he saw a regular succession of events which he could project forward as a blueprint for the future. Dutton had a complementary vision. In his panoramic sweep over the vast layers of strata peeled from the surface by erosion and in the record of ancient erosion surfaces of similar magnitude, such as are manifest in the Great Unconformity in the gorge of the Grand Canyon, he saw the burden of the past. When he peered eastward from the Aquarius Plateau, he saw a landscape of sublime desolation, the eroded buttes and crags like fallen idols, the ruins of ancient geologic empires. To give them a head and tail, as he put it, he told their life story.

III

Although it was the last mountain to be discovered on the continental United States, and the river that drained it, the Escalante, the last major stream to be discovered, the Henry Mountains never assumed the cultural significance of such other Western landscapes as the Yellowstone, which excelled it in novelties, or the Grand Canyon, whose cross-section of geologic time made it a central symbol in the search for the earth's age. Still, the Henrys are exotic enough. The debate about their structural geology still continues, though there is little disagreement about Gilbert's analysis of the Henrys' geomorphology. In 1953, for example, Charles Hunt published the *Geology and Geography of the Henry Mountains Region, Utah*, a comprehensive survey outfitted with a brilliant array of cartographic techniques in which he rejected Gilbert's interpretation of laccolith formation.[14] In 1970 Arvid Johnson brought out a spirited defense of Gilbert's method and conclusions in a book rich in the technical and conceptual apparatus of rheology, *Physical Processes in Geology*.[15] If anything, the subject promises to continue and thereby hold for Gilbert a certain place in modern geoscience.

The reason may be that earth science has abandoned the assumptions of the classical period and entered, both scientifically and metaphysically, a more modern period. A Third Age of Discovery, the exploration of space and the ocean basins, and a new Technological Revolution have altered our ideas about the earth and our philosophy of nature as radically as did the Second Age of Discovery and the Industrial Revolution in their day. As a result, the intellectual apparatus, metaphors, and broad theories of the earth sciences are changing. The new paleomagnetism, for example, is based on rhythmic oscillations rather than entropy considerations. In this context many of Gilbert's ideas seem very modern indeed.

Yet it is not quite correct to say that Gilbert anticipated modern trends. He was a pure Newtonian who sought to make geological systems analogous to mechanical systems. His impulse was conservative: to describe new marvels by analogy to old systems of knowledge, not, like Powell or Davis, to invent new sciences in response to a growing information base. Most of Gilbert's colleagues, men like T. C. Chamberlin or Powell, assumed that science itself, and scientific methodology, were undergoing an evolution not unlike that of the natural world. New theories were superior to old ones; new hypotheses, like mutations, would be selected so that those best adapted to the data would survive. Yet to Gilbert the process of science was best communicated by example, by imitating the great works of the past. Hypotheses did not resemble mutations, but a series of successive approximations in a mathematical series which gradually approached a best possible product. The thrust was basically conservative, even neoclassical: new sciences developed by imitating the successful older sciences. Even at the Henry Mountains what Gilbert the explorer saw in the strange Western landscape did not lead to new words, new techniques, or new sciences so much as to analogies of older sciences.

Gilbert's allusion to "certain allied problems in mechanics" was not a casual, parenthetical phrase. In one respect it summarized symbolically the progress of the earth sciences. Whereas James Hutton, experimenting with the analogy of the earth to a heat engine, could only imagine volcanic features as a safety valve or rupture, a feature not performing work, Gilbert, drawing on the growth of physics, could picture the laccolith as a sort of piston, an integral part of the earth's machinery, operating according to regular laws and with measurable efficiency. In another respect, the phrase tellingly synopsized the inventive process by which Gilbert became one of America's outstanding geologic thinkers. It was a process repeated throughout his long career, and one epitomized some years after the *Henry Mountains* by an episode that occurred on the High Plains. A novitiate geologist whom Gilbert was training became lost. It was past sunset and there was no wood for a signal fire. A search party had to be organized. Orienting himself by the stars Gilbert stepped off a number of paces. He turned 60° and paced off a like distance; there he turned again 60° and returned to camp having inscribed an equilateral triangle. Presumably the novice was found within that area. His approach to the terra nova of earth history, another seemingly featureless plain, was much the same. He merely transplanted old patterns of understanding into new sights, guided by a sound sense of mechanics and mathematics, and a belief in the frictionless pendulum of the stars.

Notes

1. Biographical information on Gilbert can be found in the following works: William Morris Davis, "Biographical Memoir of Grove Karl Gilbert, 1843–1918," *National Academy of Sciences, Biographical Memoirs*, 31 (Washington, 1926); W. C. Mendenhall, "Memorial of Grove Karl Gilbert," *Geological Society of America Bulletin* (1920), 31:26–64; C. Hart Merriam, "Grove Karl Gilbert, the Man," *Sierra Club Bulletin* (1919), 10, no. 4:391–399; and Stephen Pyne, "Grove Karl Gilbert: A Biography of American Geology," doctoral dissertation, Austin, University of Texas, 1976.
2. Pyne, 459–477.
3. G. K. Gilbert, "The Inculcation of Scientific Method by Example," *American Journal of Science* (1886), 3rd ser., 31:284–299. Stephen Pyne, "Methodologies for Geology: T. C. Chamberlin and G. K. Gilbert," *Isis*, 1978, 69:413–424.
4. Arthur Lovejoy, *The Great Chain of Being* (Cambridge, Harvard University Press, 1929).
5. See Arthur Eddington, *The Nature of the Physical World* (reprinted Ann Arbor, Ann Arbor Paperbacks, 1958), 36–111.
6. G. K. Gilbert, "Rhythms and Geologic Time," *Science* (1900), new ser., 11:1001–12.
7. G. K. Gilbert, *Report upon the Geology of the Henry Mountains* (Washington, 1877).
8. Ibid., 90.
9. Ibid., 91.
10. Ibid., 99.
11. Ibid., 115.
12. Ibid.
13. Ibid., 116.
14. Charles Hunt, "Geology and Geography of the Henry Mountains Region, Utah," U.S. Geological Survey, *Professional Paper 228* (Washington, 1953).
15. Arvid Johnson, *Physical Processes in Geology* (San Francisco, W. H. Freeman, 1970).

21

The Geosyncline–First Major Geological Concept "Made in America"

R. H. Dott, Jr.

"The greater the accumulation, the higher will be the mountain range." So said James Hall in publishing his concept of mountain building, which was destined to evolve into the first great American geological theory. Hall first presented his thesis in 1857 as the presidential address to the American Association for the Advancement of Science in Montreal, but it was not published until 1859.[1] Although spawned by Hall, the geosynclinal theory—American style—was named and fully nurtured by James D. Dana. But why should the geosyncline have been invented in America, which up to that time had been dependent upon Europe for its theoretical geological base? And why by Hall and Dana? These questions and an analysis of the venerable Yankee mountain-building theory seem appropriate tasks for a Bicentennial Symposium. The subject is too vast to treat fully here; I have therefore been nationalistic and concentrated upon pre-1930 American contributions.[2] Special attention is given to the geanticlines or borderlands and the role of geosynclines in overall continental evolution, for these aspects seem to have been somewhat neglected.

Before Hall and Dana, Hutton's plutonism contained a sketchy mountain-building theory, which even involved sedimentation as a precursor to uplift. And Elie de Beaumont in 1825 had proposed lateral compression and catastrophic cracking of a shrinking, cooling earth as the cause of mountains rather than the plutonists' simple vertical uplift engineered by heating, expansion, and intrusion of magma beneath. As the making of geologic maps and (especially) scaled cross-sections became widespread by mid-century, lateral compression seemed inescapable to a growing number of workers. Together with the characteristic folding of strata, the demonstration of long-distance, low-angle overthrust faulting must have been particularly compelling.[3] Though Dana was a staunch contractionist, Hall subscribed instead to large-scale vertical movements; as suggested in 1836 by British physicist J. F. W. Herschel, these movements were due to pressure and temperature changes at depth in response either to erosion or deposition at the surface. The relative importance of lateral versus vertical movements was to be debated for nearly a century.

"Made in America"

By 1850 there were many geologists of great ability working in both Europe and America, but few could have shared the exceptional egotistical drive of James Hall, and he stood almost alone in having extensive first-hand stratigraphic experience both in the Appalachian mountain belt and in the continental interior of the midwest. New York and adjacent areas constituted Hall's native ground, but his horizons were expanded westward as early as 1850–51, when he joined J. W. Foster and J. D. Whitney in their survey of the Lake Superior region. This gave Hall experience with the lower Paleozoic strata in northern Michigan and Wisconsin. Then during one of the cantankerous paleontologist's several feuds with the New York legislature, the inconvenience of suspended salary caused him to accept an invitation in 1855 to organize a geological survey of Iowa. In characteristic style, while still affiliated with Iowa as well as New York and the Geological Survey of Canada, he unabashedly accepted also an advisory role with the embryonic Wisconsin Survey in 1857. These collective experiences uniquely equipped Hall to recognize the profound stratigraphic as well as structural difference between the Paleozoic strata of the interior of the continent and their counterparts in the Appalachian Mountains. The midwestern strata, besides being almost undisturbed structurally, were only about one tenth as thick as those folded ones of the mountain ranges! This fundamental observation, not previously recorded in detail, then led him to infer a causal relation between great thickness of sedimentation and mountain making. Right or wrong, this view in one form or another has been around ever since.

Hall found Elie de Beaumont's and the Rogers brothers' lateral compression too catastrophic, for he was a strict Lyellian uniformitarian. He accepted the popular view that the interior of the earth, whether liquid or solid, was pliable enough to yield to gravitational loading to produce "depression at one point and elevation at another by the yielding mass beneath."[4] He took as his principal authority the embryonic gravitational equilibrium (isostasy) of Herschel and Babbage.[5] Hall believed that the greater thickness of strata seen in the mountain belt had caused the earth's crust at the edge of the continent to subside profoundly until it could bend downward no farther. Then the upper part of the sediment pile would become crumpled by compression as the lower part was fractured by extension. The upper layers were elevated locally by the crumpling.

> The line of greatest depression would be along the line of greatest accumulation [that is] the course of the original transporting current. By this process of subsidence . . . the diminished width of surface above, caused by this curving below, will produce wrinkles and folding of the [upper] strata. That there may be rents or fractures of the strata beneath is very probable, and into these may rush the fluid or semi-fluid matter from below, producing trap-dykes; but the folding of strata seems to me a very natural and inevitable consequence of the process of subsidence.[6]

Though Hall was ambiguous about the cause of uplift, he clearly divorced magnitude of elevation from severity of folding. In 1859 and more emphatically in 1883 he argued that the elevation of mountains must be of continental rather than local origin because the Appalachians are not significantly higher than the adjacent plateaus to the west (a circumstance almost unique to those mountains). The mechanism of continental uplift was side-stepped with the statement that it had never been his intention to present a complete theory of mountain formation. Gravitational buoyancy was implied, however, by the observation that the world's highest mountains contain younger strata than the Appalachians, which suggested that they are higher because they received greater accumulations through longer time. Hall also believed that metamorphism required a thick accumulation of sediments, and was proportional in intensity to sediment thickness.[7]

Some, like T. Sterry Hunt, N. S. Shaler, and G. L. Vose, were quickly enthralled by Hall's hypothesis, but most found it incomprehensible upon first hearing. Among notable doubters were Joseph

LeConte and Joseph Henry, but the most important was James D. Dana, who had been developing his own theory of mountain building for more than a decade. Dana's assessment was that Hall had developed a fine theory of mountain making but had left the mountains out. Hall's troughs of depression were vastly wider than deep and so amounted to insignificant scratches on the globe. Simple mathematics suggested that 40,000 feet of subsidence of the Appalachian region could have resulted in compression and elevation of but a few feet by Hall's mechanism. Dana believed that the crust has considerable strength whether or not the interior is pliable. How, for example, could the earth support high mountains if it is so sensitive to every increment of sediment? "The earth's crust would have had to yield like a film of rubber, to have sunk a foot for every added foot of accumulation over its surface; and mountains would have no standing-place."[8]

Long before Hall's AAAS address, Dana was fully committed to the prevailing view that mountains are the main zones of compression of the crust, which result from the almost universally assumed secular cooling and shrinkage of an originally molten earth whose interior was assumed to be still molten or at least plastic. The earliest statement that I have found of Dana's basic idea dates to 1846: "The greater subsidences of the oceanic parts would necessarily occasion that lateral pressure required for the rise and various foldings of the Alleganies and like regions."[9] Dana even noted in passing that the Appalachian Mountains contained "Silurian and Devonian sandstones . . . of great thickness, being five times as thick as limestone and associated deposits of the same age to the west."[10] But unlike Hall, he believed the thicker sediments to be a consequence not a cause of crustal subsidence. Warping of the crust initiated what Dana named in 1873 a *geosynclinal* or downbending and a complementary *geanticlinal* or upbending (converted by him in 1890 to *geosyncline* and *geanticline*). A newly formed geosynclinal trough provided immediately a receptacle for deposition of thick sediments whose rate of accumulation just kept pace with subsidence so as to maintain "the surface of the area near the water level."[11] Dana (like Hall) emphasized the position of mountain belts at the edges of the North American continent. Unequal early contraction of the earth presumably gave rise to the continents and ocean basins, and continuing contraction then caused bending due to lateral pressure primarily at the juncture between the two.

> The position of mountains on the borders of continents . . . is due to the fact that the oceanic areas were much the largest, and were the areas of greatest subsidence under continued general

> contraction of the globe . . . the oceanic crust had the advantage through its lower position of leverage, or more strictly speaking, of obliquely upward thrust against the borders of the continents.[12]

The bottom of the geosynclinal accumulation was presumed to be greatly weakened by heat from the interior, resulting in metamorphism and ultimate collapse of the trough and the crumpling and pressing together of the strata within it. To Dana, unlike Hall, burial alone was not sufficient to cause metamorphism, because unmetamorphosed strata lie below sequences 17,000 feet thick. He thought that structural deformation might produce some of the necessary thermal energy. As the geosynclinal became narrower by viselike compression, inevitably some of its volume was raised to form a mountain range even though its crustal basement remained deeply downwarped. Volcanoes formed where some of the fluid-like subcrustal material escaped to the surface via fractures. Thus a mountain range began as a geosynclinal and ended in upheaval as a *synclinorium*. Anticlinal folds within the synclinorium were so fractured that erosion reduced them rapidly, leaving a series of parallel synclinal mountain ridges.

Most analyses of Dana have focused only upon the geosynclinal, but of equal importance was the postulate of complementary geanticlinals warped up by viscous subcrust pushed laterally from beneath the down-warped zone. They could form on either side—the Cincinnati uplift to the west being an early example—but that on the seaward side received the greatest attention:

> . . . on the oceanic side of the progressing geosynclinal referred to, there has been generally, as the first effect of the thrust against the continental border, a progressing geanticlinal, which usually disappeared in the later history of the region . . .[13]

Although this resembles Hall's "depression at one point and elevation at another," the mechanism of formation of these complementary features and of the subsequent mountains became a matter of intense debate on both sides of the Atlantic among physicists and geologists alike.

Contraction versus Isostasy

By the 1870's the origin of mountains had become a major issue.[14] The venerable compression-due-to-contraction was subscribed to by such American notables as LeConte, Hunt (in part), Charles Whittlesey, and Whitney in addition to Dana and the British W. Mackie, J. Durham, and J. Davison. But the vertical gravitational adjustment mechanism was coming of age, and was favored by C. S. Dutton (who coined isostasy in 1889), N. S. Shaler, W. J. McGee, Hunt (in part), A. Winchell, and Europeans Osmond Fisher and V. H. Hermite. To explain gravity measurements near the Himalaya, English physicists G. B. Airy and J. H. Pratt already had argued for the necessity of a relatively mass-deficient root to support high mountain ranges. Physicists Dutton and Fisher took up the mathematical cudgel, challenged prominent British physicists Kelvin's and Mallet's estimates of earth cooling, and condemned contractional compression as hopelessly incapable of elevating mountains. The eighties became the physicists' decade, and the issue was hotly debated in allegedly definitive mathematical terms. Empirical evidence seemed to be provided by evidence of depression of the crust by the load of ice caps and upwarping or rebound due to removal of ice and the draining of large lakes.

During the 1890's isostasy became widely accepted as a real phenomenon; however, its relative importance in mountain building was by no means agreed to. Hunt was perhaps the earliest to attempt a compromise. He had strongly endorsed from the first Hall's subsidence by loading to explain great thickness of strata, but by 1861 he felt that lateral compression caused by contraction was necessary to explain the folding of strata in mountains. In the 1870's and 1880's LeConte staunchly defended lateral compression to form mountains (even though increasingly unsure of its cause), while at the same time accepting isostasy as at least a partial cause of subsidence. He seems to have coined the famous shriveling apple skin analogy for contraction of the earth. The isostasy debate, much like Kelvin's age debate, found physicists brandishing mathematics at puzzled geologists, who could see that the folds of mountains apparently demand tens of miles of lateral shortening of the earth's crust. It was not all one-sided, either, for mathematics was also employed to prove isostasy impotent for mountain making.

Finally, in the 1890's, American geologists, led by R. S. Woodward, Bailey Willis, and G. K. Gilbert, generally acknowledged that an isostatic tendency determines the overall figure of the earth. but it operates relatively quickly, and the geologic evidence indicates that structural disturbances must repeatedly upset isostatic equilibrium, which would otherwise maintain a static surface condition. Therefore, compression, presumably due to earth contraction, still must be the main cause of mountain building. Even Dana gradually accommodated, so that in the last edition of his *Manual* (1895), he adopted this same position. Moreover, he finally paid full homage to Herschel, Babbage, and Hall in conceding that a *preparatory geosyncline* is precursor to mountains, and much of the subsidence therein could be due to sediment loading. But he blurred the definition of *geosyncline* by including interior lake basins and the deep ocean basins as types. In the early twentieth century, geodecists pursued isostasy in virtual isolation from geology, but practically all geologists came to accept it in a role subordinate to lateral compression. The strength of the crust now became a central issue together with the viscosity of the upper mantle and thermal conductivity of rocks at high pressure and temperature. Mountain-building theory stalled until a better knowledge of these phenomena could be gained and the implications of radioactive heating were assimilated. This did not occur until after 1930.

The European Style

What was happening meanwhile in Europe at the turn of our century? Most European students of Alpine geology at first accepted and supported what E. Haug characterized as James Hall's Law, namely that "Mountain chains are formed on geosynclines." (1907, p. 159–160).[15] In the Alps as in the Appalachians, most Mesozoic and Cenozoic strata were found to be thicker than equivalent ones in the bordering lowlands. Profound subsidence and sedimentation seemed to have preceded upheaval of the mountains there, too. But Hall and Dana believed all geosynclinal sediments to have been shallow marine or nonmarine in origin because of the abundance in Appalachian Paleozoic strata of ripple marks, fucoids, mud cracks, and coal, which were then assumed to be impossible in deep water. Alpine geologists, however, were beginning to interpret many of their geosynclinal sediments as deep-water deposits, based in part upon analogies with the newly published descriptions of modern deep-sea sediments in the *Challenger* reports that first appeared in 1891. For example, according to Hsü, M. Neumayr suggested that radiolarian cherts were analogous to modern deep-sea radiolarian oozes, but J. Walther doubted if the cherts were deep-marine deposits. Bertrand in 1894 and 1897 pressed the deep-water interpretation, and also called attention for the first time to the characteristic progression of Alpine sedimentation from chert and shale through *flysch* (monotonous alternating sandstone and shale), all considered of deep origin, to the *molasse* (conglomerates and thick sandstones with coal) of shallow marine to non-marine origin.[16]

A parallel European development of fundamental importance was the inference that a distinctive rock succession from serpentine through greenstone or basalt to radiolarian chert—the *ophiolite sequence*—represents deep-ocean crust beneath the geosyncline. Suess in 1909 recorded ophiolites from other mountain regions, and drew comparisons with mafic igneous rock occurrences in the Atlantic.[17] The importance of the ophiolite sequence was almost entirely lost on Americans, however, for more than a quarter of a century, until H. H. Hess of Princeton began to make comparisons of American serpentine belts with those of the Caribbean island arc.[18] Even then there was a controversy over whether such belts represented intrusions and extrusions of mantle material through continental (sialic) crust during deformation of a geosyncline (favored by Americans), or whether they represented oceanic crustal basement that predated mountain building (favored by Europeans). This issue was not to be fully resolved until about 1970, by which time the results from the Deep Sea Drilling Program supported the European view.[19] No doubt the principal American preoccupation was that deep-sea rocks in a geosyncline would threaten the long-held assumption of continental permanency.

Emil Haug soon became the Dana of European geosynclinal theory. He wrote of deep-water geosynclinal sediments in 1900 as a follow-up of Bertrand's 1897 paper, and in 1907 he stated: "Usually the beds of a geosyncline are of deep water origin while the lateral beds are of shallow water. But this is not an invariable rule."[20] To Haug depth and the resulting biofacies were the most important characteristics of a geosyncline; fine, dark shales with pelagic fossils and lacking unconformities were prevalent. Also, unlike the Americans, he envisioned a deep bathyal trough symmetrically bounded on *both* sides by shelflike (continental) regions—an obvious Alpine-Mediterranean influence, as Hsü notes. He does not seem to have been very concerned about the sources of the coarser clastic sediments of the flysch and molasse, but E. Argand in 1916 conceived of elongate embryonic mountain ridges called *cordilleras* rising within the geosynclinal belt as the sources of flysch sands while the youthful Alps themselves supplied the coarser molasse.[21]

We see that the early European students of geosynclines added three important new concepts, all of which were in conflict with the American product. These were: (1) Many geosynclinal sediments are of deep-water origin; (2) The ophiolite sequence probably represents oceanic crust; (3) Major land sources of clastic sediments may lie *within* (rather than next to) the geosynclinal belt. Was it conceivable that the Appalachian and Alpine geosynclines were so completely different that two theories were required? Such was implied by the divergent evolution of thought by the early twentieth century.

Borderland Geanticlines

James Hall emphasized the location of geosynclines at the edges of continents, but as early as 1843 he recognized that Paleozoic sandstones increased in proportion eastward, thus requiring a land source east or southeast of the subsiding zone.[22] The Rogers brothers and others soon recognized independently an eastward coarsening of clastic sediments. Vague allusions to an Atlantic continent were sometimes made, but such was anathema to Dana because of his strong advocacy of permanency of oceans and continents. Instead, Dana postulated his *seaborder anticlinorium.* It was partly an inferential mechanical necessity to complement the geosynclinal, but some geologic evidence also was cited for existence of a Paleozoic borderland.

> . . . in the character of the remains of marine life, or else its absence, in the sea-border rocks, through a large part of Paleozoic and Mesozoic time, showing that a barrier of some kind existed along the sea border. In the Carboniferous rocks . . . there are almost no marine fossils. . . . It was not until the Cretaceous period that the coast was open to the ocean, through a disappearance of the geanticlinal barrier.[23]

Within a few years, T. C. Chamberlin showed just such a borderland on what was apparently the first published paleogeographic map of a large part of North America (Figure 1).[24]

Appalachian workers at first were extremely vague about the exact nature of the geanticlinal borderland. In 1890 Dana suggested that there had been several ridges of Archaean rocks (*protaxes*) that acted both as barriers between the faunas of different troughs and as sources of sediments. Because of the prevalence of crystalline rocks, then regarded as largely Pre-Paleozoic in New England and the Piedmont region, it became accepted that borderlands were upwarped masses of old, pregeosynclinal rocks. Suddenly more paleogeographic maps showing various conceptions of the borderland appeared. C. D. Walcott presented a map suggesting that the entire coastal plain had been

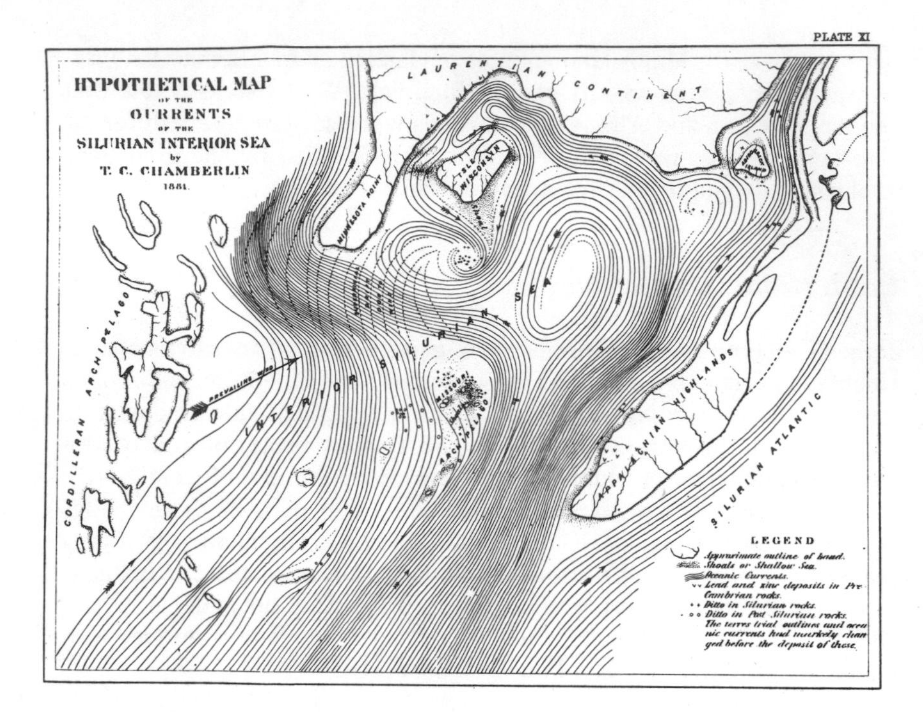

1. *Paleogeographic map of the eastern two thirds of the United States for early Paleozoic time, showing an "Archaean Appalachian Highlands stretching from the bounds of Alabama northeastward to an unknown limit" (p. 532). Ocean current flow lines, and the prevailing wind arrow, which were printed in red, were influenced by the modern Gulf Stream and prevailing westerly winds all modified by the rotation of the earth and by land obstructions (see pp. 531–538). The map and accompanying discussion related to speculations on the origin of upper Mississippi Valley lead-zinc deposits. ("Silurian" referred to the entire lower Paleozoic, reflecting the old Murchison classification; "Ordovician" had only been coined in Britain in 1879.) From* Geology of Wisconsin *(1882), 4: plate XI by T. C. Chamberlin.*

a borderland during the Cambrian (1891), and H. S. Williams published a generalized map for the Devonian (1897) on which he coined the name *Appalachia* for the land from which the Devonian Catskill clastic sediments had been derived (Figure 2).[25] The name caught on immediately, and in 1902 Bailey Willis published two colored paleogeographic maps showing Appalachia for the Cambrian and Devonian (Figure 3). The next year C. Schuchert published two Devonian maps, and in 1909 he generalized borderlands on the oceanward side of geosynclines to all American examples (Figure 4). After standing as lands for millions of years, the borderlands mysteriously faded away beneath the seas. This was a ruling hypothesis in America until the 1940's. Only then did the alternative concept of island arcs raised intermittently *within* the geosyncline begin to displace the borderland dogma.

Origin and Accretion of Continents

Dana was the first American geologist to present a complete theory of continental and ocean-basin evolution into which geosynclinal and mountain-building views were fully integrated. Most references to Dana's geosyncline have tended to ignore these broader aspects of his theory, which has resulted in an incomplete appreciation of the man's intellectual breadth. In his important 1846 paper, "Volcanoes of the Moon," he argued that the early earth must have had volcanoes over all of it, but irregular cooling led to cessation of volcanism first where continents now are situated.[26] Clearly Dana's extensive travels around the Pacific (1838–42) had convinced him of the fundamental differences between continents and ocean basins as evidenced in his early observation that oceanic islands have only volcanic and coral rock.[27] Later he observed: "Perhaps the fact that the prevailing rocks of the oceanic volcanoes are basaltic, and of the continental, andesytic and trachytic, explains how it is that the oceanic crust is made the denser."[28] He was so impressed by these distinctions that he early assumed that, once formed, continents and ocean basins were permanent fixtures in terms of both their composition and their positions. Although they might be warped and faulted, they could not disappear completely or change places.[29]

While the primitive continents were thickening, and stiffening by additions at their bases, oceanic areas presumably remained liquid much longer. When crystallization of crust between continents began, continual shrinkage of the earth as a whole guaranteed that the newer,

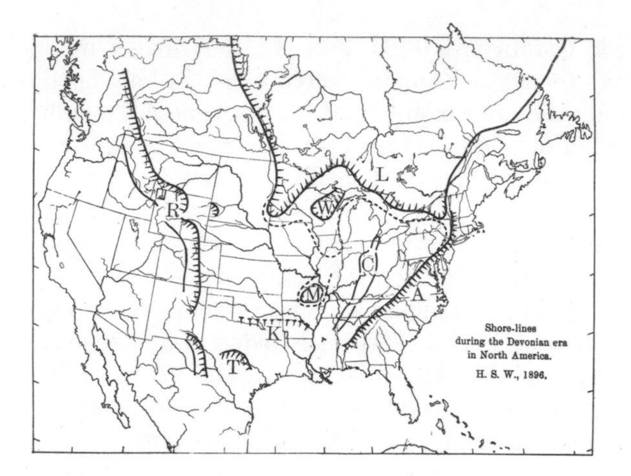

2. H. S. Williams' 1897 paleogeographic map "showing the approximate position of the Devonian intercontinental sea" on which he coined the name Appalachia (A) for the eastern Devonian borderland (pp. 395–396). The lines with fringes represent shorelines; L is Laurentia land, W the Wisconsin Peninsula, and C the Cincinnati plateau (uplift or geanticline of Dana). See note 17.

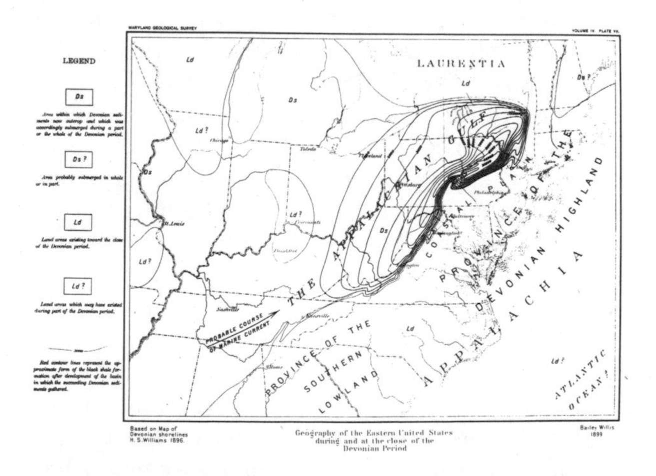

3. Paleogeographic map showing the borderland Appalachia adjacent to the central Devonian Appalachian geosyncline. The original, as well as an accompanying one for the Cambrian, was printed in color. Note the structure contours, which indicate the form of the subsiding geosyncline, and current direction similar to Chamberlin's map. From Bailey Willis, Maryland Geological Survey *(1902), 4: plate VII.*

BULL. GEOL. SOC. AM.

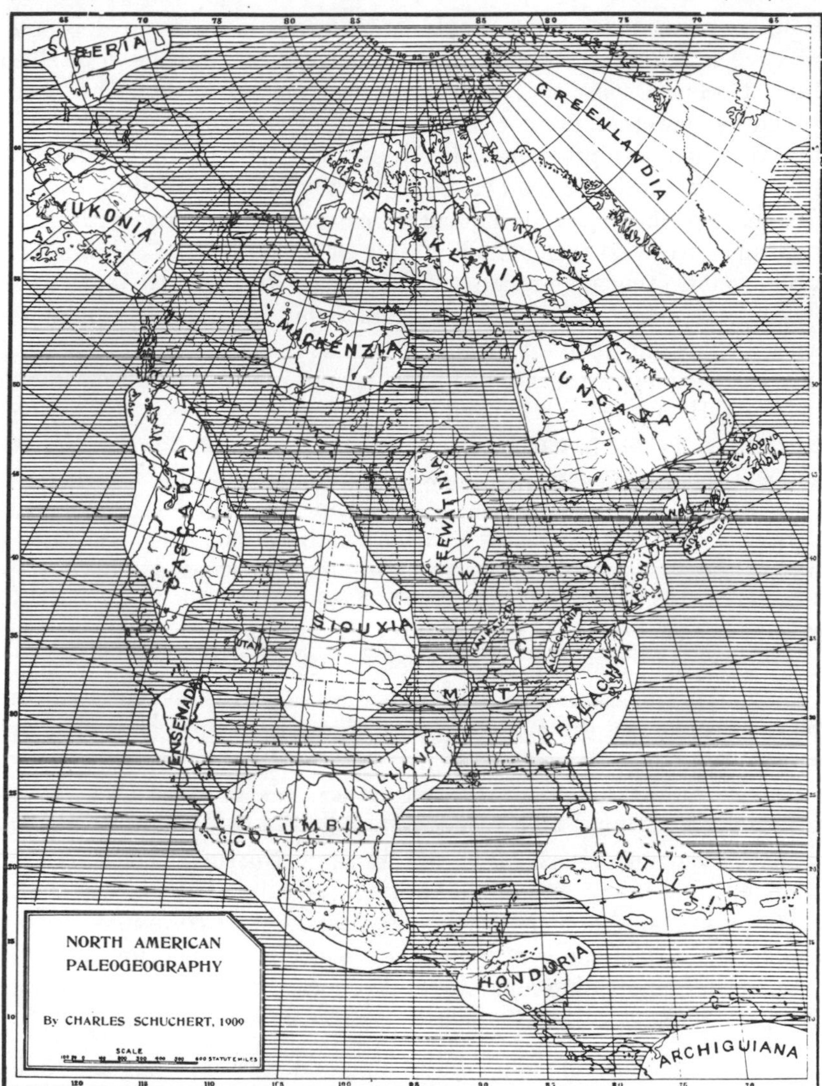

4. Charles Schuchert's borderlands for all of North America, in which the Appalachia prototype was generalized to all of the continent's geosynclines. Names (e.g. Llano, Cascadia, and Franklinia) were coined for each of these as well as for other features shown (compare Schuchert's [1923] Fig. 3 map; see note 24). This map preceded 49 paleogeographic maps of the continent. From "Paleogeography of North America," Bulletin of the Geological Society of America *(1909), 20: plate 49.*

thinner, ocean-basin crust would of necessity have been depressed relative to the thick, stiff continental blocks. As both the crystallization of oceanic crust and overall contraction proceeded, the basins would be depressed farther, and it was inferred from the rock record that they had been progressively deepening since early Paleozoic time. Conversely, continents had risen, especially in Cenozoic time.[30]

Dana thought that a fundamental "cleavage structure" with northeast and northwest trends had influenced the outlines of continents as well as most other major structural features. In 1856 and 1874 papers in the *American Journal of Science* and in his *Manual* of 1875, he presented maps showing major trends, such as the island arcs of the western Pacific and the linear mid-Pacific islands, as well as of continental margins. He argued that lateral pressure of oceanic crust against continental edges due to earth shrinkage produced mountain ranges, volcanoes, and metamorphism. Dana inferred not only that pressure was greatest from the oceans toward the continents, but also that the pressure was greatest on those continental borders facing the largest oceans, because more very high mountains and volcanoes border the Pacific. He considered "that North America should illustrate most simply and perfectly the laws of the earth's genesis. Unlike the other continents, it is bounded on all sides by ocean basins. . . . The conditions under which the lateral pressure acted were therefore the simplest possible; and the evolution was therefore regular as well as systematic" (1875, p. 748).

Dana's most profound contribution was his generalization of the roles of geosynclinals and mountain building in the evolution of continents. Although it was based chiefly upon North America, he was confident that his theory would apply universally. The essence of that theory was that after a geosynclinal had failed, it was crumpled against the stable part of the continent, the resulting synclinorium then becoming stiff, unyielding and stable. The locus of the next geosynclinal would be parallel to it and generally on the outside (oceanward):

> . . . each epoch of mountain-making ended in annexing the region upturned, thickened and solidified, to the stiffer part of the continental crust, and that consequently the geosynclinal that was afterward in progress occupied a parallel region more or less outside the former . . . (1873, p. 171).

He also postulated that as the crust thickened through time, geosynclinals would become less possible and would diminish in extent. Ever-continuing lateral pressure would be relieved more and more in the upward direction; thus younger mountain ranges should achieve greater heights.

It is abundantly clear that Dana, by no later than 1875, had a complete conception of the concentric addition of successive mountain belts that was to become the mid-twentieth-century ruling hypothesis of *continental accretion*. The accretion hypothesis eventually incorporated not only the geosyncline and mountain building but both igneous and metamorphic petrology and even ore formation into a very appealing generalization. While Suess in Europe arrived at a similar view of continental nuclei (shields or kratogens) with outward-accreting mountain belts, the theory was always most popular in North America in the form created by Dana.[31]

Dana and Hall had envisioned a definite sequential history of mountain building; subsidence and thick sedimentation were followed by crumpling and uplift of the sediments to form mountains. Then Bertrand had introduced the concept of an ophiolite-flysch-molasse sedimentary sequence or cycle. At the turn of the century, there was intense interest in igneous, metamorphic, and structural phenomena. Experimental and field evidence now indicated, for example, that all rocks could flow plastically at the pressure existing at a depth of six miles or more. Beginning with R. A. Daly in 1912, the concept of geosynclinal evolution was broadened to include supposed regular, sequential igneous, metamorphic, and even ore-forming events related to the overall evolution of mountain belts.[32] It came to be dogma that mafic rocks (ophiolites) formed first and more silicic rocks formed last in a so-called *tectonic cycle* (also known as the *geosynclinal* or *orogenic cycle*). Such a sequence fit the ideal order of crystallization from a magma, and it also reinforced the continental accretion theory, which assumed the mafic oceanic crust to be primitive and the continental crust to be a complex distillate of more silicic components concentrated from mafic parent material through time. The increasingly popular hypothesis of granitization also seemed to fit into the accretion theory. That is, if recrystallization of deformed and heated geosynclinal rocks to form granite occurred in successive, concentric mountain belts, this seemingly would provide a mechanism for enlarging the volume of so-called granitic continental crust throughout time.[33]

While stressing that mountains form only at the edges of continents and produce enlargements thereof, Dana and his contemporaries seem to have overlooked the apparent contradiction that their borderlands lay seaward of the geosyncline and contained old continental rocks, which meant that the geosyncline actually lay *within* the continent.

This seeming anomaly did not escape Charles Schuchert in 1923.[34] Schuchert subscribed to most of the Hall-Dana tradition, but he concluded that North America had been considerably larger in Pre-Paleozoic time, and that the Appalachian geosyncline had formed *within* the continent; it "has never been a part of the oceans or mediterraneans, nor does it, like the mediterraneans, *lie between continents* . . ." (p. 158; see also pp. 197–198). Instead of endorsing continental accretion, Schuchert adopted the maverick view that the continent had shrunk by two million square miles since Pre-Paleozoic time because the borderlands had been warped and fractured into the oceanic realm (p. 159). Schuchert inferred that Appalachia extended at least 250 miles east of the present shoreline, and so encompassed the entire present continental shelf, which became submerged as early as the Jurassic Period. He emphasized that, in contrast with the European model, Appalachia was not a separate continent but rather an integral part of North America. He also named several types of geosynclines, his most basic distinction being whether they were inter- or intra-continental; he denied that the ocean basins were geosynclines. In effect, Schuchert had opted for two distinct geosynclinal theories—an *American intra-continental* one (à la Schuchert) and a *European inter-continental* one.

State of the Art in 1930

Schuchert's assessment of geosynclines and geanticlines suggests a muddled state of the art paralleling the stalemate noted before in the search for the mechanism of mountain building. For every appealing generalization, there were glaring exceptions. The Principle of Simplicity simply did not seem to be working. At the time of The Great Economic Depression, there was also a depression in ideas about geosynclines and mountain making. In the early twentieth century, global structural patterns and their evolution began to receive a great deal of attention. By the turn of the century, dramatic additional evidence of profound lateral compression across mountain ranges had been amassed, particularly for the Alps. Recumbent folds, thrust faults, klippen, fensters, and especially the huge nappes had become commonplaces. Several Europeans and in America Bailey Willis had produced alpine-type folds and thrust faults in laboratory squeeze boxes, and it was estimated that Alpine structures represent shortening of a geosyncline at least 600 km wide to a mountain range only 150 km wide.

Grand unifying tectonic theories began to appear with increasing frequency, but by 1930 a point of diminishing return was passed for descriptive, comparative tectonics, and redundancy became prevalent. Isostatic excesses and analogies from glaciers led to such suggestions as outward *continental creep* (a perversion of Suess's ideas), *suboceanic spread*, and *batholithic injection* as alternatives to contraction.

It was E. Suess in his great tectonic treatise published between 1885 and 1909 who first focused geologists' attention upon the festoons of volcanic islands and their associated deep trenches or *foredeeps* on the Pacific side of Asia. He linked them genetically with the curving old mountain ranges on the continents, and reasoned that the foredeeps were downwarped as the folded mountain ranges were pushed up and outward from the continent in successively younger concentric rings.[35] Though foredeeps include the deepest zones of the ocean floor, Suess reckoned that some of them had been partially filled with sediments (e.g. the Tigris-Euphrates Valley and Persian Gulf in front of the mountains of Iraq and Iran), and others, completely silted up, were postulated in front of the Alps and Carpathians. (Confirming subsurface data did not yet exist, however.) In some cases he supposed that the mountains had overridden and concealed their foredeeps. While *Pacific-type* continental margins are characterized by arcs and foredeeps, Suess regarded *Atlantic-type* margins as quite different, because they lack the arcs. He thought the latter formed by dismemberment somehow of now-disjunctive mountain belts, which formerly extended across the site of the Atlantic basin. One school of thought postulated a former Atlantic-like continent there, which had subsided and become converted to oceanic crust, while the drifters were soon to propose the alternate explanation of continental separation.

J. Geikie in 1914 criticized Suess's foredeep concept and concluded that such troughs were not "essential accompaniments of folded mountains."[36] They are so much deeper than their complementary mountains are high that he could not believe them to be of the same age and origin. Also Geikie argued that the South American foredeep had the opposite compressional symmetry (continentward) from the western Pacific ones. Because of the many ambiguities and unanswered questions about foredeeps and mountain ranges, Geikie concluded that rather than being essential to formation of folded mountains, deep marginal trenches are a characteristic of all subsiding ocean basins. Around most of the Pacific, where sedimentation is minimal, topographic trenches are prominent, but around the Atlantic, where sedimentation is much greater, they have been filled. Both men could agree on one point, however, namely that thick geosynclinal sedimentation was not a necessary precursor to mountains. Clearly the old geosynclinal theory was a bit threadbare and starting to unravel, but in spite of Geikie's objections, it was Suess's focus upon island arcs and trenches that would ultimately rescue the geosyncline in midcentury. Indeed, it was the beginning of the island arc era—a kind of renaissance—that dictated the termination of the present paper at 1930.*

By 1910 radioactivity had been known for fourteen years, and its geologic implications were beginning to be appreciated. Heat from radioactive decay provided the eventual death knell for the thermal contraction theory of mountain building, although that venerable idea resisted final extermination at least into the thirties. T. C. Chamberlin, for example, rescued contraction for a time with his planetesimal theory, which called for large, low-density embryonic planets capable of significant *gravitational* contraction. Although for quite different reasons, a highly imaginative alternative to global contraction had appeared in 1910 in North America. This was F. B. Taylor's embryonic idea of *continental drift as a mountain building mechanism* due to the compressive drag at the leading edges of moving continents and along zones of continental collisions. Downbuckling there first created a geosyncline, which was filled with sediments, and then was crumpled and upheaved. In rapid succession came the American H. B. Baker's detailed reconstruction of the Atlantic continents (1911) and Alfred Wegener's first championing of drift (1915). It seems ironic that continental displacement had two of its earliest advocates in the United States before Wegener, yet North America was to be the bastion of permanency of continents and ocean basins. This is not the place to detail the history of continental drift; it is introduced only for chronological completeness in the evolution of mountain-building ideas.[37]

Early advocates of continental drift called upon various mysterious forces to drive their vagrant land masses and thus also to form geosynclines and mountains. During the 1920's radioactive dating advanced, and the thermal energy released by decay finally began to attract the interest of a few geologists. One of these was the Scot Arthur Holmes, who in a rather obscure 1931 paper in the *Transactions of the Geological Society of Glasgow* dared not only to endorse continental drift, but also to suggest a new mechanism for it, by thermal convection in the mantle. Convection has been the favorite driving mechanism ever since for mountain building and drift.

*R. H. Dott, Jr., "Tectonics and Sedimentation a Century Later," *Earth-Science Reviews* (1978), 14:1–34.

Post-Mortem

Through its long history the geosyncline has experienced extensive metamorphoses. It commonly has been proclaimed as one of geology's most important unifying concepts, but lately has been scoffed at as either dead or so profoundly altered by plate tectonics as to be an anachronism. A century of laborious, frustrating, yet exciting work by untold numbers of geologists and geophysicists was necessary to sift through the clutter and thresh out wheat from chaff concerning mountains and crustal evolution. Whatever its future may be, the Yankee geosyncline has played an important role in this great intellectual harvest.

It is interesting to return to the speculation about why the geosyncline and Dana's general theory of continental accretion were born in America rather than Europe. Such explanations as the homogeneity of language over an entire continent, the large size of individual countries in America, and the rapid exploration in the late nineteenth century of the vast western regions with their excellent exposures and youthful structures have all been offered to me as explanations; none are very satisfying, however. More significant may be the fact that the United States was literally born in the greater Appalachian Mountain belt; all of the original states touched a part of the system. Furthermore, the mountains themselves were barriers to westward expansion, which would have made them a focal point of great interest for Americans from the start. There was growing interest in mountains in Europe, as well, by the mid-nineteenth century, but the Appalachian Mountains seemed simpler to Europeans than their own. For example, it was much easier to trace thin strata of the continental interior into their thicker mountain-range counterparts here than in Europe. Yet another factor may have been the early studies of the Pre-Paleozoic complex of the southern Canadian shield in the Great Lakes region, which were stimulated by vast ore reserves. Because Sir William Logan and his associates in Canada were very active contributors to nineteenth-century American geological organizations, their growing insights into the old rocks in the core of the continent were quickly disseminated to workers like Hall (who actually worked with Logan) and Dana. Knowledge of Pre-Paleozoic rocks lagged slightly in Europe, and this difference may have given Americans a broader historical perspective on crustal history. Specifically, knowledge of the very old rocks stimulated an early interest in questions about the nature of the earliest continental crust and of the relations of granites and metamorphism to continental evolution.

Dana was correct in regarding the structural symmetry of North America as simpler than that of most continents. His concentric accretion seemed to fit very well here, while the structure of other continents conversely provided evidence that suggested other histories —ultimately continental drift. This difference of geology probably explains in large measure the bitter American opposition to drift; it did not seem needed here. In general, collective experience has influenced most major theories, as is evidenced so clearly by the contrast between the Alpine and the American definitions of a geosyncline. Progress in geology for the past one hundred years has been to a considerable extent proportional to the shedding of the constraints of provincial field experience.

It is clear from the nineteenth-century literature that scientific communication between Europe and North America, as well as within America, was excellent. Indeed, it seems to have been better in some respects then than now, especially between physics and geology. It was only natural that geology would flower in a rapidly expanding young nation, and the record leaves no doubt that North American geology had matured, and was able not only to keep pace but to offer leadership even in theoretical geology. Hall, Dana, Dutton, and many of their contemporaries were as good as the best anywhere. America had in Dana, particularly, a truly exceptional intellectual giant. He was a prolific writer whose textbooks especially would help to shape American geology for the coming century.

Notes

1. James Hall, *Description and Figures of the Organic Remains of the Lower Helderberg Group and the Oriskany Sandstone*, New York Geological Survey, Natural History of New York, Pt. 6, Paleontology (1859), 3, pt. 1:96. A brief statement of the Paleozoic rock-thickness contrast between the midwest and Appalachians was contained in J. Hall and J. D. Whitney, *Report on the Geological Survey of the State of Iowa* (Des Moines, State of Iowa, 1858), 1, pt. 1:35–44. The text of the original lecture was finally published as "Contributions to the Geological History of the American Continent," *Proceedings of the American Association for the Advancement of Science* (1883), 31:29–69. J. M. Clarke's biography, *James Hall of Albany* (Albany, 1923), 565 pp., is a valuable source of background on Hall's career. An unpublished Ph.D. dissertation by D. E. Mayo, "The Development of the Idea of the Geosyncline"

(University of Oklahoma, 1968), 286 pp. (available on microfilm), provides a more extensive account of Hall's work and its reception, with more complete referencing than is possible here.

2. Other recent historical treatments of the geosyncline and related tectonic matters include J. Auboin, *Geosynclines* (Amsterdam, Elzevier, 1965), 335 pp.; P. J. Wyllie, *The Dynamic Earth* (New York, Wiley, 1971), 416 pp.; A. Hallman, *A Revolution in the Earth Sciences* (Oxford, Clarendon Press, 1973), 127 pp.; K. J. Hsü, "The Odyssey of Geosyncline," in *Evolving Concepts in Sedimentology*, ed. R. N. Ginsburg (Baltimore, The Johns Hopkins University Press, 1973), 66–92; and R. H. Dott, Jr., "The Geosynclinal Concept," in *Modern and Ancient Geosynclinal Sedimentation*, ed. R. H. Dott, Jr., and R. H. Shaver, Society of Economic Mineralogists and Paleontologists, Special Pub. 19 (1974), 1–13. These provide especially more treatment of European ideas and/or peripheral structural and geophysical issues. Mayo (1968; above, note 1) provides a thoroughly documented treatment of the early geosynclinal concept, especially in America.
3. An especially important pioneer work was H. D. Rogers and W. B. Rogers, "Mountain Building Forces Exemplified by the Appalachian System," *Transactions of American Geologists and Naturalists for 1840–1842* 1:474–531. The Rogers brothers documented an asymmetry of folds, which was interpreted as due to westward push of Appalachian strata over a fluid subcrust; they coined *anticlinal*, *synclinal* and *monoclinal mountains* in this paper. The Rogers' work was known in Europe, but similar cross sections showing spectacular folds of the Jura Mountains and of southwestern England also appeared about the same time. A thrust fault was recognized as early as 1826 near Dresden, Germany, where C. S. Weiss found ancient granite lying flat upon Cretaceous strata with a faulted contact (A. Rothpletz, "Die Uberschiebungen und ihre methodische erforschung," *VIe Congr. Internat. (1894), Compte Rendu* (Lausanne, 1897), 6:161–177. H. D. and H. B. Rogers also had recognized overthrust faults in the Appalachian Mountains by 1842.
4. Hall, *Description and Figures*, 88.
5. Hall acknowledged the ideas of Herschel expressed in two 1836 letters quoted by Charles Babbage in *The Ninth Bridgewater Treatise: A Fragment* (London, John Murray, 1837), 207, 216–217. Herschel argued that there is no internal driving force. Rather, an increase of pressure due to sedimentation causes subsidence of the surface, and yielding beneath will flow laterally to produce an adja-

cent elevation. Secondly, he postulated subterranean isothermal surfaces roughly parallel to surface topography. Where thick sedimentation occurred, these surfaces would rise, heat the mass, and either heave it up or crack it to form volcanoes. In suggesting an equilibrium of both pressure and of temperature, Herschel clearly anticipated the principle of isostasy; the influence of his two letters on nineteenth-century geology was profound. Europeans William Hopkins and F. D. R. de Montlosier also contributed to Hall's notion of complementary uplift and subsidence on a continental or oceanic scale, and Charles Lyell influenced Hall's conception of gradualism of sedimentation by ocean currents and of deformation and uplift (Lyell visited Hall during both of his trips to America). See Clarke's and Mayo's works (above, note 1) for further background.

6. Hall, 70, 73.
7. Hall rejected any need of igneous intrusions or of folding for metamorphism. He reckoned that deep subsidence would result in a rise of isogeotherms, which would be sufficient to produce metamorphism. He also was favorable toward a suggestion by both Lyell and T. Sterry Hunt that vapors released from the interior by chemical reactions contributed to metamorphism as well as producing earthquakes and volcanoes. Hunt, especially, thought it but a small step from metamorphism to vulcanism. Hall was aware that William Logan had found in Canada fragments of preexisting sedimentary rocks preserving their original characteristics within metamorphic terranes; his recognition that many so-called Primary crystalline rocks in the Appalachians were in reality metamorphosed Paleozoic ones was prophetic.
8. J. D. Dana, *Manual of Geology* (2nd ed., New York, Ivison, Blakeman, Taylor, 1875), 748–749. This edition contains a more concise statement of Dana's concepts (735–754) than his famous, nearly contemporaneous paper "On some Results of the Earth's Contraction from Cooling, Including a Discussion of the Origin of Mountains, and the Nature of the Earth's Interior," *American Journal of Science* (1873), 3rd ser., 5:423–443, 443, 474–475; 6:6–14, 104–115, 161–172, 381–382. His earliest paper on continental evolution was "Volcanoes of the Moon," *American Journal of Science* (1846, 2nd ser., 2:335–355. His presidential address to the American Association for the Advancement of Science, which preceded Hall's by two years, was published as "On the Plan of Development in Geologic History of North America," *American Journal of Science* (1856), 2nd ser., 22:305–335. Three papers in

1890 dealt with his latest ideas about geosynclines; see, for example, "Archaean Axes of Eastern North America," *American Journal of Science* (1890), 39:378–383. Dana published many other papers touching on the subject, especially in the *American Journal of Science* (of which he was editor), but the above ones, together with the 4th edition of the *Manual* (1895), provide full insight into the evolution of his ideas.

9. See Dana's 1846 paper (above, note 8), 355.
10. Dana's 1856 paper, 344.
11. Dana, *Manual*, 748.
12. Dana's 1873 paper, 170–171.
13. Ibid., 171.
14. Relevant literature on the isostasy debate is too vast to cite fully; Mayo (above, note 1) has documented it well. The most important references include the following: C. S. Dutton, "A Criticism upon the Contractional Hypothesis," *American Journal of Science* (1874), 3rd ser., 8:113–123; "On Some of the Greater Problems of Physical Geology," *Bulletin of the Philosophical Society of Washington* (1889), 11:51–64; J. LeConte, "On the Structure and Origin of Mountains, with Special Reference to Recent Objections to the Contractional Theory," *American Journal of Science* (1878), 3rd ser., 16:95–112; Osmond Fisher, *Physics of the Earth's Crust*, 1st ed. 1881, 2nd ed. 1889 (London, Macmillan); R. S. Woodward, "The Mathematical Theories of the Earth," *Annual Report of Board of Regents of the Smithsonian Institution to July, 1890* (Washington, 1891); B. Willis, "The Mechanics of Appalachian Structure," *Geology, Part II, Thirteenth Annual Report of the U.S. Geological Survey to the Secretary of Interior, 1891–92*, ed. J. W. Powell (Washington, 1893), 211–281.
15. See below, note 20, for the Haug citation. There were important exceptions, however. For examples, Eduard Suess was not convinced that so-called geosynclinal sediments, which he regarded simply as pelagic deposits, were linked with mountains (*Die Entstehung der Alpen*, Wien, W. Braumuller, 1875, 168 pp.); and J. Geikie disbelieved emphatically that thick sediments were necessary precursors to mountains (*Mountains—Their Origin, Growth and Decay*, New York, Van Nostrand, 1914, 311 pp.).
16. Hsü (above, note 2) cites M. Neumayr, *Erdgeschichte* (Leipzig, Bibliographisches Inst., 1875), 1:364, and J. Walther, "Lebensweise fossiler Meeresthiere," *Zeitschr. Deutsch. Geol. Ges.* (1897), 49:209–273, which were not available to me. M. Bertrand gave an oral presentation in 1894, which was published as "Structure des

Alpes Francaises et recurrenee de certains faciès sedimentaires," *VIe Congr. Géol. Internat. (1894), Compte Rendu* (Lausanne, 1897), 161–177.

17. The most classic paper is G. Steinmann, "Die geologische Bedeutung der Tiefseeabsätze und der ophiolithischen Massengesteine," *Naturf. Ber.* (1905), 16:44–65. According to Auboin (above, note 2), however, D. Pantanelli anticipated Steinmann with his paper "I diaspiri: della Toscana e loro fossili," *Reale Accad. Nazl. Lincei.* (1883), 7:13–14, which I have not seen. E. Suess, *Das Antlitz der Erde* (Leipzig, Freytag, 1909), pt. 2, 3:789 pp.
18. H. H. Hess, "Gravity Anomalies and Island Arc Structure with Particular Reference to the West Indies," *Proceedings of the American Philosophical Society* (1938), 79:91–93.
19. There may be an interesting postscript to this success story, however, in light of a recent challenge by Americans R. L. Folk and E. F. McBride, "Possible Pedogenic Origin of Ligurian Ophicalcite: A Mesozoic Calcified Serpentinite," *Geology* (1976), 4:327–332. These authors reinterpret *some* of the classic European ophiolites to show evidence of formation at shallow depths and in some cases even to show emergence above sea level.
20. E. Haug, *Traité de Géologie* (Paris, Colin, 1907), 1: p. 159, Cf. "Les géosynclinaux et les aires continentales," *Soc. Geol. France Bull.* (1900), 28:617–711.
21. E. Argand, "Sur l'Arc des Alpes Occidentales," *Ecologae Geol. Helvetiae* (1916), 14:145–191.
22. James Hall, *Geology of New York, Part IV, Comprising the Survey of the Fourth Geological District* (Albany, State of New York, 1843).
23. Dana, *Manual of Geology* (1875), 752.
24. This little-known map coupled with the fact that Chamberlin was a midwesterner who had not himself worked on Appalachian rocks provides a measure of how widely known and discussed were the ideas of Hall, Dana, and the Rogers brothers by the 1880's.
25. C. D. Walcott, "Correlation Papers—Cambrian," *U.S. Geological Survey, Bulletin 81*, Pl. III; H. S. Williams, "On the southern Devonian formations," *American Journal of Science* (1897), 4th ser., 3:393–403.
26. See above, note 8.
27. There is a striking similarity between Dana's travels and research by ship around South America and across the Pacific with Darwin's slightly earlier (1831–32) experiences in *H.M.S. Beagle*. Both expeditions had similar economic goals for their respective govern-

ments, both followed similar routes, and both men did geologic and biologic research of great significance.

28. *Manual of Geology* (1895), 379.
29. The inference of permanency seems only natural (if not inevitable) as a first guess in 1846, for it accords with the Principle of Simplicity. Later, isostasy provided supporting physical evidence. But the clinging to permanency, especially in America, for a whole century in the face of so much contrary evidence (see the symposium *Theory of Continental Drift*, Tulsa, American Association of Petroleum Geologists, 1928) prompted Alexander DuToit's assessment in *Our Wandering Continents* (Edinburgh, Oliver and Boyd, 1937) that "Geology has not wholly recovered from the cramping influence of the Uniformitarian doctrine of Lyell" (p. 7). He also characterized drift "as the unescapable deduction from the wealth of geological evidence available to the unfettered mind" (p. 21).
30. Dana, 1846, 1873, 1875, 1895.
31. Examples of subsequent support and amplification of the basic concept include A. W. Grabau, "Migration of geosynclines," *Bulletin of the Geological Society of America* (abs.) (1919), 30:87; R. T. Chamberlin, "Significance of the framework of continents," *Journal of Geology* (1924), 32:545–574; M. Kay, "North American Geosynclines," *Geological Society of America*, Memoir (1951), 48:143 pp.
32. R. A. Daly, "Geology of the North American Cordillera at the 49th Parallel," *Geological Survey of Canada*, Memoir (1912), Pt. 2, 38:547–857; F. Kossmat, "Die mediterranen Kettengebirge in ihrer Berichtigung zum Gleichgenichtszustande der Erdrinde," *Sachsische Akad. Wiss.* (1921), Abh., Math-Phys. Kl. 38, 46–68; E. Kraus, "Der orogene Zyklus und seine Studien," *Centralbl. fur Mineralogie* (1927), Abt. B, 216–233; A. Knopf, "The Geosynclinal Theory," *Geol. Soc. America*, Bulletin (1948), 57:649–670.
33. Extreme granitization contained a crucial fallacy, namely that continents could hardly increase in size if all granite were derived by recrystallization only of geosynclinal sediments, no matter how many times the material was eroded, deposited, recrystallized and re-eroded in what amounted to a closed system. Additions of new volcanic material from the mantle, however, if later granitized, could correct this shortcoming.
34. Charles Schuchert, "Sites and Nature of the North American Geosynclines," *Geological Society of America*, Bulletin (1923), 34:151–230. This, his presidential address, was a keynote paper for a Symposium on the Structure and History of Mountains and the Causes

of their Development, which included papers by seven other prominent geologists (pp. 231–400). Schuchert's classification of geosynclines and other tectonic elements foreshadowed M. Kay's more elaborate classification twenty years later. This paper by Schuchert may well have been a model for Kay's famous memoir (above, note 31). That memoir clearly shows a strong Dana influence.

35. E. Suess, *Das Antlitz der Erde* (Wien, Tempsky, 1885), 1:778; (1888), 2:703; (1901), 3, pt. 1:508; (1909), 3, pt. 2:789. It was translated into French as *La Face de la terre* (Paris, Colin, 1897–1918), and into English as *The Face of the Earth* (Oxford, Clarendon, 1904–24). O. Ampferer and E. Argand in Europe and B. Willis, A. C. Lawson, and W. H. Hobbs in America were among those who subscribed to underthrusting (later called *subduction*) of oceanic beneath continental slabs, as opposed to Suess's view of continental material thrust out over oceanic.
36. J. Geikie, p. 226.
37. For the history of drift theory, see Dutoit, 1937; Wyllie, 1971; Hallam, 1973; as well as U. B. Marvin, *Continental Drift* (Washington, Smithsonian Institution, 1973) and R. H. Dott, Jr., and R. L. Batten, *Evolution of the Earth* (2nd ed., New York, McGraw-Hill, 1976).

22

Geodesy and the Earth Sciences in the Philosophy of C. S. Peirce

Val Dusek

This is an investigation of the relationship between Charles Sanders Peirce's researches for the U.S. Coast and Geodetic Survey and his general philosophy. It is not a history of his work for the Survey, which has been treated in articles by Victor Lenzen[1] and Carolyn Eisele,[2] and will be treated in the massive biography by Max H. Fisch. Nor is it an investigation of his institutional and political connections with the Survey, which have been investigated by Manning.[3] This is, rather, an investigation of ways in which the philosophy of the founder of pragmatism and America's greatest philosopher may have been moulded by the particular sorts of professional scientific work in which he engaged.

Peirce, perhaps the one truly universal mind that nineteenth-century America produced, has been characterized as something of a geological phenomenon himself. Bertrand Russell, confusing his vulcanology somewhat, called him "a volcano spouting vast masses of rock of which some, on examination, turn out to be nuggets of pure gold."[4] Although Peirce was known in his own day primarily for his scientific work undertaken for the Survey, from the time of his death until recently he has been known mainly as a philosopher, historian of science, and mathematical logician. During the last decade historians have investigated his scientific work, and the Coast and Geodetic Survey named a ship after him to recognize his contributions. But so far there has not been an extended investigation of the relation between the type of research he did and his philosophy.

After an intense, disciplined, early-childhood education by his father in mathematics, puzzles, and sensory acuity (including wine-tasting) Peirce had an undistinguished record at Harvard. After graduation he became an aide to the Coast and Geodetic Survey on July 1, 1861. He worked for a year and returned to Harvard to receive an advanced degree in chemistry. He studied classification with Louis Agassiz for six months, claiming that this later aided his work on the classification of the sciences.[5] He mentioned that the one science of which he was most ignorant was geology.[6]

On July 1, 1867, Peirce was made an assistant in the Survey. He was in charge of the Coast Survey Office in 1872. He continued as an assistant until he resigned on December 31, 1891. During the 1880's, changes in administration and Peirce's always blunt and outspoken character made his work difficult.[7] In 1887 he partially retired to Milford, Pennsylvania, hoping his students would follow him there (they did not).

Peirce's tenure with the Survey almost precisely corresponds to the period of his work on pragmatism and the philosophy of science. After 1891 he turned more to metaphysical speculations. He worked in a variety of scientific capacities. He was computer for the almanac in 1873 and was special assistant in gravity research 1884–91. He made stellar observations, compared modern results with ancient star catalogues, invented an improved pendulum, discovered errors in European pendulum measurements of gravity, improved calculations of the shape of the earth, and measured the meter's length in terms of the wavelength of light, among other activities. He was well-known in Europe for his gravimetric researches and his work in metrology.[8] He invented a new method of map projection using elliptic integrals which has finally found use in maps of airline routes. (Carolyn Eisele has described this).[9]

There are a number of influences traceable from Peirce's science to his philosophy. Some of these are very direct and easily documented. Peirce used examples from his work with pendulums in his papers on pragmatism.[10] He several times mentions stellar observations (on occasion to ridicule Comte's use of the example of the chemistry of the stars as an unknowable datum).[11] Also two of Peirce's papers stand intermediate between his strictly scientific work and his philosophy of science. One, "On the Theory of Errors of Observation," and the other, "A Note on the Economy of Research," are practical outcomes of his work on metrology and his part in the socially organized research of the Survey.[12] The latter paper Peirce claimed predated Ernst Mach's popularization of the criterion of economy in evaluating scientific theories. Peirce's paper is indeed far more advanced than Mach's work, in that he considers the economics of a sequence of research decisions and actually works out the mathematical economics for the "operations research" involved.[13]

There are also more general influences of Peirce's research activity on his philosophy. One of these noted by Eisele is the generalization of the notion of mapping in Peirce's philosophy.[14] Peirce did practical work on map projection. He also used graphical methods in logic in a highly innovative way ("logical graphs").[15] Indeed, Peirce is unique among modern philosophers in the extent to which he recognizes the possibilities of mapping techniques to represent philosophical notions which philosophers such as Hegel, Bergson, and Nietzsche claim are "non-spatializable." Presumably his experience with a wide variety of mapping techniques did not lead him to identify "spatializing thought" solely with Euclidean descriptive geometry.[16]

Another area where influences from the earth sciences may be surmised, but wherein the connections are much more hypothetical, is in Peirce's cosmology, on which he labored toward the end of his life in retirement. One of the major principles of his philosophy is "synechism"—continuity. Commentators generally attribute this idea to the influence of mathematical studies and of Darwinism on Peirce's thought,[17] but it is perhaps significant that Chauncey Wright, Peirce's philosophical mentor, from whom Peirce absorbed evolutionary doctrine, was influenced by Lyell and by uniformitarianism.[18]

Although Peirce's references to geology and mineralogy are not numerous (appearing primarily in his classifications of the sciences in Vol. 1, of the *Collected Works*, in his use of classification and nomenclature of rocks in his theory of signs and his letters to Lady Welby, and in "The Century's Great Men of Science,"),[19] there is one mineralogical metaphor that is central to his cosmology: crystallization. This could come from his chemical studies, of course, but he uses the metaphor of an extremely gradual and long-term process of crystallization from disordered heterogeneity to ordered homogeneity to describe the overall evolution of the universe.[20]

However, the most significant and general area of influence of

Peirce's scientific work on his philosophy is, I believe, in the area of his pragmatism and philosophy of science. Commentators on Peirce all emphasize the centrality of Peirce's scientific training to his thought in general, but they often fail to point out that the particular sorts of scientific work in which Peirce engaged makes him a rarity, indeed in many ways unique even for a scientifically trained philosopher. Even today, although scientifically trained philosophers are less of a rarity than in Peirce's day, few philosophers of science have background in the *observational* and *experimental* sciences. Most twentieth-century philosophers with science backgrounds or eminent scientists who have written on philosophy have been mathematicians or theoretical physicists. Few experimental physicists (with the notable exception of Bridgman, whose operationalism has considerable resemblance to Peirce's more sophisticated treatment of "How to Make Our Ideas Clear")[21] have reported at length their philosophical reflections on their work.

Among major philosophers since Aristotle who have philosophized on the sciences, most have been mathematicians insofar as they had any training in the subject-matters they discuss. A notable exception is Kant, who lectured on physical geography and published papers on volcanoes and earthquakes (and who was the modern philosopher most closely studied by Peirce).[22] Peirce's one major deficit in his scientific experience, however, was his lack of work in an historical natural science.

Although Peirce was a geodesist trained in mathematics and physics, he was not a geologist. Although he said that geology was the science in which he was least knowledgeable, given the scope of his knowledge in mathematics, physics, chemistry, and biological classification, this was *comparative* ignorance, not absolute. Also, although Peirce was a specialist in geodesy, a branch of physics, he was in frequent personal contact with geologists in his work in the Survey. Thus if he was not trained in the historical science geology, he was a professional in a science closely allied to it in his professional work.

Peirce, although trained in and creative in pure mathematics, (the bulk of his mathematical manuscripts are only now being published)[23] was professionally an experimentalist and observational scientist and was known primarily in his own lifetime for his observations and experimental apparatus. He emphasized the difference between his own "laboratory" mind and the "seminary" minds that dominated the philosophy of his day (and perhaps ours).[24] Indeed Peirce, through his study of the logical works of the scholastics (especially Duns Scotus) was well aware that the logical and semantic precision which often pass today for "scientific philosophy" were in the Middle Ages compatible with subject matters and modes of thought at odds with that of experimental science.[25]

Indeed Peirce's long and patient work in observational and experimental science sets his philosophy of science apart not only from that of the seminary men of his own day but also from the recently popular philosophers of science of our own. Karl Popper, Norwood Russell Hanson, Paul Feyerabend, and others have tended to emphasize the huge conceptual shifts, scientific revolutions, and metaphysical aspects of science rather than aspects which Peirce emphasized. Peirce's philosophy of science was neither a narrow psychological empiricism (as in David Hume or John Stuart Mill) nor a study in broad, conflicting ideologies or world-views (as is much post-Koyré, post-Kuhn philosophizing about science). I believe that Peirce's experience in accurate recording of positions of stars, correction of pendulum data, etc. gave him a much more balanced view of the scientific enterprise than the view of those whose science is learned from reading only the major conceptual innovators in science, such as Galileo and Einstein, and not from experience with less spectacular cumulative research.

Peirce was a "mathematical physicist" in his services for the Survey, but he was not a "theoretical physicist." His application of the principles of physics was to the determination of the irregular, heterogeneous, given structure of the earth, and to the determination of the deviations of the actual shape of the earth from calculations thereof based on various idealizing assumptions. I think this may account for the emphasis on experiment and observation in his philosophy which distinguishes it from the philosophies of science of twentieth-century mathematicians and theoretical physicists such as Alfred North Whitehead, Werner Heisenberg, and Eugene Wigner, parts of whose philosophies Peirce anticipates.[26]

Peirce's area of concentration, geodesy, was a particularly apt one in which to gain a balanced view of science during the late nineteenth century. Unlike much of descriptive biology it was able to utilize the laws and techniques of exact physics. But unlike, say, theoretical celestial mechanics, its connections with detailed and constantly novel observations was close. Thus Peirce was led neither to the British empiricist error of treating science as a simple, mechanical, automatic outcome of numerous observations, nor to the rationalist, Platonist error of believing that science could be done a priori by mathematical reasoning, with no observation. He appreciated the power and almost mystical prescience of mathematical thought without underestimating the role of brute fact (what he called "secondness") and of unexpected and disorienting observations. Poincaré in contrast was mainly active in an area of theoretical celestial mechanics which was more concerned with working out the mathematics than with novel observations, and this may account in part for the latter's tendency to treat the observational, factual component of mechanics as largely a matter of definition and convention.[27]

Peirce's participation in science was as part of an organized community of scientists, each contributing a small part to the overall advance of research. His work in applying known laws of physics to the determination of the shape of the earth was by no means trivial and involved energy and effort in evaluating and criticizing data and in constructing and even inventing apparatus. This, I think, gave Peirce a more balanced view of both "revolutionary" and "normal" science (to use the current jargon) than have many of our contemporaries. He recognized the feats of ingenuity involved in even apparently humdrum "normal" science and realized that, although cumulative, it was not the product of simple inductions. On the other hand (for instance, in his studies of Kepler) he did not overemphasize the irrationality and discontinuity of even major scientific breakthroughs.[28] In his notion of "abduction," or the logic of discovery, which is neither induction nor deduction, he showed the rational element in scientific discoveries which from the viewpoint of simple inductivism would seem arbitrary and inexplicable.[29] Unlike Mill and Bertrand Russell, who treated Kepler's work as a simple outcome of mechanical inductions and calculations, and Kuhn, Popper, et al., who treat the major scientific revolutions as rationally inexplicable or as major discontinuous leaps of thought and history, Peirce did justice to the common elements of both normal and revolutionary science: the insight and ingenuity involved in routine research and the elements of abductive logic and continuity in even the major theoretical shifts. Peirce's own research experience (along with his studies of the history of science) allowed him to avoid the excesses of both simplistic inductivism and crude falsificationism or irrational overreaction to the failures of inductivism and Popperianism as accounts of science.

Peirce's work was with extremely accurate measurements to seven or eight decimal places. He worked and wrote on the theory of errors of observation and noted that:

> infallibility in scientific matters seems to me irresistibly comical . . . In those sciences of measurement which are least subject

> to error—metrology, geodesy, and metrical astronomy—no man of self-respect ever now states his result without affixing to it a probable error; and if this practice is not followed in other sciences it is because in those the probable errors are too vast to be estimated.[30]

The sciences referred to are, of course, those in which Peirce was active. The very "romance of the next decimal place" of late nineteenth-century physical science made Peirce aware of the fallible nature of science (his "falliblism") and of the inaccuracy inherent in all measurement. Peirce generalized this finding methodologically in his falliblism and self-correcting method of induction. He also generalized it metaphysically in his view that there is "real vagueness" in the structure of the universe, that it is not really made of separate objects neatly distinguishable from one another in all cases. Unlike Laplace, he attributed the probable errors not to us alone but ultimately to the nature of reality. In "The Architecture of Theories" he makes the suggestion (along with an argument for the non-Euclidean nature of astronomical space based on probable-error considerations) that processes at the atomic level are indeterminate (1891).[31] He believed that nature itself did not allow measurements beyond a certain accuracy.

Peirce also extrapolated his scientific work on accuracy of measurement to his theory of truth. He treats truth as a regulative ideal of inquiry which, approached ever more accurately, better and better approximated, is never fully reached (thereby justifying falliblism, and the self-corrective methods and at the same time the progress of science and of the scientific community).[32] His idea here is an application of ideas of Kant to an area where Kant did not apply them, but it is also a generalization of his work on measurement.

Peirce also defined scientific truth in terms of the long-term evolution of the scientific community and its self-correction.[33] Indeed, his emphasis on the social aspect of science and the role of the scientific community in the definition of scientific truth (as that upon which the scientific community will agree in the long run) presages developments in the philosophy of science three quarters of a century later.[34] Here there is a clear biographical connection. Peirce was associated with a large-scale, government-financed, scientific research organization which gave him knowledge "from within" of the structure of a scientific community.[35] Indeed his association with the Survey was perhaps his longest and often his only institutional tie. Peirce never gained tenure at a university.[36] He was persona non grata at Harvard after divorcing his first wife, who was well connected there. He returned to Harvard to lecture briefly only through the efforts of his friend and benefactor William James. He left Johns Hopkins apparently because of a scandal. He was not a part of the community of philosophers. His logical ideas were too advanced to be appreciated even by friends or students like William James and John Dewey. James called his early articles "exceedingly bold, subtle, and incomprehensible," and his Lowell Lectures "Flashes of brilliant light relieved against Cimmerian darkness." Dewey said that his presence in Peirce's class at Johns Hopkins was merely that of "two ships that pass in the night."[37] Peirce finally retired to relative isolation with his second wife during his last decades. His association with the Survey was his one constant institutional tie during his middle decades. He idealized the scientific community and made it central to his philosophical conceptions but has very little to say about general social philosophy.[38]

In summary, Peirce's professional work for the Coast and Geodetic Survey influenced his philosophy in a number of ways: in his technical methodological work, in his use of mapping, in his pragmatist emphasis on experimentalism, in his treatment of the nature of scientific truth, in his treatment of the logic of scientific discovery and the cumulative nature of science, in his emphasis on the role of the scientific community, and perhaps in his metaphysics.

Notes

1. Victor F. Lenzen, "Charles S. Peirce and die Europäische Gradmessung," *Actes du Xme Congres International d'Histoire des Sciences* (Paris, 1964) 781–783. This volume will subsequently be referred to as "Ithaca," since the Congress was held there and the volume has that title.
   ———, "Charles S. Peirce as Astronomer," Edward C. Moore and Richard S. Robin, eds., *Studies in the Philosophy of Charles Sanders Peirce, Second Series* (Amherst, Mass., University of Massachusetts Press, 1964), 33–50.
   ———, "The Contributions of Charles S. Peirce to Metrology," *Proceedings of the American Philosophical Association* (1965), 109:29–46.
   ———, "Charles S. Peirce as Mathematical Geodesist," *Transactions of the Charles Peirce Society* (1972), 8:90–105.
   ———, "Charles S. Peirce as Mathematical Physicist," *Transactions of the Charles Peirce Society* (1975), 11:159–160.
   For a bibliography of further works of Lenzen on Peirce, see "Vic-

tor F. Lenzen (1890–1975)," *Transactions of the Charles Peirce Society*, ibid., 195–196.

2. Carolyn Eisele, "Charles Sanders Peirce," *Dictionary of Scientific Biography*, ed. C. C. Gillispie (New York, Scribner, 1970), X, 482–488.
   ———, "The Quincuncial Map-Projection of Charles S. Peirce," *Ithaca*, 2:687.
   ———, "Charles S. Peirce and the Problem of Map Projection," *Proceedings of the American Philosophical Society* (1963), 107: 299–306.
   ———, "Charles S. Peirce, Nineteenth-Century Man of Science," *Scripta Mathematica* (1959), 24:305–324.
   ———, "The Charles S. Peirce—Simon Newcomb Correspondence," *Proceedings of the American Philosophical Society* (1957), 101, no. 5:410–433.
3. Thomas G. Manning, "Peirce, The Coast Survey, and the Politics of Cleveland Democracy," *Transactions of the Charles Peirce Society* (1975), 11:186–193.
   ———, *The U.S. Coast and Geodetic Survey*, forthcoming.
4. Israel Shenker, "A Thinker's Thinker is Honored Belatedly," *The New York Times* (12 October 1976), 39.
5. Charles Sanders Peirce's Letter to Lady Welby, 14 March 1909, in Philip P. Wiener, ed., *Charles S. Peirce: Selected Writings* (New York, Dover Publications, 1966), 417. Paul Weiss, "Biography of Charles Sanders Peirce," in Richard J. Bernstein, ed., *Perspectives on Peirce* (New Haven, Yale University Press, 1965), 1–12.
6. Charles S. Peirce, *The New Elements of Mathematics*, ed. Carolyn Eisele (Atlantic Highlands, N.J., Humanities Press, 1976), Vol. 4.
7. Manning, "Peirce" (note 3, above).
8. Lenzen, "Contributions" (note 1, above).
   ———, "Die Europäische Gradmessung" (note 1, above).
9. Eisele, "The Quincuncial Map-Projection" (note 2, above).
10. *Collected Papers of Charles Sanders Peirce*, ed. Charles Hartshorne and Paul Weiss (Cambridge, Mass., Harvard University Press, 1934), 5: pars. 403, 404, 407. Further references to Peirce will be in the form of volume number and paragraph.
    Lenzen, "Astronomer" (note 1, above), 33–34.
11. *Papers*, 6:556.
12. Charles Sanders Peirce, "Note on the Theory of the Economy of Research," in *Report of the Superintendent of the United States Coast Survey Showing the Progress of the Work for the Fiscal Year Ending with June 1876* (Washington, Government Printing Office,

1879), reprinted in W. E. Cushen, "C. S. Peirce on Benefit-Cost Analysis of Scientific Activity," *Operations Research* (1967), 14: 641–648.
13. Ibid.
14. Nicholas Rescher, "Peirce and the Economy of Research," *Philosophy of Science* (1976), 43:71–98.
15. Don Davis Roberts, "The Existential Graphs and Natural Deduction," in Moore and Robin, 109–121, and Don Davis Roberts, *The Existential Graphs of Charles S. Peirce* (Atlantic Highlands, N.J., Humanities Press, 1972).
16. Peirce, *The New Elements of Mathematics*, 4, "Abstracts of Eight Lectures [Topological Basis of Philosophy of Continuity]," esp. 146–147.
17. Philip P. Wiener, *Evolution and the Founders of Pragmatism* (New York, Harper and Row, 1965). Charles Hartshorne, "Peirce's 'One Great Discovery' His Most Serious Mistake," in Moore and Robin.
18. Wiener, *Evolution*, 7.
19. *Papers*, 1:266, 260; 2:121; 5:578. Wiener, *Selected Writings*, 149–150, 273, 409–410.
20. *Papers*, 6:33.
21. *Papers*, 5:388–410, originally *Popular Science Monthly* (1878), 12:286–302.
22. Charles Sanders Peirce, "Entry in the Harvard Class Book of 1859," in Wiener, *Selected Writings*, xxiii–xxiv. *Papers*, 1:4.
23. Eisele, "Correspondence" (note 2, above).
24. *Papers*, 1:126–129.
25. Charles S. Peirce, "The Laws of Nature and Hume's Argument against Miracles," in Wiener, *Selected Writings*, 289–321. "Critical Review of Berkeley's Idealism," ibid., 73–88, esp. 76–79.
26. Victor Lowe, "Peirce and Whitehead as Metaphysicians," in Moore and Robin, 430–454. Werner Heinsenberg, *Physics and Philosophy* (New York, Harper and Row, 1958), chapter 8. Eugene Wigner, "On the Unreasonable Effectiveness of Mathematics in the Natural Sciences," *Symmetries and Reflections* (Bloomington, Indiana University Press, 1967), 222–237.
27. Henri Poincaré, *Science and Hypothesis* (New York, Dover Books, 1952), xxii, xxvi, 91–97.
28. Charles Sanders Peirce, "Kepler," chapter 14 of Lowell Lectures on the History of Science, in Wiener, *Selected Writings*, 250–256. *Papers*, 1:72–74, 2:96–97.
29. *Papers*, 5:171–172; 5:197; 6:525.

30. Ibid., 1:9.
31. Ibid., 6:29, 11.
32. Ibid., 5:407, 565.
33. Ibid.
34. Michael Polanyi, *Personal Knowledge* (Chicago, University of Chicago Press, 1958). Thomas Kuhn, *The Structure of Scientific Revolutions* (Chicago, University of Chicago Press, 1962).
35. Manning (above, note 3).
36. Weiss, "Biography."
37. W. B. Gallie, *Peirce and Pragmatism* (New York, Dover Books, 1966), 55. John Dewey, personal communication to George Gentry. William James, *Pragmatism* (New York, Longmans, 1970), 5.
38. Rulon Wells, "Charles Peirce as American," in Bernstein, *Perspectives*, 13–41.

Agassiz' Later, Private Thoughts on Evolution: His Marginalia in Haeckel's Natürliche Schöpfungsgeschichte *(1868)*

Stephen Jay Gould

The aphorism that new theories triumph only when their opponents die has been attributed to so many prominent scientists that it must be essentially valid. The success of evolutionary theory was rapid and unbounded in the years following 1859. Ten years after the *Origin*, only a handful of prominent opponents remained, none more notable or more uncompromising than our own Louis Agassiz, America's foremost naturalist and Director of the Museum of Comparative Zoology at Harvard. But Agassiz, after an initial flurry of opposition in professional journals (for example, his 1860 review of the *Origin* in the *American Journal of Science*), chose largely to keep his own council for the rest of his life. In his later years he confined his attacks to popular articles (most famous is his posthumous work "Evolution and Permanence of Type" in the *Atlantic Monthly*, 1874). Thus, though we understand his intransigence, we do not really know how Agassiz reacted—if he reacted at all—to the professional literature of evolutionists during the late 1860's. Did he retreat to unconsidered dogmatism, tinged with theology? Did he simply lose his nerve or come secretly to doubt his own convictions?

Two years ago I discovered on the open stacks of Harvard's Museum of Comparative Zoology library Agassiz' copy of the first edition (1868) of Ernst Haeckel's *Natürliche Schöpfungsgeschichte* (*The Natural History of Creation* in later English translations)—according to Nordenskiöld (1929), the most influential book ever written in support of evolution (in terms of convincing large numbers of people). The book is crammed full of Agassiz' penciled marginalia, in German script.* In the context of his withdrawal from the debate among professionals, these marginalia form an important document in the history of evolutionary thought—for they represent Agassiz' longest and most coherent reaction, in later life, to the arguments of a prominent evolutionist. When we realize, in addition, how intensely Agassiz disliked Haeckel —more for his arrogance, materialism, strident antitheism, and flippant treatment of scientific evidence than for the content of his views (see Lurie, 1960), these marginalia assume added interest, if only because we are all voyeurs at heart. I present here only an abstract and some preliminary thoughts to the more extensive work that these documents demand.

I think that these marginalia immediately dispel both a common myth about Agassiz himself, and a general interpretation, often made by Whiggish historians, of opposition to triumphant, modern theories —i.e., that continued opposition to "truth," long after the graceful retreat of most original opponents, can only be ascribed to blind dogmatism, prejudice, superstition, and an unwillingness to consider new evidence. I wish to make four general observations on the Agassiz marginalia: all display the seriousness and cogency of his alternative world view. Evolution was too powerful a theory for the Western world to resist in the midnineteenth century: it explained too much information, previously unsynthesized; it resolved too many anomalies; and it accorded too well with the social and political views of expanding imperial and industrial nations. But it was not, like a mathematical proof or a theorem in physics, a firm, deductive argument that could only be opposed by theistic dogmatism. As a world view, embedded in metaphor and analogy, it remained weak in several areas—notably in the empirics of the fossil record, Agassiz' forte. Moreover, evolution

*Their transcription runs to 40 pages, and I give special thanks to my secretary, Agnes Pilot, who (*Gott sei Dank*) learned to read these mysterious squiggles as a girl in Germany.

in the hands of so slippery (and often downright dishonest) a debater as Haeckel offered much to be attacked.

1. The basic observation that Agassiz read every page of a long book so carefully should lay to rest the idea that he had suffered a *crise de confiance* before advancing truth, and had abandoned the serious consideration of evolutionary thought.

2. Agassiz' commentary deals very little with the extensive metaphysical sections of Haeckel's book. Agassiz presents very few theistic defenses against Haeckel's explicit materialism (though he does express his disgust at Haeckel's flippant treatment of subtle arguments), but comments primarily upon Haeckel's empirical claims and his statements about purely biological theories.

3. The prevailing theme of Agassiz' critique is not the metaphysical defense of creation as true a priori, but a consistent, empirical literalism directed against the speculative inferences of evolutionists, mainly concerning the geological record of fossils. This theme represents the continuing concern of Agassiz' entire career:

i) As a disciple of Cuvier and a convert from earlier flirtation with Oken's *Naturphilosophie*, Agassiz had long upheld the view that the geological record, indeed, all scientific evidence, should be read literally: do not interpolate between sudden stratigraphic breaks the gradual transition that uniformitarianism requires and that Lyell provided with his explicitly antiempirical claim that the task of science (as opposed to common sense) must be to impose good theories upon literal evidence. (What saved Lyell from dogmatism was not an empirical methodology, but a willingness to alter or abandon theory should the literal evidence resist his imposition too strongly.)

ii) As Rudwick (1972) and Hooykaas (1970) have shown in opposition to conventional history, catastrophists like Agassiz were the empirical literalists of nineteenth-century geology, while the uniformitarians interpolated their notions of gradual change into a geological record that spoke (and still speaks) loudly for rapid transitions.

Thus, for example, Agassiz writes (p. 47) "Ist heut zu Tage noch wahr" next to Haeckel's claim that newly discovered transitional forms are destroying an old belief in the sharp distinction of successive faunas. He defends catastrophism as good science (p. 103) by denying Lyell's claim that it puts the past outside the sphere of scientific investigation: catastrophic events, he writes, do not imply unknowable causes without modern analogs: "No one has ever denied that relationship [between past and present causes of change], in the same sense as the storm and good weather stand together; but uproar and peace in nature are not the same thing." He emphasizes the continued absence of transitional forms in the fossil record as evidence against evolution: missing links cannot be assumed a priori: "Why have such ancestors not been found, since we now have collected so many fossil vertebrates?" (p. 506).

4. Although Agassiz does cite some arguments against the mechanism of Darwinian theory (p. 132 on the missing explanation of variation, later supplied by genetics), he treats the geological record as his primary target and directs almost all his criticism against Haeckel's attempted construction of phyletic trees from the evidence of fossils and living embryos: How can such an orderly story as the geological history of fossils arise from a planless mechanism like natural selection (p. 131)? Why, if gradual, divergent evolution occurs, are the earliest members of a major group as distinct one from the other as the last representatives (p. 526—"Compare the beginning of each paleontological sequence with its end. The first trilobites are as different from each other as the last; also, the crinoids, etc. etc."). Why, if natural selection leads to progress, are the earliest representatives of a phylum as (or more) complex than their "descendants" (p. 424, on Haeckel's "convenient" noncitation of trilobites)? Interspersed among these general criticisms are Agassiz' copious criticisms of specific phylogenies, particularly of Haeckel's hypothetical ancestors from the Precambrian mists (inferred, according to the theory of recapitulation, from early stages of modern embryos).

Beyond these general arguments, Agassiz focuses on Haeckel's own techniques as a consummate but tricky (and often dishonest) debater. Agassiz' most frequent comments express his disgust at Haeckel's method of presenting truth *ex cathedra* without any citation of evidence. Over and over again he cites variants of the German proverb: "Behaupten ist nicht Beweisen" (to assert is not to prove). Agassiz directs his strongest anger and sharpest irony, all abundantly justified, against Haeckel's dishonest methods of debate:

i) He accuses Haeckel of rendering the antievolutionists as absurd caricatures to establish straw men for attack. Often, he states correctly, the arguments cited in evolutionary perspective are those developed not by Haeckel (as implied in the book), but by the very men he attacks. Thus Haeckel uses Owen's classification of fossil vertebrates to attack a caricature of Owen's position. But nothing arouses Agassiz' fury more than Haeckel's appropriation (without credit and in an opposing context) of Agassiz' own cardinal argument that geological sequences and embryological stages run in parallel. "Das ist mein Resultat!" he writes (p. 290). "This method is not the author's, but was first used in my *Poissons fossiles*" (p. 318). In addition he castigates Haeckel for including among his supporters older scientists who were not advocates of evolution. He, himself, had heard Oken lecture as a student; and the great *Naturphilosoph* did not, as Haeckel claimed, advocate the transmutation of species: "I heard Oken discourse on this subject. He did not mean that man arose from the next lowest form, but directly from primordial substance [*Urschleim*]" (p. 79). In his angriest passage he blasts Haeckel for a caricature of creationism based on supposedly conflicting data actually discovered before Haeckel's birth by those he attacks (pp. 236–237): "Wir sind ja alle so dumm! Wir sind ja alle so unwissend! Oh sancta simplicitas. Dies ist durch unsere Methode erreicht worden. Unverschämt—für einen der nicht geboren war, als die Embryologie ihre jetzige Entfaltung erreicht hatte."

ii) He directs his most withering scorn at Haeckel's continual citation of pure speculation as fact. "Haeckel ist der Apostel" he writes (p. 62). He summarized his disgust in a final comment on Haeckel's phylogenies, written boldly under the last paragraph of the book: "Gegeben im Jahre 1 der neuen Weltordnung. E. Haeckel."

iii) He noticed, as his 1874 *Atlantic Monthly* article did later in print, that Haeckel often doctored diagrams to suit his arguments. He recognizes, for example, that Haeckel supports his important belief in the early similarity of all vertebrate embryos simply by repeating the same figure for turtles, dogs, and humans: "Where did these figures come from? There is nothing like this in our entire literature. This identity is not true" (p. 248). "Naturally [they are so similar], because these figures were not drawn from nature, but copied one from the other. Detestable [*abscheulich*]" (p. 249).

Yet through Agassiz' pugnacity, cogency, and righteous anger, we also sense a sadness and growing despair—that from such flimsy evidence and dishonest argument, a new consensus continued to flourish and grow. Agassiz knew that the tide of history was flowing against him and that in the end (despite his attachment to empiricism) a decision for or against evolution rested on differing philosophical predisposition, rather than upon the clear testimony of his beloved fossils. For he wrote on the title page, in summary of his anger and sadness:

> Their entire doctrine is naked affirmation throughout. Of the proof that we demand in scientific investigation, they provide nothing. They believe that the entire story [*das Ganze*] cannot be read in any other way, and they demand that all scientists accept their opinion as proof. A few years ago such arrogance would have

> simply appeared laughable; it seems to me that this should still be so. I have, as honestly as anyone else, sought the connections among these phenomena [*Dinge*]. They themselves, and Darwin before them, have acknowledged that my own work provided, in part, the facts upon which they base many of their conclusions. I am therefore, in their eyes, fully justified to speak about these matters. Let us therefore consider their objections to my conception of these issues. In the end, their opinions are derived purely from their philosophical position. They laugh at my opinion as gross anthropomorphism . . . I confess that, after long consideration of their moneran and ascidian phylogeny of the human race, I would rather stick with the old views than profess such error.

"God or monerans [spontaneously generated, simple forms]," he writes in another place (p. 539), "I prefer God."

References

Agassiz, L., 1860. "Professor Agassiz on the Origin of Species," *American Journal of Science*, 30:142–154.

Agassiz, L., 1874. "Evolution and Permanence of Type," *Atlantic Monthly*, 33:92–101.

Haeckel, E., 1868. *Natürliche Schöpfungsgeschichte*, Berlin, G. Reimer, 568 pp.

His, W., 1874. *Unsere Körperform und das physiologische Problem ihrer Entstehung*, Leipzig, F.C.W. Vogel, 224 pp.

Hooykaas, R., 1970. *Catastrophism in Geology*, Meded. Konink. Nederl. Akad. Wetens. Letterkunde, 33 (No. 7), 50 pp.

Lurie, E., 1960. *Louis Agassiz: A Life in Science*, Chicago, University of Chicago Press, 449 pp.

Nordenskiöld, E., 1929. *The History of Biology*, London, Kegan Paul, Trench, Trubners, 629 pp.

Rudwick, M., 1972. *The Meaning of Fossils*, Amsterdam, Elsevier, 287 pp.

VIII

Natural Science in a Post-Industrial World: The American Themes, II

100th Anniversary of Observations on Petroleum Geology in the U.S.A. by Dmitriy I. Mendeleyev

Eugene A. Alexandrov

Dmitriy Ivanovich Mendeleyev arrived in the United States during the summer of 1876. He was 42 years old at that time and was already well known for his discovery of the periodic law of the elements his outstanding achievement in chemistry. In addition to being a professor at the Institute of Technology and at the University of St. Petersburg, he was also a consultant for the Russian petroleum industry, which, as in the United States, was in its incipient stage of development. Mendeleyev was commissioned to visit the United States, particularly Pennsylvania, to become acquainted with the organization of the American petroleum industry. Chemistry was his main field of interest, but judging by his publications, he also showed interest in physics, economics, law, oceanography, metrology, aeronautics, mineralogy, coal deposits, petroleum geology, and even psychology.

Mendeleyev's visit to the United States coincided with the centennial celebration of the American Revolution. In the nontechnical part of his report about his journey (Mendeleyev, 1877) he makes keen observations on many aspects of American life in addition to describing the colorful celebrations on the 4th of July in Philadelphia and the great exhibition of all the marvels of technology of that time. Mendeleyev traveled from Philadelphia to Pittsburgh and then on to Karns City, Parker, Millertown, Oil Creek, and Titusville. From there he went to Buffalo and Niagara Falls—which fascinated him greatly—and then back to New York for the return trip to Russia.

As a result, Mendeleyev had traveled through that part of the oil-producing province in North America which stretched along the Appalachians. Most of the oil at that time was produced around Karns City and Parker, while natural gas was found mainly in the area around Millerstown. Mendeleyev made a thorough study of all available information on the physical and chemical properties of American crude oil, the methods of its production since its discovery by Seneca Indians, the methods of refining and transportation, including the construction of pipe lines, and taxation. During his trip he examined outcrops and the stratigraphic successions of formations. He studied the description of Appalachian geology compiled for him in Russia by E. O. Romanovsky (Mendeleyev, 1877). Lastly, he made correlations between geological conditions in the Caucasus and in Pennsylvania, and drew conclusions about the geological conditions that control the formation of oil pools.

After his trip to the oil fields of Pennsylvania, Mendeleyev began to ponder the problem of the origin of petroleum. He believed that a correct understanding of the conditions that contribute to the generation of petroleum would assist exploration for new oil fields. In 1876 the dominant hypothesis was that petroleum was generated from organic relics. Mendeleyev suggested that petroleum in Canada and Pennsylvania, where it occurs in Silurian and Devonian sediments, respectively, formed at depth, in formations containing no organic relics. He rejected the possibility that petroleum had migrated downward from Pennsylvanian sediments containing coal deposits, which were considered by some contemporaries as a possible source of oil. Mendeleyev noted that all oil fields with which he became familiar, both in Russia and in America, were in the vicinity of mountain ranges—in foredeeps. The second important feature stressed by Mendeleyev and American petroleum geologists was the trends along which the oil fields occur. These trends represent very broad arcs with very large radii. The directions of the trends (lineaments) closely follow those of the mountain ranges, as demonstrated by the Alleghenies in Pennsylvania, the northern Caucasus, and the Baku region. Mendeleyev as-

1. D. I. Mendeleyev and his assistant, V. Gemilian, at Niagara Falls.

sumed the existence of regional open faults in foredeeps. He thought that despite slumping and partial closure of these faults, they probably penetrated deeply into heavy rocks, in which petroleum might have been generated.

Starting from the assumption that petroleum has a deeper source than is traditionally believed, Mendeleyev began to consider inorganic chemical processes and mineral substances that could produce hydrocarbons. He turned to an examination of the origin of the earth and the resulting distribution of metals in it. Iron, being abundant at the center of the planet, attracted his attention as one of the metals capable of combining with carbon. The existence of carbides in meteorites and some terrestrial basalts encouraged Mendeleyev to suggest that iron carbides occur in sufficient quantity deep below the earth's crust. Since open faults form in foredeeps of mountain ranges contemporaneously with orogenies, water could penetrate along them into rocks containing carbides of iron and other metals. These metals in carbides would oxidize, using oxygen of the water, while hydrogen was liberated, and partially combine with carbon to form hydrocarbons. Therefore, the age of petroleum corresponds to the age of the orogeny. He recommended that oil wells should be drilled along a trend parallel to that of a mountain range.

Mendeleyev called his theory the "mineral hypothesis" for the origin of petroleum. He supported it with laboratory experiments, but his ideas were generally not accepted. Geologists did not agree that deep faults could remain open. They argued that lithostatic pressure would force these faults to close and cut off any upward or downward flow of fluids. The existence of carbides in sufficient amounts was also considered questionable. On the other hand, progress in petroleum geology yielded much evidence in support of the organic origin of petroleum, including microscopic relics of organic life in crude oil. But when Mendeleyev himself looked for relics under the microscope he could not find them. His hypothesis was almost completely forgotten until fairly recently, when it was suddenly reborn. New protagonists of the theory of an inorganic origin of petroleum have appeared, who followed in Mendeleyev's footsteps. Their ideas were synthesized by Porfir'yev in 1974.

References

1. Mendeleyev, D. I., 1877. *Neftyanaya promshlennost v severo-amerikanskom shtate Pensil'vanii i na Kavkaze*, (*The Petroleum Industry in the North American State of Pennsylvania and in the Caucasus*) St. Petersburg, 304 pp.
2. Porfir'yev, V. B., "Inorganic Origin of Petroleum," American Association of Petroleum Geologists, *Bulletin*, 1974, 58:3–33.

The Influence of Geology on American Literature and Thought

Dennis R. Dean

The rise of geological awareness in America simultaneously affected aspects of thought and expression that we commonly regard as humanistic—literature, art, philosophy, and theology in particular. Though I mention all of them, this short paper has space only for literature, and even then is reduced to naming especially significant examples. Despite its limitations, however, the literary record presented here is a faithful guide to what, throughout its history, the nation as a whole was thinking about geology, the subject matter and utility of which have been differently regarded at different times.[1]

The earliest descriptions of American landscape, beginning with Columbus' enthusiastic raptures about the "miracle" of Hispaniola, were deliberate attempts on the part of the explorers to attract European finance, and to save their own reputations, by stressing the availability of gold; Coronado's expedition of 1540, the most obvious enactment of Spanish desires, helped to give geological exploration in the American West a uniquely mythological basis that endured. On the Eastern seaboard, among northern European settlements, the richness sought tended to be less imaginative, and students of natural history emerged early. Thomas Hariot's *A Brief and True Report on the*

New Found Land of Virginia (1588) is the first such analysis to include significant geological details. George Sandys, America's first author of belles lettres, came to Virginia during the 1620's and translated most of Ovid's *Metamorphoses*, adding to its notes original geological observations of his own. A more general survey of natural history, designed to encourage colonization, was William Wood's *New England's Prospect* (1634), which combined verse and prose. Still another early work, the *Discoveries* of John Lederer (1672), includes Virginia geology and a prophetic look over the Appalachians into the Middle West. By the end of the seventeenth century, ours was no longer an unknown land.[2]

It was only reasonable that the first interest shown in American landscape should be utilitarian. Since aesthetic responses satisfy needs less tangible than hunger, we scarcely find American landscape poetry before the eighteenth century. A description of the Connecticut Valley, perhaps the earliest clear example, was versified by Roger Wolcott in 1725. Four years afterward an associate of Benjamin Franklin's named James Ralph published an imitation of Milton that includes powerful Andean scenes, the first high mountains in our literature.[3] Authentically American landscapes came even later in art, the first prominent painter of them being Ralph Earl (1789), whom we also associate with the Connecticut Valley. In the meantime, prose descriptions were flourishing, including well known ones by Peter Kalm, Jonathan Carver, John and William Bartram, and Thomas Jefferson.

Although empiricism is fundamental to modern geology, theories (as historians well know) do not necessarily depend upon it. Our Puritan ancestors, for instance, regularly allude to the Biblical ideas of the deluge (as in poems by Anne Bradstreet, 1650) and a volcanic Judgment Day (as in Michael Wigglesworth's *Day of Doom*, 1662, stanza 15). Earthquakes and other geological disasters were instances of divine wrath. Two particularly forceful earthquakes struck New England during the eighteenth century, however, and between them virtually created American geology, both in science and in literature. The first, on Sunday, 29 October 1727, was the subject of more than twenty published sermons by such leading clergymen as Cotton and Samuel Mather. Although most interpreted the tremors of the earthquake as divine warnings, Thomas Price and others were willing to discuss its natural causes also. The Rev. Mather Byles, who versified such contemporary sermon topics as "The God of Tempest and Earthquake" (1727) and "The Conflagration" (1729), is our earliest geological poet. That second earthquake, of 18 November 1755, called forth additional sermons, including one by Byles, and Jeremiah Newland's extended poem *Earthquakes Improved* (1755). But explanation now was increasingly naturalistic; with John Winthrop's *Lecture on Earthquakes* (1755) and Samuel Williams' later conjectures (1785), America had unmistakably developed geological theories of its own.[4]

Theorizing independent of the Bible began in America with the Enlightenment, during which the personal letter was cultivated as a literary form. Accordingly, there are sustained attempts in the letters of Benjamin Franklin, especially during the 1760's. Not long afterward (in *The Rising Glory of America*, 1772) young Philip Freneau published the first extended geological theorizing in American verse.[5] *The Age of Reason* (1794), Thomas Paine's deistic attack upon the Bible, helped to make all such theorizing unpopular for a few years, but Samuel Miller's extensive compilation of geological theories in his *Brief Retrospect of the Eighteenth Century* (1803) demonstrated the irrepressible vigor of the new science, as did *The Columbiad* (1807), an epic poem by Paine's friend Joel Barlow. Geological passages in that remarkable work, while supporting the deluge, also stress the immense age of the earth (Book IX), demonstrating clearly that Genesis was in trouble. Among much else, Barlow and other poets had also discovered the mammoth—one of American geology's most influential contributions to the arts, as our dinosaurs would be later—and the whole problem of extinction now arose, to plague Thomas Jefferson especially.

Washington Irving is the first major American author to stress in imaginative contexts the destructive effects of time; in *Rip Van Winkle* (1819) and elsewhere, Uniformitarianism is assumed. The Italian portions of Irving's European journal (1805) are frequently geological—they include descriptions of Vesuvius and Etna—and we know that he was conversant with European developments of the science. From Henry Brevoort, for example, Irving received glowing descriptions of John Playfair and the other Huttonians of Edinburgh,[6] whom Benjamin Silliman (*A Journal of Travels*, 1810) would also visit.

James Gates Percival and Henry R. Schoolcraft made what now seem minor contributions to both geology and literature during the 1820's. As poets, Schoolcraft (*The Rise of the West*, 1827) is more geological, but Percival stood higher in contemporary esteem. He was soon overshadowed, however, by William Cullen Bryant, who, having ridiculed Jefferson's fascination with mammoth bones in an early poem (1808), went on to write (in "A Hymn to the Sea," 1842) the finest geological passage yet seen in American verse:

> These restless surges eat away the shores
> Of earth's old continents: the fertile plain

Welters in shallows, headlands crumble down,
And the tide drifts the sea-sand in the streets
Of the drowned city.

Simultaneously, however, God's creative powers are equally at work, bidding the "fires, / That smoulder under ocean, heave on high / The new-made mountains . . ." Hutton would have been pleased.

Bryant, who learned some of his natural history from Amos Eaton, was a close friend of the painter Thomas Cole, whose landscapes are geologically meticulous. Also associated with Bryant was the literary figure James K. Paulding, whose novel *Westward Ho!* (1832) borrows descriptions of the Mississippi Valley, shows insight into fluvial geology, and depicts the New Madrid earthquake of 1811–12 (chapter 14)—perhaps the earliest geological disaster in American fiction. One of Paulding's personal letters (1846) unfairly disparages the great Geological Survey of New York; another (1847) discusses Hutton, Werner, and some specimens with fellow New Yorker Martin Van Buren, whose secretary of the navy Paulding was.[7] That self-taught naturalist Constantine S. Rafinesque published in 1836 *The World, or Instability*, a book-length scientific poem advocating organic evolution. Fossils, Lamarck, and Cuvier are prominent. Finally, this is as good a place as any to mention the geological fantasies of Edgar Allan Poe, including "The City in the Sea," "Dream-Land," "Ulalume," *The Fall of the House of Usher*, *A Descent into the Maelstrom*, and one of the oddest geological landscapes in American literature, chapters 20ff. of *The Narrative of Arthur Gordon Pym* (1838), Poe's only novel. Like "MS Found in a Bottle" (1833), *Pym* was influenced by the bizarre geological theory which John Symmes, its originator, had himself novelized in *Symzonia: Voyage of Discovery* (1820, by "Captain Adam Seaborn"). According to Symmes, the earth is open at the poles.

Landscape description had been common in American novels ever since *Wieland* (1798) by Charles Brockden Brown, who is better known to historians of geology for having translated C. F. Volney's *View of the Soil and Climate of the United States of America* (1804), but a particularly impressive accomplishment was the Leatherstocking novels of James Fenimore Cooper (1823–41), which popularized the scenery of the nearby West. Meanwhile, painters of the Hudson River School, like Cole and later Frederick Church, gave visual impact to geological analyses.

Geology was especially influential in America during the mid-nineteenth century, between the publication of Lyell's first volume in 1830 and Darwin's *Origin* in 1859. Although Lyell's visits in 1841 and 1845 were high points, there had been lectures, discussions, pamphlets, and books on geology in America much earlier, and we may turn to popular magazines in order to find out why. In *The Knickerbocker* for April 1834, Samuel L. Metcalf's argument for "The Interest and Importance of Scientific Geology as a Subject for Study" is a tribute to Lyell:

> Philosophers have recently accomplished for geology, what Sir Isaac Newton did for astronomy—they have reduced it to a system of principles—so that more useful knowledge of the earth's structure may now be acquired from a dozen lectures, than formerly by the labor of a life-time. The prejudices which have hitherto retarded its progress are rapidly giving way; and it is at the present time, employing the talents of many of the most distinguished philosophers of Europe and this country. It is, indeed, the fashionable science of the day, and may be said to form a necessary part of practical and ornamental education.

The usually sedate *New England Magazine* (for April 1835) described Bostonians as "geologically mad," and George Ticknor's visits to British geologists that year corroborate the charge. But the danger to religious orthodoxy that geology represented was now equally apparent. In May 1836 *The Knickerbocker* reviewed no fewer than four books on "Geology and Revealed Religion." As an attempt at reconciliation, Edward Hitchcock's *The Religion of Geology* (1851) was by no means the first of its kind.[8]

A major response by American religion to the encroachments of science was Transcendentalism, according to which God's revelations of himself to man through nature are continuous, for all newly discovered scientific facts have spiritual significance. The intellectuals who believed this, some of America's finest, not merely tolerated but welcomed advances in science, and it was primarily this optimistic accommodation that made geology—the modernizer of religion—so exciting to the 1830's.

We see this very clearly in the writings of Ralph Waldo Emerson, whose mature interest in geology began about the time that his search for religious truth led him to renounce the pulpit and become a lay philosopher. In 1836, under the influence of Lyell, he noted that "Geology teaches in a very impressive manner the value of facts and the laws of our learning from nature . . . No leaps, no magic,—eternal tranquil procession of old, familiar laws, the wildest convulsions never overstepping the calculable powers of the agents, the earthquake and geyser as perfect results of known laws as the rosebud and the hatching of a robin's egg." Besides affirming the stability of nature, geology teaches us to be objective about time. As Emerson wrote in 1867, "Geology, a science of forty or fifty summers, has had the effect to throw an air of novelty and mushroom speed over entire history. The oldest empires,—what we called venerable antiquity,—now that we have true measures of duration, show like creations of yesterday." Some of his most striking metaphors are also geological. "Every new writer," he suggested in 1863, "is only a new crater of an old volcano." More than forty geologists and geological works are mentioned in his writings, with Louis Agassiz and C. T. Jackson (his brother-in-law) especially prominent.[9]

Jackson's geological reports also influenced Henry D. Thoreau (particularly in *The Maine Woods*, 1864), who has geological passages throughout his works, including the chapter in *Walden* (1854) on "Spring." Other members of Emerson's circle showing the impress of geological concepts are Jones Very, "The Glacial Marks on Our Hills" and other poems; James Russell Lowell, "The Progress of the World" (essay, 1886); Oliver Wendell Holmes, who thought New England's three most interesting geological features to be its drifted boulders, fossil footprints, and trilobites; and Henry Wadsworth Longfellow, whose "Footprints on the sands of time" derive from Edward Hitchcock and the Connecticut Valley. Geological moments in the reclusive poetry of Emily Dickinson were fostered by Hitchcock's teaching at Amherst.[10]

The period between Lyell and Darwin also saw geology especially prominent in fiction, but this is not to say that the science was always treated with respect. For example, a little-known novel by James Fenimore Cooper called *The Monikins* (1835) is a thoroughgoing satire upon pre-Darwinian theories of evolution, the reconstruction of fossil animals, and various geological controversies (chapters 11, 12, 16). Even more strikingly, his novel of ideas *The Crater* (1847) [see below with Stevenson] depends for its plot upon Leopold Von Buch's Craters of Elevation theory, which Cooper probably found in Lyell. A considerably less serious work was Cornelius Mathews' *Behemoth, a Legend of the Moundbuilders* (1839), in which Indians attempt to defend themselves against an invading mastodon. Henry Hirst's narrative poem, *The Coming of the Mammoth* (1845), is similar.

Of all the major American novelists, none utilizes a wider variety of accurate geological information than Herman Melville, particularly in *Mardi* (1849), another novel of ideas like *The Crater*. Chapter 132 takes us to the Isle of Fossils, where we learn about fossil bird tracks, the formation of volcanic coral atolls, and "the celebrated sandwich system," which turns out to be an elaborate satire upon stratigraphy.

Moby-Dick (1851), often regarded as America's finest novel, is even more geological. The whale himself is called a "salt-sea Mastodon," and other geological references include fossil footprints, strata, vulcanism, glacial scratches (Agassiz is mentioned), and earthquakes. In chapter 104, "The Fossil Whale," Melville presents his "credentials as a geologist." Two of his earlier novels, *Typee* and *Omoo*, describe volcanic scenery, but Melville's finest use of geological landscape appears in *The Encantadas* (1854), a thoughtful reaction to the Galapagoes Islands.

Like so many other mid-nineteenth century American writers interested in geology, Melville lived in Massachusetts. What is in some respects the final statement of this school appears in Oliver Wendell Holmes's novel *Elsie Venner*, which began its serial publication in December 1859, a few weeks after Darwin's *Origin of Species* appeared. (In chapter 7 Holmes refers explicitly to Darwin and the "struggle for life"; he is the first American novelist to do so.) There are several references to New England geology, including bird tracks and trilobites, and the novel ends with a geological catastrophe. But its most interesting feature is a debate between a doctor of medicine (as Holmes was) and a doctor of theology. "'Doctor!' the physician said, emphatically, 'science is knowledge . . . So Geology *proves* a certain succession of events, and the best Christian in the world must make the earth's history square with it'" (chapter 22). Eventually, John W. Draper (1875), Andrew D. White (1896), and many others would be outrightly militant, but it was already clear to Holmes in 1860 that the Transcendentalist compromise was over—and so was geology's foremost period as a literary influence.

Following Darwin, writers were often skeptical about religion, as Melville and Holmes had been earlier. Mark Twain managed to be skeptical about everything except his ability to make us laugh, but he is frequently geological in his usual offhand manner. We ascend Vesuvius and visit Pompeii in *The Innocents Abroad* (1869). *Roughing It* (1872) takes us through the Rocky Mountains, the canyon country of Utah, the Great American desert, and Mono Lake (especially geological) to Virginia City and its mines. Half a dozen chapters at the end deal with Hawaiian volcanoes. Twain satirizes paleontological fervor, and perhaps the early rivalry of Cope and Marsh, in "Some Learned Fables" (*Sketches Old and New*, 1875) intended primarily for children; animals send out scientific expeditions to study the remains of man. *A Tramp Abroad* (1880), concerning the Alps, includes a chapter on glacial movement; *Life on the Mississippi* (1883) has a geological beginning and reliable insights about our greatest river; "Fenimore

Cooper's Literary Offenses" (1895), on the other hand, include an ignorance of fluvial geology. Finally, Twain's *Following the Equator* (1897), the product of a triumphant world tour, tells us about geological features in Australia, New Zealand, and South Africa. Several of his later, private works ridicule the Biblical story of man.

With geology's cultural hey-day now in the past, writers became increasingly more interested in biological (especially human) evolution, or tended to ignore science altogether. Fewer major writers are now involved, and it is easier to speak of schools and trends. One such trend, already evident in Twain, is regionalism. During the latter nineteenth century, most of America's geological literature comes from the West. There are, for example, a great many books about mining and the gold rushes: Bayard Taylor, *Eldorado* (1850); Edward Buffum, *Six Months in the Gold Mines* (1850); Frank Marryat, *Mountains and Molehills* (1855); Franklin Langworthy, *Scenery of the Plains, Mountains and Mines* (1855); Henry Villard (1860, the Pike's Peak gold rush); William Wright De Quille (1876, the Comstock lode); and others later on. Robert Louis Stevenson's well known sketch *The Silvarado Squatters* (1883) includes Mount St. Helena and the Calistoga volcanic region in California, but the sentimental mining camp stories of Bret Harte were even more influential. Three of his Western poems are geological and amusing.[11]

Besides regionalism, a second and related trend was romantic naturalism, which continued in literature and art that close observation and appreciative acceptance of nature common among both scientists and laymen prior to Darwin. Important forerunners were Gilbert White, William Bartram, the poet Wordsworth, and Alexander von Humboldt, whose *Personal Narrative* of South American travels and synthetic *Kosmos* were regarded as models of observation and theory. For American writers, however, the greatest of their predecessors was Thoreau. Among his many followers, John Burroughs popularized geology in *Time and Change* (1912); Joseph Wood Krutch proved worthy of his subject in *Grand Canyon* (1958); and Edward Abbey was knowledgeably responsive in *Desert Solitaire* (1968). Of all the writers one might name, the greatest to follow the hermit of Walden was John Muir, who fought and won geological disputes with J. D. Whitney and Clarence King regarding the origin of Yosemite—all three will forever be associated with the High Sierras—and whose name has become almost synonymous with the wise conservation of nature. We owe many of our national parks to the dedication, persistence, and vision of men like him.

Clarence King, like Muir and some other geologists, was an author in his own right, and certainly a character—in the excellent biography of him by Thurman Wilkins (1958); in Wallace Stegner's mining era novel *Angle of Repose* (1971), with its geological title; in Stegner's *Beyond the Hundredth Meridian* (1954), a biography of John Wesley Powell; in the collected Clarence King *Memoirs* (1904) by distinguished men of his generation; and especially in our finest autobiography, *The Education of Henry Adams* (1907), which is also a tribute to Louis Agassiz, whose Harvard University course "on the Glacial Period and Paleontology . . . had more influence on [Adams'] curiosity than the rest of the college instruction altogether." As a result, the *Education* is one of the most geological of major American literary works, for Adams had not only Agassiz and King to ponder, but also Darwin and Lyell (chapters 15, 20, and 26). Eventually, he rejects the latter's Uniformitarianism, on grounds that now seem shrewd rather than eccentric. Other geologists Adams refers to are T. H. Huxley, Raphael Pumpelly, Roderick Murchison, S. F. Emmons, J. D. Whitney, J. D. Hague, James Croll, Archibald Geikie, and Eduard Suess. Inevitably, King's U.S. Geological Exploration of the 40th Parallel is emphasized.

A third important literary trend in the latter nineteenth century, besides regionalism and romantic naturalism, was what we more properly call naturalism—the recognition that man is powerless before the immense destructive capabilities of nature. Bret Harte's "Lessons from the Earthquake" (1868) is an early instance, Adams' *Autobiography* provides the best discussion, and Jack London's novels are sufficient for examples. The recurring theme of personal survival is primarily Darwinian, but disastrous Caribbean eruptions in 1902 and the San Francisco earthquake of 1906 reminded everyone that geological forces could not be disregarded. Once it had become clear that science rather than theology was to be the guiding light of the twentieth century, it was common to wonder where science was taking us. Some who tried to answer that question were professional philosophers, some were would-be pundits, and others preferred to express themselves through images in fiction.

Within the newly flourishing genre of science-fiction, cavemen struggling for survival became our heroes in novels by Stanley Waterloo (*The Story of Ab*, 1903), Jack London (*Before Adam*, 1907), Mary Johnston (*The Wanderers*, 1917), and even Vardis Fisher (*The Testament of Man* series, 1940's). Worldwide cataclysms were popular, as in Sidney Fowler Wright's *Deluge* (1928) and *Dawn* (1929). Perhaps our most characteristic expression of anxiety regarding science, however, was the famous Scopes trial of 1925, during which geological and evolutionary theories became subject to the law—a controversy not yet ended. The transcript of that trial, in which the agnostic Clarence Darrow cross-examined his Fundamentalist opponent, William Jennings Bryan, is literature itself and has also given rise to a play and a movie (both called *Inherit the Wind*), as well as to biographies, essays, journalism (H. L. Mencken), and cartoons.

Potential opposition notwithstanding, geology and evolution soon became common in the movies also. Because earthquakes and eruptions effectively climaxed often silly screenplays, Atlantis, Pompeii, and other unlucky places were regularly done in, and prehistoric men and monsters jerked through rubbery forests with varying degrees of success and chronological authenticity. D. W. Griffith produced a version of *Man's Genesis* as early as 1912; Conan Doyle's novel of that year became *The Lost World* (1924), with Wallace Berry and ingenious dinosaurs animated by Willis O'Brien; to be followed by Victor Mature in *One Million B.C.* (1939). *The Lost World* was then remade in 1960, with Michael Rennie, as was *One Million Years B.C.* in 1966, with Raquel Welch; *When Dinosaurs Ruled the Earth* (1970) and *The Land That Time Forgot* (1974, based upon Edgar Rice Burroughs) were further tries. But the genre's one authentic masterpiece remains *Fantasia* (1939), which coordinates a scientifically respectable sequence of the evolution of life with passages from Stravinski's *Rites of Spring* and stunning visual imagery. Disney's geological consultant, Roy Chapman Andrews, had recently published nontechnical accounts of his paleontological expeditions to Mongolia.

In turning to popular culture for a time, I do not wish to suggest that geology in modern arts has reduced itself to triviality. On the contrary, its perspective has become so much a part of the way we view landscape and our planet's past that specific influences become harder to trace as they become less self-conscious. Several twentieth-century novelists, for instance, use geological understanding regularly. Willa Cather, in *O Pioneers* (1913), *Death Comes for the Archbishop* (1927), and the central episode of *The Professor's House* (1925), seems to me the outstanding example. Two other recent but noteworthy instances were both inspired by Hawaii. James Michener's novel of that name (1959) has a geological beginning; and in Peter Gilman's *Diamond Head* (1960) lava flows from an eruption of Mauna Loa are slowed by bombing them, the idea of a fictitious vulcanologist obviously derived from Thomas Jaggar. Chapter 1 of Robert M. Pirsig's *Zen and the Art of Motorcycle Maintenance* (1974) includes an extended fluvial metaphor, while the second and third of Michener's *Centennial* (1974) are straightforwardly geological. One of America's best short stories in 1976 was John Updike's "The Man Who Loved Extinct Mammals" (*New Yorker*).[12] Among our playwrights, geology is conspicuous in Thornton Wilder's *The Skin of Our Teeth* (1942), with its Ice Age references. Drama and (to a lesser extent) the novel are, however, social forms of literature, and geology—having carried the day—is no longer the current problem that it was in the last century. For this reason, we are more likely to find it strongly emphasized in minor, old-fashioned, or eccentric writers, our major novelists and playwrights having gone on to other things.

Fortunately, this is not the case in poetry, a more individual literary form in which the imagistic power of geology and its concepts continues to be appreciated. From Walt Whitman, our most revolutionary nineteenth-century poet, came the message that poetry should celebrate the present, by including science. Accordingly, there are geological references in many of his works (*Song of Myself*, sections 19, 20, 23, 31, 44, and 45, for example), and his successors have continued the tradition, often with wonderful results. Robert Frost's "Fire and Ice" (1923), the most famous geological poem in American literature, makes psychological metaphors of the Biblical conflagration and Buffon's theory of refrigeration (perhaps as revived by T. C. Chamberlin in *The Origin of the Earth*, 1916). In "Once by the Pacific" (1928) Frost similarly extends the concept of erosion. Two minor poets of the 1920's, Raymond Holden (*Granite and Alabaster*, 1922) and Wilfred Earl Chase (*Poems*, Madison, Wisc., 1928), though geological to good effect, are outdone by the coastal geology of Conrad Aiken ("Senlin," 1918) and Robinson Jeffers (*Tamar*, "Songs of the Dead Men," "To the House," "To the Rock," "Continent's End," all 1924; "Granite and Cypress" and "The Treasure," both 1925; "Tor House," 1928). Coastal scenery also appears in the poems of George Santayana ("Cape Cod," 1923). Geology abounds in Edwin Thomas Whiffen, *Cosmogony: An Evolution-Epic* (1927), especially Book XVI.

An impressive series of poets utilized geology during the 1930's, beginning with Hart Crane in *The Bridge* (1930). Conrad Aiken followed with *Preludes for Memnon* (1931, esp. XXII and XXXII) and *Time in the Rock* (1936). Thomas Hornby Ferrill's *Westering* (1934) includes the very geological "Time of Mountains." Two other individual poems of note are Robert Penn Warren's "History" (1935) and Lola Ridge's "Via Ignis" (1935, section 9). W. H. Auden, the decade's greatest geological poet, utilizes his knowledge with consistent brilliance in *Look, Stranger!* (1936), *Letters from Iceland* (1937), and later works. As a child, Auden had landscape fantasies (limestone with lead mines) and wanted to be a mining engineer.

The Soldier (1944) and *The Kid* (1947), two longish poems by Conrad Aiken, both have geological beginnings. Following the war, three geological poets indebted to Whitman did interesting work: John Gould Fletcher, *The Burning Mountain* (1946); John Theobald, *The Earthquake and Other Poems* (1948); and Charles Erskine Scott Wood, *Collected Poems* (1949, esp. "Thoughts in Harney Valley"). Theobald's title poem, written in 1945, includes a reference to continental drift (p. 19). Robert Frost's "Directive" (1947, glacial scratches), Conrad Aiken's "Summer" (1949), and Kenneth Rexroth's "Lyell's Hypothesis Again" (1949), all of them fine poems, were soon joined by a distinguished series from Auden, including *The Age of Anxiety*, *Bucolics*, "In Praise of Limestone," "Ischia," "Not in Baedeker," "Ode to Gaea," and "A Walk After Dark." Cosmic pessimism, science, and superb poetry combine to form Robinson Jeffers' visions of *The Beginning and the End* (1954).

In *A Range of Poems* (1966), Gary Snyder collects a decade's worth of poetry often geological, including "Milton by Firelight," "Riprap," and "Through the Smoke Hole" ("a Permian reef of algae"), among others also having Western or Japanese locales. Randall Jarrell's final volume, *The Lost World* (1966), turns that old dinosaur movie into a poignant metaphor. W. D. Snodgrass, "Planting a Magnolia" (dinosaurs) and Gary Snyder, "A Walk" (glaciated landscape) were both written in 1968, the same year that Dudley Randall recalled "bones of dinosaurs buried under the shale of eras" in "Roses and Revolutions." Judith Johnson Sherwin's *Uranium Poems* (1969) make poetic use of radioactive ores, and there are "underwater lava flows" in Gary Snyder's "Burning Island" (*Regarding Wave*, 1970). Philip Booth (*Margins*, 1970) explores New England geology. One of the most successful users of science in contemporary poetry is A. R. Ammons; among his geologically knowledgeable *Collected Poems* (1972), "Requiem," "Possibility Along a Line of Difference" (fault), "Risks and Possibilities," "Terrain," "Expressions of Sea Level," "The Constant," "Corson's Inlet," and "Delaware Water Gap" are all especially fine. Ammons' *Sphere* (1974) likewise contains geological imagery. "Unpredictable But Providential," in W. H. Auden's final book (*Thank You, Fog*, 1974), includes a striking image of continental drift. Joan Johnson's poem "Continental Drift" (1976) is less concrete.[13] The last complete volume of poetry I shall mention, Gary Snyder's *Turtle Island* (1974), stresses ecology, Turtle Island being the Earth itself. Among its poems, "Straight-Creek—Great Burn" includes a geosyncline, and "What Happened Here Before," a poem organized by geological time, deals with both the deposition and the consequences of California gold, as if the propaganda of Columbus, with which this survey began, had finally and ironically come true.

Though these later poets are less familiar to us than some others, their use of geology is fully as informed, and perhaps even more sophisticated, than that of nineteenth-century figures. If geology in literature tends to be a little rarified now, it is frequently more brilliant, and of greater worth as art. In general, American writers throughout our history show themselves to be as knowledgeable about geology and as responsive to it as one might reasonably expect. Although the minutiae of an increasingly complex science seem unlikely material for verbal art, literary metaphors are often surprisingly technical. Clearly, those writers who have benefited from geology repay the science by popularizing it, and by fostering a sense of wonder toward its finds. As our nation begins yet another century, we have no reason whatsoever to suppose that this fruitful relationship is over. Already, new geological theories, and discoveries on this and other planets, await their poet.

Notes

1. Robert E. Spiller, et al., *Literary History of the United States* (New York, Macmillan, 1955) is a basic reference; for geologists, see chapter 53. Particularly useful histories of ideas include William M. Smallwood, *Natural History and the American Mind* (New York, Columbia University Press, 1941); Hans Huth, *Nature and the American: Three Centuries of Changing Attitudes* (Lincoln, University of Nebraska Press, 1972); Dirk J. Struik, *Yankee Science in the Making* (New York, Collier Books, 1962); Henry Nashe Smith, *Virgin Land: The American West as Symbol and Myth* (Cambridge, Mass., Harvard University Press, 1970); and John C. Greene, *The Death of Adam* (Ames, Iowa State University Press, 1959). George P. Merrill's *The First One Hundred Years of American Geology* (New Haven, Yale University Press, 1924; rep. 1969), seriously outdated now, is supplemented by George W. White, *Essays on History of Geology* (New York, Arno Press, 1977) and Robert M. Hazen and M. H. Hazen, *Bibliography of American-Published Geology: 1669 to 1850*. Microform 4, iv (Boulder, Col., Geological Society of America, 1976). Scholarship specifically relating geology and American literature is extremely scarce. About the only articles are Elizabeth S. Foster, "Melville and Geology," *American Literature* (1945), 27:50–65; Harold H.

Scudder, "Cooper's *The Crater*," *American Literature* (1947), 19: 109–126; Donald K. Ringe, "William Cullen Bryant and the Science of Geology," *American Literature*, 26 (1955), 507–514; and my own "Hitchcock's Dinosaur Tracks," *American Quarterly* (1969), 21: 639–644. In the present essay I am also indebted to Harry Hayden Clark (on Cooper) and Archie Lamont. A brief survey of the influence of geology upon American art might begin with C. W. Peale ("Exhuming the Mastodon"), Cole, Durand, Crapsey, Gifford, Church, Bierstadt, and Moran (who was associated with Hayden's Survey) in The Metropolitan Museum of Art, *19th-Century America: Paintings and Sculpture* (New York, 1970).

2. Except in a very few libraries, one can only read about the original editions of these and other famous books, but later versions generally exist (consult the *National Union Catalog*), while reprints, facsimiles, and film reproductions continue to appear. There is no point in citing specific, soon-to-be-outdated editions for well known literary classics that college libraries have or can obtain in a variety of forms. Occasionally, I allude to writers or works that prior researchers (note 1 above, especially Spiller) have discussed in more detail. Otherwise, author, title, and date will sufficiently identify a literary work. Shorter pieces by the famous can easily be found in their collected editions, and often in anthologies. I comment in my remaining notes only when less obvious guidance is required.
3. Wolcott, Ralph, and Byles (mentioned below) are available in Samuel Kettel, ed., *Specimens of American Poetry*, 3 vols. (1829; rep. 1967), which also includes James Gates Percival. For other passages of geological interest, see 1:312–318; 2:104–106 (the mammoth), 182–184, 190–191, 226; and 3:115.
4. For 1755 see Charles Edwin Clark, "Science, Reason, and an Angry God: The Literature of an Earthquake," *New England Quarterly* (1965), 38:340–362. Hazen (note 1) lists the sermons individually.
5. See, for example, Franklin's letters of 16 July 1747 (on springs); 7 May 1760 (on the sea's salt); 20 Sept 1761 (on rivers); 31 Jan 1768 (mammoths at Big Bone Lick); 22 Sept 1782 (geological observations); and 31 May 1788 (theory of the earth). The geological portion of Freneau's poem comprises lines 46–84.
6. *Letters of Henry Brevoort to Washington Irving*, ed. George S. Hillman (1918), 86.
7. Paulding, *Letters* (1962), 423, 456–459.
8. Useful articles include Conrad Wright, "The Religion of Geology," *New England Quarterly* (1941), 14:335–358; and Stanley M. Gural-

nick, "Geology and Religion before Darwin: The Case of Edward Hitchcock," *Isis* (1972), 63:529–543. See also Frank L. Mott, *A History of American Magazines, 1741–1850* (1966), 446–447.

9. I quote Emerson's *Journals* (1909–1914), 4:129; *Works* (1890–1899), 8:202; and *Journals*, 9:555–556. Newer editions are in progress.
10. This paragraph, like the previous one, derives from my own published research; see note 1, above.
11. The three are "The Society upon the Stanislaus," "To the Pliocene Skull," and "A Geological Madrigal"; see *Writings* (rep. 1966), 12:132–133, 268–269, 279–280. His prose "Lessons from the Earthquake," mentioned below, is in 20:162–164.
12. It was reprinted in *The Best American Short Stories: 1976*, ed. Martha Foley (1976), 333–339.
13. Modern poems can be harder to locate than 19th century ones. Collected editions, for example, are complete only if posthumous. Several of the poems I name individually, however—by Frost, Crane ("The River"), Warren, Rexroth, Auden, Randall, Jarrell, Snodgrass, and Ammons—can be found in *The Norton Anthology of Modern Poetry* (1973), ed. Richard Ellmann and Robert O'Clair. Joan Johnson's "Continental Drift" is in *Literature* (an anthology), ed. Joseph K. Davis and others (1977).

North American Vertebrate Paleontology, 1776–1976

Joseph T. Gregory

The year 1776 was not one of importance in the study of fossil bones. Mastodon remains, discovered along the Ohio River in 1739, had puzzled European scholars and prompted several collecting expeditions prior to the Revolutionary War. There was still confusion between the fossils which today are known as mammoth and mastodon. These fossils played a significant role in establishing the concept of extinct animals and faunas in the history of the earth.[1] But one can scarcely speak of scientific study of vertebrate fossils in America prior to 1799, when Caspar Wistar described remains of the ground sloth, *Megalonyx*, obtained by Thomas Jefferson from a cave in what is now West Virginia.

Until about 1850 the study of fossil vertebrates in America was limited to infrequent descriptions of such remains by physicians interested in natural history, and to the collection and exhibition of such large and spectacular skeletons as the mastodon and Eocene whale, *Basilosaurus*. Between 1850 and the late 1880's the western exploration of the continent was accompanied by numerous discoveries of prehistoric novelties and the rapid development of this science, largely at the hands of Joseph Leidy, E. D. Cope, J. W. Dawson, O. C. Marsh, and J. S. Newberry. A stratigraphic sequence of vertebrate faunas was determined and correlated with those of Europe; fossil horses and other vertebrates were used as evidence for the evolutionary transformation of organisms; attempts were made to derive "laws of evolution" from the fossil record, particularly the Neolamarckian doctrines of Cope. Discoveries of dinosaurs and spectacularly large mammals such as titanotheres kindled public interest in the field.

This interest led to the development of public exhibits of vertebrate fossils at several large museums with financial resources for collecting material suitable for exhibition. The growth after 1890 of numerous centers for the study of fossil vertebrates and the more complete specimens which became available provided means for review and critical evaluation of the work of the previous half century by a new and larger group of specialists. Extensive faunal revisions consolidated and organized the results of the early discoveries. Unfortunately, these were carried out in isolation from the developing science of genetics and followed an essentially pre-evolutionary typological systematic philosophy. Field work stressed more precise determination of the stratigraphic level of specimens in an ever finer stratigraphical framework; sedimentological analyses of fossiliferous deposits improved interpretations of past environments; vertebrate paleontology was probably more closely identified with geological sciences than at any time before or since. Matthew's essay "Climate and Evolution" (1915) reoriented biogeographical thought. Artistic attempts (as by C. R. Knight) to recreate the appearance in life of the myriad of extinct animals of the past led to anatomical researches and restorations of the musculature of various extinct forms, which in turn prompted discussions of the functional significance of morphological changes.

Transition from this early twentieth-century phase to modern work was gradual. George Simpson began analyzing large collections of fossils as samples of former animal populations, and gradually adopted for fossil materials the biological species concepts developed by vertebrate zoologists and geneticists. This led, by 1950, to a synthesis of genetics with systematic biology and paleontology into a new, essentially Darwinian, evolutionary biology, and to renewed emphasis by paleontologists on the biological aspects of their material. Besides the new systematics, functional approaches to the meanings of morphological changes have provided insight into the adaptive significance of crucial steps in vertebrate phylogeny. New collecting techniques supplied large samples of small vertebrate fossils in place of small samples of large animals; much new material of previously unstudied groups such as rabbits, rodents, lizards, and frogs was discovered. This more comprehensive record of the life of past epochs has encouraged attempts to investigate paleoecology by applying ecological principles to the population dynamics, food chains, and community evolution of former faunas. Attempts to infer ancient community structure from fossil assemblages have been made with increasing frequency and sophistication. Continued refinement of biostratigraphic classifications combined with radiometric dating of fossil-bearing deposits has improved intercontinental correlations. The widespread acceptance of continental drift and development of plate tectonic theories in the late 1960's has created much interest in reassessing the distributional patterns of terrestrial vertebrates and revision of biogeographic theories.

Published histories of vertebrate paleontology concentrate too often upon the adventures of the bone hunters.[2] The rivalry of Marsh and Cope and the fanciful battles between their collecting parties; expeditions of students from eastern colleges, armed to the teeth (it really is remarkable that none were accidentally shot by their comrades); the awesome size of the dinosaurs and the remarkable transformation of "little eohippus" into the modern horse—all these make interesting reading but give a woefully incomplete picture of the development of this subject and its contributions to related branches of science. If the following discussion slights these more dramatic events, it is hoped that calling attention to matters less well known will lead to a better understanding of the significance of the studies of fossil vertebrates.

From the Beginnings of the Republic until 1846.[3]

At the same time that Cuvier was beginning his studies of fossil vertebrates in France which were to establish vertebrate paleontology as a discipline based upon comparative anatomy, Caspar Wistar recognized the relationship of the large clawed foot which Thomas Jefferson had obtained from a cave in western Virginia to the sloths. A few years later Charles Willson Peale collected two fairly complete mastodon skeletons from Newburgh on the Hudson, New York, and displayed them in his museum in Philadelphia along with gigantic bison skulls from Kentucky. Some of Peale's drawings of these bones were published by Cuvier in his memoir on the mastodon.[4] The Lewis and Clark expedition across the continent in 1804–1806 brought back the jaw of a Cretaceous fish (*Saurocephalus* Harlan) and reports of other fossil bones from the upper Missouri River.[5] Around 1840 Dr. Albert Koch collected other mastodon remains from various localities in Missouri and displayed them in St. Louis and elsewhere. Later he brought together bones of several individuals of the large Eocene whales of the

Gulf coastal region in a composite monster which he displayed about the United States in a traveling show.[6] Most of the fossils that came to the attention of scientists before 1850 were Pleistocene mammals. More ancient fossils included reptilian and fish remains from the greensands of New Jersey and birdlike tracks in the red sandstones of the Connecticut Valley, which also yielded a few bones and fishes. Most of a large reptile skull from the upper Missouri River region went to Europe. Sir William Logan found tracks of land animals in the Coal Measures of Horton Bluff, Nova Scotia, in 1841, the first fossil evidence of air breathing vertebrates as early as the Carboniferous ever found.

The chief American students of vertebrate fossils during the early nineteenth century were Wistar, Harlan, DeKay, Morton, and Mitchell, with contributions near the end of the period by E. Hitchcock and the Redfields. Excellent accounts of these early developments are given by Simpson (1942) and Gerstner (1967), to which interesting details are added in Howard (1975).

The Heroic Period. 1846–1891. Westward Expansion and Paleontological Discoveries in the Missouri River Basin and Rocky Mountains.[7]

Although fossil bones had been reported from the western interior of North America as early as 1796, they attracted little scientific attention until 1846, when Dr. Hiram Prout of St. Louis, Missouri, published information on a large fossil jaw brought by trappers from an area known as the "Badlands." The publicity on Prout's large fossil led to scientific expeditions to the White River Badlands, by John Evans in 1849, by Thaddeus Culbertson in 1850, and by F. B. Meek and F. V. Hayden in 1853. Fossil bones collected by these explorers were sent to Dr. Joseph Leidy, who had already published a note concerning the evidence afforded by fossils for the existence of horses in North America prior to their reintroduction by Europeans in the sixteenth century. Leidy rapidly became the authority on vertebrate fossils and was able to arrange with Spencer F. Baird of the Smithsonian to have all such specimens obtained by government surveys referred to him for study and identification.

Thus was inaugurated the intensive exploration for, and collection and description of, fossil bones from western America.[8] Except for the interlude of the Civil War, this activity continued with increasing tempo until about 1890, and with changing aims and methods to the present day. Three figures dominated the field in the later nineteenth century: Joseph Leidy, E. D. Cope, and O. C. Marsh. Less well known are the important contributions of J. W. Dawson and J. S. Newberry.

Joseph Leidy, physician and talented artist, professor of anatomy at the University of Pennsylvania, published impressive research on freshwater microorganisms as well as in vertebrate paleontology. He supported the idea of transmutation of species well before the publication of Darwin's *Origin*. His morphological descriptions of fossils have been praised repeatedly for their clarity and accuracy, but his work has been unjustly criticized for the absence of phylogenetic hypotheses or theoretical discussions. His writings portray the life of the past far more than merely describing the broken fragments of bones sent back by early explorers. Leidy made known the wealth of fossils of the White River Badlands (Oligocene in modern parlance), of the overlying later Tertiary (mainly Miocene and some Pliocene), and a portion of the middle Eocene Bridger fauna of Wyoming. He published the first notice of dinosaur remains from this continent (from the Judith River beds of Montana), and shortly thereafter described the skeleton of a duck-billed dinosaur from New Jersey (*Hadrosaurus*). Leidy's other paleontological contributions ranged from Carboniferous fishes to Pleistocene edentates and giant bison. From 1847 until 1866 he worked alone on all material from the western region. After the strident entry of Marsh and Cope into the field, he turned to other areas for fossil materials.[9]

Othniel Charles Marsh, wealthy by inheritance, was trained as a geologist and paleontologist. He held a Yale professorship without salary or classes, and for many years was paleontologist for the U.S. Geological Survey. Imperious and impatient of obstruction or contradiction, he had a flair for recognizing a significant point or feature among a myriad of details. His brief diagnoses of fragmentary fossils usually stress important peculiarities of the specimens. He ably capitalized upon the sensational features of such discoveries as Cretaceous birds with teeth, the strange horned ungulates of the Eocene, which he called *Dinocerata* (horned dinosaurs), and the evolution of the horse.[10]

Edward Drinker Cope, who like Marsh inherited wealth, studied briefly under Leidy and also worked under Baird on the herpetological and ichthyological collections of the Smithsonian. His systematic work in paleontology was comparable to that of Marsh but far broader in its scope, including fishes and amphibians as well as the higher vertebrates. He is equally noted for prolific work on recent ichthyology and herpetology. Cope paid greater attention than Marsh to the stratigraphic distribution of fossils and to their use for correlation with Europe. Most of all, he constantly engaged in speculation and generalization from his data. His papers are filled with phylogenetic diagrams and comments concerning the relationships of genera and families and the evolutionary origins of these and of higher categories up to the phylogeny of the vertebrate classes. The origins of mechanical adaptations in the limbs and feet of mammals and of the grinding and shearing mechanisms of teeth were of special interest to him, and around these he formulated his Neolamarkian theories of the causes of evolutionary change.[11]

John Strong Newberry, a collector of Coal Measures plant fossils from an early age, studied medicine but soon abandoned that field for geology. He took part in several western explorations and surveys in the 1850's, and became professor of geology at Columbia University in 1866; he also directed the Ohio Geological Survey for many years. Better known as a paleobotanist, he was nevertheless the first serious student of fossil fishes of this continent, especially those of the Paleozoic. He made known the large armored placoderms of the Devonian of Ohio, the multitude of shark teeth in the Mississippian limestones of the Mississippi Valley, and fishes of the Ohio Coal Measures. His essay on the succession and relationships of Paleozoic fishes, in the first report of the Geological Survey of Ohio, is an excellent statement of the status of this subject in the later nineteenth century. Newberry's work shows appreciation of the factors of growth and individual variation in systematics, and his comparison of European and American fish faunas takes into account the environmental influences upon distribution as well as age factors. He shows more appreciation of fossils as remains of living organisms than any of his contemporaries except Leidy.[12]

J. William Dawson of Pictou, Nova Scotia, later Principal of McGill University, discovered the earliest known terrestrial vertebrate fossils in the erect tree stumps at Joggins, Nova Scotia, in 1851. He established the Order Microsauria for some of these in 1863, many of which have indeed proven to be early reptiles. His contributions to the description and interpretation of Carboniferous vertebrates continued until the end of the century.[13]

Perhaps the major achievement of the nineteenth-century vertebrate paleontologists was making known the wealth and variety of life of past epochs of the history of the earth. Hope of discovery of wholly new kinds of extinct animals certainly motivated the describers of the material, and lay behind such incidents as the 1872 race for priority in the description of Uintatheres from the Eocene of Wyoming and the similar contest over Jurassic dinosaurs in 1877 (below, notes 27, 30).

Of no less importance was establishing the sequence of terrestrial faunas in North America, especially in the Cenozoic. For the most part this was done by straightforward application of the principle of superposition; particular assemblages of fossils were collected from formations that were being mapped and described by geologists. The paleontologists contributed their interpretations of the ages of the fossils. Leidy at first compared the fossils of the White River Badlands to those of the late Eocene of France (Cuvier's paleotherian fauna), but in 1857 he advanced strong arguments for considering them Miocene (at that time the term Oligocene was little used). When fossils were discovered at Fort Bridger, Wyoming, in 1869, he immediately recognized their greater antiquity and probable Eocene age.

Marsh generally used the stratigraphic classification of Hayden, but he applied the generic names of characteristic fossils to the "beds" which produced the bones he studied. His major contribution to stratigraphic paleontology is interesting chiefly for an early discussion of the Miocene-Pliocene boundary which Marsh chose to place at an unconformity in Nebraska. Cope also used Hayden's terminology in his earlier work on Tertiary mammals. In 1879 he presented a classification of the vertebrate faunas of North America in which various levels were labeled with generic names and correlated with faunal stages of France recognized by Gervais (1859). Both Marsh and Cope discussed the fossil content of the various units (which were rather thick, lithic formations or groups) and the implications of the various genera for their geologic age. Cope was responsible for recognition and description of the early Eocene (Wasatch) and Paleocene (Puerco) mammalian faunas and their correlation with the Suessonian and Thanetian stages of Europe.[14]

Leidy recognized dinosaurs among the reptilian remains that Hayden had collected from lignitic clays near the mouth of the Judith River in Montana. He compared them with those from the Wealden of England. Hayden had observed lignitic deposits (the Fort Union) above the highest marine Cretaceous in the Dakotas, and was inclined to consider the Judith River deposit of similar age early Tertiary, on the basis of angiospermous plants. Leidy quietly noted the disagreement and left its explanation to Hayden.[15]

The long and at times bitter dispute over the position of the Cretaceous-Tertiary boundary in the High Plains region, known as the Laramie Problem, arose over the presence of horned dinosaurs and other reptiles which Cope and other vertebrate paleontologists insisted had not lived beyond the Cretaceous, in the same formation with floras that Lesquereux, Newberry, and later generations of paleobotanists asserted were of Tertiary type. Arguments over the possible survival of dinosaurs into the Tertiary lasted well into the present century and were apparently settled only when paleobotanists Erling Dorf and Roland W. Brown showed in the late 1930's that the floras of the Lance and Denver formations, associated with dinosaurs and Mesozoic mammals, were distinguishable from those of the immediately overlying Fort Union and equivalent formations, which by then were known to contain Paleocene mammals.[16]

Marsh at first regarded the dinosaurs of the Morrison Formation (his *Atlantosaurus* beds) as equivalent to the Wealden of England, but soon changed his determination to Jurassic, without stating reasons for either view. Vertebrate fossils played a major role in successive determinations of the age of the Newark Group of red beds along the Atlantic Coast as Permian (New Red Sandstone), early Jurassic, and finally Triassic. The fishes were identified with *Palaeoniscus* of the German Copper Slate in 1819 by Alexandre Brongniart, with early Jurassic forms by W. C. Redfield (1853–1857), and as Triassic by Newberry (1888). Discovery of the skull and other bones of a reptile similar to "Belodon" of the German Triassic in North Carolina by E. Emmons in 1856, and of teeth of similar reptiles in Pennsylvania about the same time, led to the Triassic age determination of Cope in 1866 and 1869. Cope also identified Triassic strata in New Mexico (1874) and Texas (1892) on the basis of fossil reptiles and labyrinthodonts, and inferred the Permian age of red beds in north Texas and eastern Illinois from the character of the fossils.

Newberry pointed out similarities in the fossil fishes of the Devonian and Carboniferous rocks of the Mississippi Valley and in Ohio with those of Europe (1973), but I find little evidence that the age determinations were based upon the fishes; rather the opposite: the age of the fossil fishes was determined by the associated invertebrates.

Fossil vertebrates and evolutionary change. At the beginning of the nineteenth century Georges Cuvier compared fossil vertebrates from Jurassic, Cretaceous, Eocene, and Pleistocene deposits and found no evidence of transition between their respective species and no basis for inferring that later faunas were descendants of the earlier ones. Half a century later Charles Darwin considered the absence of transitional fossils between various species a major obstacle to the theory of evolution, and developed his arguments about the incompleteness of the geological record to answer this objection. Although numerous fossiliferous deposits had closed the large gaps in the record known to Cuvier by that time, T. H. Huxley, in 1862, found little paleontological support for evolution. By 1870, however, several examples of progressive modifications of the structure of successive representatives of various families of mammals had been worked out by L. Rütimeyer, A. Gaudry, and Huxley. The latter summarized these in his presidential address to the Geological Society of London, in which he proposed a paleontological genealogy of the horse consisting of a series of fossils from successive Tertiary deposits representing several stages in the ancestry of the modern horses.[17]

In America Joseph Leidy had noted the abundance and variety of fossil horses from the Tertiary deposits of the west with much interest; by 1869 he had described 6 genera and 19 species of these fossils. He did not attempt to arrange these in any genealogical sequence, but his descriptions stress the intermediate or transitional features of *Anchitherium*, *Parahippus*, and *Protohippus* between "adjacent" genera.[18] O. C. Marsh became involved in the study of fossil horses through the specimens he hastily collected at a railroad water stop in western Nebraska in 1868. By 1874 he had assembled sufficient material to propose that horses had evolved in North America rather than in Europe, and he succeeded in convincing Huxley of this when the latter visited America for a lecture tour. Marsh's horse series, augmented by many side branches, remains the backbone of the vertebrate paleontologist's most frequently cited demonstration of evolutionary change.[19] Leidy also pointed out the greater similarity of the early Tertiary members of various orders of animals to one another than is seen among their later representatives.

Marsh contributed to the documentation of evolutionary change by his discovery of Cretaceous birds with teeth[19] and his investigations of the evolution of brains from the endocranial casts of fossils.[20] Of the toothed birds, Darwin wrote to Marsh: "Your work on these old birds and on the many fossil animals of N. America has afforded the best support for the theory of evolution, which has appeared within the last 20 years." Marsh proposed some "general laws of brain growth," especially that of a "general increase in brain size all during the Tertiary." These views were widely accepted for many years. His statements concerning the relative enlargement of the cerebrum, increase in convolutions, and reduction of the olfactory lobes have stood the test of time.

Cope speculated more widely than Marsh on the causes of evolutionary change, and marshaled paleontological evidence concerning the relationships of the larger systematic groups of vertebrates. In 1867 and 1871 he pointed out anatomical similarities of dinosaurs to birds, anticipating Huxley's view on this matter.[21] The convergence of diverse orders of ungulate mammals toward a common structure represented by *Phenacodus* of the early Eocene was illustrated in 1883. The course of evolutionary change among all vertebrate classes was discussed in detail in 1885.[22] In 1892 he presented the first strong argument for derivation of tetrapods (land vertebrates) from rhipidistian fishes rather than from dipnoan lung fishes.[23] He first recognized the relationship of theriodont (mammal-like) and pelycosaurian reptiles to mammals, and the primitive position of the cotylosaurs rather than rhynchocephalians to other reptilian orders.[24] Numerous diagrams of mammalian lineages accompanied his papers after 1880. Marsh rarely published diagrams of phylogenetic relationships.

Cope argued that "the prevalence of non-adaptive characters, in animals, proves the inadequacy of hypotheses which ascribe the survival of types to their superior adaptation to the environment."[25] With Alphaeus Hyatt, the student of ammonites, he stressed the role of acceleration and retardation of development of various parts of the organism in the origin of new characters. Mechanical forces imposed by the environment further modify growth (kinetogenesis) in ways that explain the progressive adaptive modifications in the joints of the feet and limbs of ungulates and in the teeth of various kinds of mammals. Fossil representatives of several mammalian lineages were illustrated to support these arguments. Though kinetogenesis has long been discarded as a causal factor in evolution, the mechanical principles of limb action and joint construction worked out by Cope are basic to modern functional anatomy.

One of the most important of Cope's observations and generalizations was the derivation of diverse types of mammalian molar teeth from a primitive, 4-cusped (1874) and ultimately 3-cusped (1884) pattern. This principle, with the addition of Osborn's theory of the origin of the triangular pattern by rotation of the basal secondary cusps of theriodont reptiles and Jurassic triconodont mammals from their initial linear alignment, is known as the Tritubercular Theory of mammalian teeth and forms the basis for a universally used system of tooth-cusp nomenclature. The theory has been extremely influential in the twentieth-century studies of mammalian phylogeny. Although Osborn's explanation of the origin of the tritubercular pattern is no longer accepted, it underlies the continuing efforts to understand this complex process by J. L. Wortman, J. W. Gidley, W. K. Gregory, G. G. Simpson, P. M. Butler, B. Patterson, A. W. Crompton, and others.[26]

Dinosaurs. Leidy first recognized dinosaurs among the isolated teeth that F. V. Hayden collected near the Judith River in 1855. His study of the skeleton of *Hadrosaurus* from New Jersey in 1859 led to the first realization that some dinosaurs walked bipedally on their rear legs rather than sprawled like a lizard. The race between Cope and Marsh for priority in describing the dinosaurs from several rich Jurassic deposits in Colorado and Wyoming that had been opened in 1877 led to a rapid succession of discoveries of the diversity and peculiarities of these reptiles. Most of the current terms for various kinds of dinosaurs are derived from Marsh's work of this period, which resulted in illustrations of complete skeletons of these reptiles.[27]

G. M. Dawson found dinosaur bones in the late Cretaceous of Saskatchewan in 1874; in ensuing years he and his assistants, particularly J. B. Tyrrell and T. C. Weston, located the principal fossiliferous areas along the Red Deer River in Alberta.[28] Their earliest finds were described by Cope, but before the end of the century L. M. Lambe of the Geological Survey of Canada had become a leader in the research on Cretaceous reptiles. The major investigation of the Alberta dinosaur beds occurred during the first quarter of the twentieth century. Horned dinosaurs were first recognized from fossils found in eastern Wyoming by J. B. Hatcher in the late 1880's.

Mesozoic mammals had been known in England since 1824. In the course of collecting dinosaurs at Como Bluff, Wyoming, in 1878 a tiny jaw and tooth was found, and the following year a rich pocket of Jurassic mammal remains. Cretaceous mammals were found with the horned dinosaurs in Wyoming in 1889.

Devonian fishes close to the ancestry of land vertebrates, discovered at Escauminac Bay, Quebec, about 1879 and described by J. F. Whiteaves of the Geological Survey of Canada, provided the basis for various of Cope's theories concerning vertebrate phylogeny, including his 1892 paper concerning tetrapod origins.

During the nineteenth century the National Museum of Mexico acquired various collections of fossil vertebrates, as did the Mexican School of Mines (later the Geological Institute) under Prof. Antonio del Castillo. Richard Owen in London described (1870) late Cenozoic fossil horses and llamas from the Valley of Mexico from material sent him by Prof. del Castillo in 1866; Falconer published a note on mastodons from Mexico in 1868. Cope described the mammalian fauna preserved in the Mexican museums in 1884.[29]

The second half of the nineteenth century was a time when fossil collecting in the western United States and Canada was both arduous and adventurous. It was a time when almost any specimen one found had a good possibility of belonging to a previously unknown kind of animal. Leidy established this by thorough and cautious analysis of the earliest collections. His followers applied it as a working principle, sometimes with less than due caution. Nevertheless, the discoveries of this era greatly enlarged the scope of the panorama of life of past geologic ages. Numerous new orders of both reptiles and mammals were discovered. An impressive start was made toward documenting the ancestry of many families of mammals.

Leidy and, more especially, Cope and Marsh were the first professional vertebrate paleontologists in America. In 1880 they were joined by H. F. Osborn and W. B. Scott, who were appointed to the Princeton faculty that year following graduate study in Germany. Scott and Osborn had been instrumental in organizing a Princeton student expedition to the Rocky Mountains region in 1877, patterned somewhat on the earlier Yale expeditions, and in the following years Scott regularly took students west to collect. Osborn left Princeton in 1891 to establish the Vertebrate Paleontology Division of the American Museum of Natural History.

The pioneer period closed with the sensational airing of the longstanding feud between Cope and Marsh on the pages of the New York *Herald* in January 1890.[30] One result of this scandal, which also involved Director Powell of the U.S. Geological Survey, was the termination of federal appropriations for paleontological research and a consequent relocation of several paleontologists. The many new faces on the scene in the 1890's belonged to leaders of the early twentieth century.

The Classic Period. Paleontology in the Growth of Major Exhibition Museums, 1891–1929

The appointment of Henry Fairfield Osborn as Curator of Vertebrate Paleontology of the American Museum and Professor of Zoology at Columbia University opened a new era of rapid development of public museum displays of vertebrate fossils and the accompanying growth of research collections both at museums and at several universities. During Marsh's lifetime the collections at Yale were used strictly for research, and no fossil skeletons were mounted for exhibition. Cope's collections were kept at his residence in Philadelphia and sold after his death to the American Museum. After 1872 the U.S. National Museum had the skeleton of an "Irish Elk" and some of Ward's casts on display.

Transfer of the portion of the Marsh collection made with government funds to it between 1886 and 1899 made it the largest study collection in the country. Vertebrate fossils were displayed in the Museum of the Geological Survey of Canada by 1883. Harvard had important collections of fossil vertebrates, especially from Europe, for teaching and research purposes, but no active programs in the field. The American Museum in New York was soon followed by others, and by 1915 a dozen more museums exhibited fossil vertebrates.

Osborn soon brought together a professional team with the object of assembling for public exhibition several series of skeletons of animals from successive stratigraphic levels to demonstrate evolutionary change. This required a different kind of collecting from gathering up weathered specimens from the surface of the ground that had supplied the nineteenth-century pioneers with most of their riches. Techniques for holding fragile fossils together during the long journey from fossil quarry to museum had been developed by the collectors employed by Cope and Marsh, notably J. B. Hatcher, C. H. Sternberg, and J. L. Wortman. These methods were further perfected by younger collectors too numerous to mention in the highly successful campaigns of the early twentieth century. New localities were discovered, especially in the Great Basin and Pacific Coast regions, including Devonian fish deposits at Beartooth Butte, Wyoming, and Water Canyon, Utah. Excavation of the Rancho la Brea tar pits near Los Angeles, at Dinosaur National Monument, Utah, and late Tertiary quarries near Agate, Nebraska, were especially significant. The late Cretaceous deposits of western Canada, which had been discovered by G. M. Dawson in 1874, were exploited by L. M. Lambe, Barnum Brown, and the Sternbergs to produce the spectacular dinosaur exhibits at Ottawa, Toronto, and New York. Collecting and displaying large fossil skeletons was costly business, possible only with the financial support of such men as Andrew Carnegie, Marshall Field, and J. P. Morgan, or appropriations of national governments. About 1920 Childs Frick began collecting Tertiary mammals on a scale far surpassing all previous efforts, and amassed material for study by generations of paleontologists yet to come.[31]

In Mexico, M. Villada and A. del Castillo described more Tertiary and Quaternary mammals, the latter in collaboration with W. Freudenberg. W. Soergel published on Mexican fossil elephants sent to him by del Castillo. C. Burkhardt listed fossil fishes found in various Mesozoic formations of Mexico in his comprehensive survey of these rocks.

The turn of the century also saw the beginning of exploration of foreign fields by paleontologists from American institutions. Ameghino's outstanding discoveries in Patagonia prompted expeditions to that region from Princeton, the American Museum, Amherst, and the Field Museum. India, Egypt, and South Africa were sampled, and the much publicized Central Asiatic Expeditions of the American Museum opened new fields for both Cretaceous and Tertiary vertebrates in the Gobi Desert, which added significantly to knowledge of the diversity and geographic distribution of land animals; important data are still being derived from these collections.

Biostratigraphy. Not only was far more complete material of many species collected from the now well known localities, but the new generation of collectors took pains to record more detailed information about localities and stratigraphic levels. Questions of the appropriate subdivision of the Tertiary Epochs on the basis of vertebrate faunas, correlation of North American and European stratigraphic units by vertebrate fossils, and related problems received attention. Under Osborn's direction, W. D. Matthew prepared faunal lists of all described North American Tertiary faunas, which he and Osborn then arranged in a series of "Life Zones"—assemblage zones, in modern parlance. In 1924 Matthew introduced a modified scheme of zonation based upon stages in the evolution of the horse, successive faunal zones being designated by the names of characteristic genera of the equid lineage. These classificatory schemes culminated in the "Wood Committee" report of 1941, which established a series of "North American Provincial Land Mammal Ages" characterized by composite lists of genera but also defined as the faunas of particular lithologic formations, a duality that led to difficulties in establishing boundaries between successive "Ages." In spite of these defects it has been widely adopted even outside of North America, and has served as a model for similar classifications on other continents.[32]

Systematics. The new and better collections of this period provided the basis for many systematic revisions and phylogenetic studies. Phylogeny was perhaps the major preoccupation and emphasis of vertebrate paleontologists during the first third at least of the present century. Their systematics was still strictly typological, in spite of the strong evolutionary beliefs of all these workers. Greater attention was given to intraspecific variation and especially to stratigraphic occurrence in defining taxa, and the more complete specimens enabled many of the names that had been established upon single teeth or other fragments to be associated and synonymized. The leading students of submammalian vertebrate fossils came to regard the genus as the operational taxonomic unit.

A new classification of reptiles, based upon the structure of the temporal region of the skull, emerged from the work of Cope, Baur, Case, and Williston in the 1890's and rapidly replaced older arrangements that had emphasized limb structure. Osborn published a comprehensive review in 1903, and his ideas were further developed by Williston, in whose hands essentially the modern arrangement appeared by 1917. Concurrently, William K. Gregory led the way to a consensus on the nomenclature of reptilian skull bones.[33]

Two supposed laws of evolution, Irreversibility and Orthogenesis, conspicuously influenced phylogenetic thinking during the early twentieth century. The Law of Irreversibility, proposed by the Belgian Louis Dollo, stated that in the course of evolution an organism never returned exactly to the condition of one of its ancestors. Dollo had especially in mind the loss of skeletal elements during evolution, and also ways in which secondarily aquatic reptiles and mammals differed from their adaptive counterparts among the fishes. Orthogenesis, the idea that organisms evolve in certain directions determined by "internal" causes and independent of adaptiveness or selection, combined with the concept of irreversibility, led to a highly skeptical attitude toward fossils as possible ancestors to other, later forms. Any feature that was not on the straight line between a given later animal and its presumed ancestor was considered to bar that fossil from a place in that particular lineage. The result was the type of phylogenetic diagram—so abundant in the works of H. F. Osborn, for example—in which a separate line, springing from the unknown ancestry below the base of the segment of phylogeny being considered, led to practically every known taxon shown on the chart. Instead of a family tree, what Jepsen has called the "candlewood ocotillo" picture of phylogeny developed.[34]

This period also saw conflict between many paleontologists, schooled in the Neolamarkian ideas of Cope and Hyatt, and the new school of geneticists and experimental zoologists over the mechanism and possible cause of evolutionary change. Osborn, Matthew, and Gregory contended that the fossil record of Tertiary mammals demonstrated continuous unidirectional evolutionary change rather than the discontinuous saltations of DeVries and T. H. Morgan.[35]

Biogeography, which had run to extremes of building "land bridges" to explain disjunct distributions in the late nineteenth century, was strongly influenced by W. D. Matthew's "Climate and Evolution," in which he showed that the distribution of most families of mammals could be explained on the basis of the fossil records of their origins in the northern hemisphere and outward migration along lines of present land regions or over recently submerged shallow seas. Appearing almost at the same time as Wegener's book on drifting continents, it provided arguments for advocates of fixed continents and oceans in their early debates with drifters. Important additional contributions to Matthewsian biogeography were George Myers' study of the dispersal of freshwater fishes and Simpson's analyses of faunal resemblance, migration routes, and the history of the South American fauna.[36]

Functional morphology, which in the nineteenth century was limited to studies of the adaptive features of dentitions for plant or animal diet, and of limbs and joints for various types of locomotion, was further developed by W. K. Gregory and his students, who studied the occlusal relationships of mammalian teeth, restored the musculature of dinosaurs and other extinct vertebrates, and sought the functional meaning of morphological differences.[37]

E. C. Case outlined the problems of paleoecological inference and attempted a restoration of the environments of late Paleozoic tetrapods.[38] A qualitative attempt to evaluate the environments of Tertiary mammalian faunas by analogy with the habitats of the closest living relatives of various members of the fossil assemblages was used by J. C. Merriam and others. At the very beginning of the century Matthew and Hatcher showed that the nature of the sediments as well as the composition of the vertebrate faunas conflicted with the long accepted notion that the Tertiary deposits of the western United States had been laid down in freshwater lakes.[39]

The Development of Present-Day Approaches, 1930–1976

Changing approaches to the study of fossil vertebrates are noticeable in the work of the generation of paleontologists led by G. G. Simpson and A. S. Romer, who began to work in the 1920's. By the middle of the twentieth century radical changes had taken place in the goals, methods, and many of the basic assumptions of the field. Aside from important pioneering studies, these were not evident until after the end of World War II, when the new concepts spread rapidly. Probably the greatest revolution was the adoption of the "New Systematics" including the application of a "biological species concept" to fossils and most recently a strong interest in the cladistic taxonomic philosophy of Hennig. Even more abrupt has been the swing of historical biogeography to the plate tectonics bandwagon and reassessment of patterns of dispersal of land animals in terms of the continental drift theory.[40] Paleoecological problems, such as the changes in ecosystems through time, have been investigated from numerous approaches in spite of formidable difficulties of interpretation of fossil assemblages in terms of biotic communities.

New techniques of collecting and preparing fossils have opened new fields and problems for investigation. The greatest impact has come from concentrating small fossils in the field by washing and screening large volumes of sediment, a method developed by C. W. Hibbard in the 1930's.[41] Hosts of rodents, insectivores, lizards, amphibians, and other small vertebrates have been added to faunal lists that were previously limited to relatively large animals. Many of these small forms are sensitive environmental indicators. Other laboratory tools, particularly air-abrasive and air-dent machines, and the judicious use of acids for removing tightly adherent matrices from the fossils have vastly improved the quality of specimens for morphological study. Radiometric measurement of the age of fossiliferous deposits has provided an independent test of the reliability of paleontological correlations, and led to some readjustments in ideas about the length of various epochs and stages, particularly in the Cenozoic. The firmer chronology provides an improved basis for studies of evolutionary rates.[42]

Increased attention to morphological detail as more and better material became available has led to specialization on limited systematic groups. Noteworthy examples include the studies of fossil birds by Wetmore, Howard, and Brodkorb; of lizards, snakes, turtles, and amphibians by Auffenberg, Holman, Estes, and others; of rodents by A. E. Wood, R. Wilson and their followers; and of fishes by Denison, Schaeffer, and R. W. Miller.

Active study of fossil vertebrates in Mexico was renewed after World War II by A. R. V. Arellano, M. Maldonado-Koerdell, and, more recently, I. Ferrusquia V. Additional exploration in Mexico by the Smithsonian and other museums in the United States has brought to light mid-Tertiary vertebrates, and Cretaceous dinosaurs and mammals in Baja California. Parties from the United States have also tested sites in Honduras and Panama. Following leads from Colombian petroleum geologists, R. A. Stirton obtained an important collection of mid-Tertiary vertebrates from Colombia, the first from tropical South America; he later succeeded in finding the first extensive Tertiary mammalian faunas in Australia and stimulated development of the field on that continent.

Post World War II research on mammals has emphasized the importance of basicranial characters, especially indications of the arterial circulation in the ear region, in addition to tooth morphology for interpreting phylogeny. Late Cretaceous mammals are now known from abundant specimens, and their study is modifying long-held ideas concerning the relationships of mammalian orders. Students of reptiles have sought unsuccessfully for osteological features that could be correlated with the sauropsid-theropsid dichotomy, and have questioned the value of temporal arches as a basis for reptilian classification. Ostrom's study of *Archaeopteryx* has demonstrated a closer relationship between birds and dinosaurs than had previously been held. American students have been reluctant to accept various suggestions of polyphyletic origins for the tetrapod classes advanced by European paleontologists, and the viewpoint that these taxa are merely grades of organization. Disagreement has been especially strong and prolonged on the question of the derivation of amphibians from fishes.[43]

The New Systematics. This seems an appropriate term for the changes in the attitudes of vertebrate paleontologists toward taxonomic problems and the resulting modifications of their systematic practices that have occurred with increasing tempo since the 1930's and especially following the end of World War II. Inextricably interwoven with the changes in systematic concepts and procedures were the clarification of many problems in the mechanism of evolutionary change. The combination of Darwinian Natural Selection, Mendelian genetics elaborated by studies of *Drosophila*, mathematical models demonstrating the efficacy of selection in modifying the genetic constitution of species, and the polytypic species concepts of ornithologists led to what has been termed the Synthetic Theory of Evolution.

Most important was the application to fossils of the biological species concept, developed during the 1930's by T. Dobzhansky and E. Mayr. G. G. Simpson began to treat collections of fossil mammals as statistical samples of the once-living populations from which they came, and published an explicit statement of the reasoning in 1943. Since 1950 this approach has prevailed in most paleontological systematics.[44] The large samples of small mammals obtained by underwater-screening techniques lend themselves particularly well to statistical treatment, measurements of appropriate morphological features serving as the "characters" for systematic analysis. With the availability of electronic computers since the late 1960's, this type of study has been extended to various methods of multivariate analysis. The impact of more abundant material is seen in studies such as the analysis by Gingerich of changes in the dentition of early condylarths and primates, taken from numerous closely spaced stratigraphic levels within a formation, which provide a far more detailed documentation of the gradual nature of morphological changes in populations evolving through time than was possible with the formation-by-formation sampling of W. D. Matthew's generation. A currently debated question is how far Mayr's model of allopatric speciation applies to all evolution. The cladistic technique of character analysis is used increasingly in paleontological systematics.[45]

Vertebrate paleontologists have shown less interest in the methods of numerical taxonomy, which are also part of the revolution in systematic biology. But a by-product of the increasing use of statistical methods in systematics has been the adoption of quantitative techniques in the study of population dynamics of extinct species, analysis of fossil assemblages for ecological associations, study of growth patterns, and more complex interactions of morphological configurations and their functional significance in evolving populations.

Functional Morphology. Evolutionary changes in form shown by sequences of fossils have been analyzed in terms of their functional significance and then tested against models of the operation of selective forces as understood by geneticists. In a qualitative fashion this type of analysis had its beginning in the work of Cope and especially W. Kowalewsky in the 1870's, and is displayed in Matthew's study of the evolution of creodonts, horses, and other mammals. To the earlier mechanical analyses of bone-muscle systems have been added a variety of techniques for the study of locomotor and masticatory movements of living animals by photography, X-ray cinematography, and simultaneously recorded neuromuscular activity, which have vastly increased our understanding of how various anatomical structures function.[47] Kinematic analyses of the role of fins in swimming fishes have helped explain the body forms of such extinct vertebrates as ostracoderms. Problems currently receiving attention in the functional field include tooth occlusion and the origin and early evolution of mammalian dentitions, the posture of various dinosaurs, and the evolution of aggressive behavior implied by the thick skulls of dinocephalians, pachycephalosaurid dinosaurs, and titanotheres.[48] Related to such mechanical studies are attempts to determine whether dinosaurs and flying reptiles were warm blooded. Paleoneurology has refined functional interpretations through increasingly detailed determinations of the function of the cortical areas of the brain delimited by sulci preserved on endocranial casts of mammals.[49]

Detailed morphological studies by reconstruction from serial sections of fossils, a technique originated by Sollas about the beginning of this century and highly developed by the Stockholm school of morphologists, has been little used in America, although it has yielded significant results in the study of the evolution of mammal-like reptile skulls and a few other scattered instances.[50]

Paleoecology. A qualitative history of Pleistocene climatic changes on the High Plains was worked out by C. W. Hibbard by comparing the present geographic distributions of the small mammals from successive deposits. He was able to demonstrate the alternation of boreal and warm temperate conditions which he correlated with glacial and interglacial stages of the Pleistocene. Similar qualitative reasoning has been used by Webb in a study of environmental change from Late Miocene into the early Pleistocene.[51]

Modern paleoecology attempts to reconstruct parts at least of ancient ecosystems and then to identify changes in these which may be related to environmental conditions or possibly to observed evolutionary changes in some of the taxa of the community. E. C. Olson has analyzed the Permian terrestrial faunas of North America and Russia for the various community associations that compose them and related these to different habitats implied by the entombing sediments. By observing changes in composition of these associations over appreciable stratigraphic intervals, he has been able to relate them to environmental changes suggested by sedimentological and floral data. A particularly interesting result has been recognition of the steps by which the typical terrestrial ecosystem developed from an aquatic one.[52]

Early attempts to relate the accumulation of vertebrate fossils to catastrophic mortality of modern animals by J. Walther and J. Weigelt were followed by a general analysis of the processes involved in burial and preservation of land vertebrates by Efremov, who coined the term Taphonomy for this aspect of paleontology. Recent investigations by Behrensmeyer, Dodson, and Voorhies, have stressed the effects of stream transportation on bone assemblages.[53] Shotwell attempted to separate the remains of animals from different habitats or communities that had become mixed in deposits of transported bones by the relative completeness of different taxa. He has applied this technique with some success to late Tertiary assemblages and at the same time raised a storm of criticism of the assumptions involved.[54] Zangerl and Richardson have provided a thorough taphonomic and paleoecologic analysis of the accumulation of fossiliferous carbonaceous marine shales.[55]

Causes of major episodes of extinction remain an interesting enigma. The impressive list of coincidences between times of extinction of large

Pleistocene mammals and the first appearance of man in various regions has not convinced advocates of the hypothesis of extinction due to climatically controlled ecological shifts that man was the key factor. Less well understood are the Cretaceous and Permian extinctions, for which both climatic and catastrophic events have been suggested.[56]

Fossil Man. Reports of fossil human bones in North America go back at least to 1805, when Gibbs announced the discovery of a skeleton near the Chesapeake. Albert Koch claimed that implements were associated with the mastodon bones he excavated in the Ozark highlands of Missouri in 1839–1840. Far better documented was the human pelvis found below gravels containing mastodon bones on an island in the Mississippi River near Natchez, Mississippi, reported by M. W. Dickeson in 1845. Various other finds have been catalogued by Sellards.[57]

Critical investigations by W. H. Holmes in the 1890's and by Aleš Hrdlička in the present century showed that neither paleolithic artifacts nor skeletal remains of premodern man occurred in North America. Hrdlička insisted that man was a recent arrival on this continent, and when E. H. Sellards found human bones associated with a Pleistocene fauna at Vero, Florida, in 1916–1917, a long controversy ensued. After discoveries of stone projectile points embedded in skeletons of extinct bison and mammoth at Folsom, New Mexico, near Fort Collins, Colorado, and especially near Clovis, New Mexico, between 1926 and 1939 by J. D. Figgins, E. B. Howard, F. H. H. Roberts, Jr., and others, the arrival of man on this continent prior to the end of the Pleistocene was acknowledged. A number of anthropologists have advanced arguments for the presence of man in America from 24,000 to 40,000 years ago, but thus far radiocarbon age determinations do not support a greater antiquity than about 14,000 years before the present.[58]

American paleontologists have become active in studies of primitive man on other continents. Barnum Brown collected a number of ape jaws from the Siwalik Hills of India in 1922–1924, and H. de Terra and G. E. Lewis found other important Tertiary hominids there in 1934. Since 1960 Elwyn Simons has discovered numerous early Tertiary primates in the Fayum basin of Egypt. From about the same period a number of American institutions have been participating in excavations at Olduvai, Lake Rudolph, and the Omo regions in East Africa, which have produced increasing numbers of primitive human fossils from well-dated early Pleistocene deposits.[59] These endeavors have involved teams of specialists in diverse aspects of anthropology, archaeology, paleontology, and various geological disciplines. They demonstrate the effectiveness of interdisciplinary cooperation in the investigation of complex problems.

The century from 1846 to 1946 witnessed the development of vertebrate paleontology in the United States and Canada from occasional description of curious fossil bones to a highly organized and internally specialized science carried on by a body of professionals which grew exponentially during this period. The episode of rapid discovery and initial description of novelties between 1850 and the 1880's was followed by inventory and systematic revision during the first half of the present century, accompanied by considerable theoretical development. The late 1930's saw the beginnings of modern approaches to systematic paleontology. Since the end of World War II the field has been actively involved in new developments in evolutionary theory and systematics, and at the same time shifted emphasis from descriptive and phylogenetic studies to paleoecology, functional morphology, and mathematical modeling, which now form the frontiers of the subject. Combining biostratigraphic data with radiometric and magnetic-reversal age determinations has significantly improved the accuracy of the paleontologists' time scale and the precision of correlations by fossil vertebrates; with increasing frequency the age of fossils or of fossiliferous deposits is stated in years rather than the old scale of geologic epochs or stages. Criteria for reconstructing distributional histories developed on the basis of Matthew's biogeographical theses have facilitated reinterpretation of historical biogeography in terms of the concept of drifting continents.

Vertebrate paleontology from the outset has been an international science, and the regional focus of this brief account has led to omission of many extremely important developments in the field. With increasing orientation toward problems of the evolution of function, paleoecology, and biogeography rather than the description of local collections of fossils, international travel and sharing of collections have become common.

Notes

1. John C. Greene, *The Death of Adam* (Ames, Iowa, Iowa State Univ. Press, 1959), chapter 4.

2. R. S. Lull, "On the Development of Vertebrate Paleontology," *Amer. Jour. Sci.* (1918), ser. 4, 46:192–221. A. S. Romer, "Vertebrate Paleontology," in *Geology 1888–1938, Fiftieth Anniversary Volume, Geological Society of America* (1941), 107–135; "Vertebrate Paleontology, 1908–1958," *Jour. Paleontology* (1959), 33, no. 5:915–925. C. W. Gilmore, "History of Vertebrate Paleontology in the U.S. National Museum," U.S. National Museum, *Proc.* (1941), 90:305–377. J. C. Merriam, "An Outline of Progress in Paleontological Research on the Pacific Coast," Univ. California Dept. Geological Sciences, *Bull.* (1921), 12:237–266. C. Stock, "Progress in Paleontological Research on the Pacific Coast, 1917–1944," Geol. Soc. America, *Bull.* (1946), 57:319–354. C. L. Camp et al., "Paleontology in the West," *Journal of the West*, (1969), 8, no. 2:165–290. Popular accounts in R. H. Howard, *The Dawnseekers* (New York, Harcourt Brace Jovanovich, 1975), and H. Wendt, *Before the Deluge* (Garden City, Doubleday, 1968).
3. G. G. Simpson, "The Beginnings of Vertebrate Paleontology in North America," Amer. Philos. Soc., *Proc.* (1942), 86, no. 1:130–188. Patsy A. Gerstner, "The 'Philadelphia School' of Paleontology, 1825–1845," New Haven, Conn., Dissertation, Yale University, 1967.
4. Rembrandt Peale, *An Historical Disquisition on the Mammoth . . .* (London, 1803), vi + 91 pp. G. Cuvier, "Sur le grand Mastodonte . . ." Museum National d'Histoire Naturelle, Paris, *Ann. 8* (1806), 270–312, pls. 49–56.
5. Details are given in Simpson (1942), 168–172. An earlier report of fossil bones seen in Nebraska by James MacKay in 1796 was published in France. Cited by M. R. Voorhies, *Univ. Wyoming Contrib. Geology, Sp. Paper* (1969), 1:5. A nearly complete skull of a mosasaur collected along the Missouri River by trappers prior to 1835 was purchased by Prince Maximilian of Wied-Neuwied, who took it to Europe, where it was described by Goldfuss (1845). The missing tip of the rostrum of this skull was obtained somehow by Richard Harlan of Philadelphia, who described it as *Ichthyosaurus missouriensis*.
6. Koch's scattered publications are listed by O. P. Hay, "Bibliography and Catalogue of the Fossil Vertebrata of North America," *U.S. Geol. Surv. Bull.* (1902), 179:126–127. See also Simpson (1942), 168; J. D. Dana, *Amer. Jour. Sci.* (1875), ser. 3, 9:335–346. R. Bruce McMillan, "Man and Mastodon: A Review of Koch's 1840 Pomme de Terre Expeditions," in *Prehistoric Man and His En-*

vironments: A Case Study in the Ozark Highland, ed. W. R. Wood and R. B. McMillan (New York, Academic Press, 1976), 288 pp.

7. Government surveys provided substantial public support for vertebrate paleontology during the second half of the nineteenth century. Cope and Marsh also expended most of their personal fortunes on the acquisition of fossils. The public airing of the Cope-Marsh-Powell feud in the New York *Herald* of January 12 and 19, 1890, which resulted in Congress closing the purse to paleontologists, coincided closely with the beginning of the ensuing phase of support by public museums and private philanthropy.
8. Leidy gave an excellent contemporary account in the introduction to his "Extinct Mammalian Fauna of Dakota and Nebraska," Academy of Natural Sciences of Philadelphia, *Journal* (1869), 2nd ser., 7:23–27. Thaddeus Culbertson, "Journal of an Expedition to the Mauvaises Terres and the Upper Missouri in 1850," ed. J. F. McDermott, Smithsonian Inst., Bur. American Ethnology, *Bull.* (1952), Vol. 147. D. D. Owen, *Report of a Geological Survey of Wisconsin, Iowa and Minnesota and Incidentally of a Portion of Nebraska Territory* (Philadelphia, 1852).
9. H. F. Osborn, "Biographical Memoir of Joseph Leidy, 1823–1891," National Academy of Sciences, *Biographical Memoirs* (1913), 7:335–396. E. S. Morse, H. S. Jennings, and W. B. Scott, "The Joseph Leidy Centenary," *Scientific Monthly* (1924) 18:422–439.
10. Charles Schuchert and C. M. Le Vene, *O.C. Marsh, Pioneer in Paleontology*, (New Haven, Yale University Press, 1940).
11. H. F. Osborn, *Cope, Master Naturalist* (Princeton, N.J., Princeton University Press, 1931).
12. J. F. Kemp, "Memorial of J. S. Newberry," Geol. Soc. Amer., *Bull.* (1893), 4:393–406.
13. Lawrence M. Lambe, "Progress of Vertebrate Paleontology in Canada," Royal Soc. Canada, Trans. (1904), 2nd ser, 10, Sec. IV: 13–56. Lull, 1918 (note 2 above), 202. F. D. Adams, "Memoir of Sir J. William Dawson," Geol. Soc. America, *Bull.* (1900), 11:550–557.
14. O. C. Marsh, "Introduction and Succession of Vertebrate Life in America," *Popular Science Monthly*, March and April, 1878. E. D. Cope, "The Relations of the Horizons of Extinct Vertebrata of Europe and North America," U.S. Geol. Geogr. Surv. Terr., *Bull.* (1879–80), 5:33–54. Paul Gervais, *Zoologie et Paléontologie Françaises . . .* (2nd ed., Paris, 1859).
15. J. Leidy, "Notice of Remains of Extinct Reptiles and Fishes, Discovered by Dr. F. V. Hayden in the Bad Lands of the Judith River, Nebraska Territory," Acad. Nat. Sci. Philadelphia, *Proc.* (1856),

8:72–73. See K. M. Waagé, "Deciphering the Basic Sedimentary Structure of the Cretaceous System in the Western Interior," Geol. Assoc. Canada, *Sp. Paper* (1975), 13:55–81, esp. 63–64.

16. G. P. Merrill, *The First One Hundred Years of American Geology* (New Haven, Yale University Press, 1924), 579–593. Erling Dorf, "Flora of the Lance Formation at Its Type Locality," Carnegie Inst. Washington *Publ.* (1942), 508:88–92. Ronald W. Brown, "Paleocene Flora of the Rocky Mountains and Great Plains," *U.S. Geol. Surv. Prof. Paper* (1962), 375:3–11.
17. Geol. Soc. London, *Quart. Jour.* (1870), 26:xlii–liii.
18. Leidy, 1869 (note 8 above); O. C. Marsh, "Fossil Horses in America," *Amer. Naturalist*, (1874), 8:288–294. G. G. Simpson, *Horses* (New York, Oxford Univ. Press, 1951).
19. O. C. Marsh, "Odontornithes: A Monograph on the Extinct Toothed Birds of North America," Geological Exploration of the Fortieth Parallel, *Report* (1880), 7:xv + 201, 34 pls. Darwin's letter quoted in Schuchert and Le Vene, *O.C. Marsh, Pioneer in Paleontology* (note 4 above), 246–247.
20. O. C. Marsh, "Dinocerata: A Monograph of an Extinct Order of Mammals," *U.S. Geol. Surv. Mon.* (1886), 10:56–67. Marsh commented on the small size of brains in Eocene mammals in 1874. Cope pointed out that E. Lartet had made the same observations in 1868. See Tilly Edinger, "Die fossilen Gehirne," *Ergebnisse der Anatomie und Entwicklungsgeschichte*, Munich (1929), 28:1–249.
21. E. D. Cope, Acad. Nat. Sciences Philadelphia, *Proc.* (1867), 19: 234–235; "On the Reptilian Orders *Pythonomorpha* and *Streptosauria*," Boston Natural History Soc., *Proc.* (1869), 12:250–266. T. H. Huxley, "Further Evidence of the Affinity between the Dinosaurian Reptiles and Birds," Geol. Soc. London, *Quart. Jour., Proc.*, 1869 (1870), 26:12–31, esp. 22–25.
22. E. D. Cope, *On the Primitive Types of the Orders of Mammalia Educabilia* (privately printed? 1873); "Note on the Phylogeny of the *Vertebrata*," *Amer. Naturalist* (1884), 18:1255–1257; "On the Evolution of the Vertebrata, Progressive and Retrogressive," *Amer. Naturalist* (1885), 19:140–148, 234–247, 341–353.
23. E. D. Cope, "On the Phylogeny of the *Vertebrata*," Amer. Philos. Soc., *Proc.* (1892), 30:278–281; see Bobb Schaeffer, "The Evolution of Concepts Related to the Origin of the Amphibia," *Systematic Zoology* (1965), 14:115–118.
24. E. D. Cope, "The Relations between the Theromorphous Reptiles and the Monotreme Mammals," Amer. Assoc. Adv. Sci., *Proc.*, 1884 (1885), 33:471–482.

25. E. D. Cope, *Origin of the Fittest* (New York, Appleton, 1887), vii, 106. Cope's views on causes of evolution are stated at length in "The Origin of Genera," Acad. Nat. Sci. Philadelphia, *Proc.* (1868), 20:242–300, esp. 290.
26. E. D. Cope, "On the homologies and origin of the types of molar teeth of Mammalia Educabilia," Acad. Nat. Sci. Philadelphia, *Jour.* (1874), 8:71–89. "On the Evolution of Vertebrates, Progressive and Retrogressive," *Amer. Naturalist* (1885), 19:345–353. A detailed history and bibliography is in W. K. Gregory, "A Half-Century of Trituberculy . . . ," Amer. Philos. Soc., *Proc.* (1934), 73:169–317.
27. Details may be found in the biographies of Marsh and Cope (notes 10 and 11 above); also in E. H. Colbert, *Men and Dinosaurs* (New York, Dutton, 1968); J. H. Ostrom and J. S. McIntosh, *Marsh's Dinosaurs* (New Haven, Yale Univ. Press, 1966); and C. H. Sternberg, *The Life of a Fossil Hunter* (New York, Holt, 1909).
28. L. S. Russell, "Dinosaur Hunting in Western Canada," Royal Ontario Museum, Life Sciences, *Contrib.* (1966), 70:37 pp. C. H. Sternberg, *Hunting Dinosaurs in the Bad Lands of the Red Deer River, Alberta, Canada: A Sequel to The Life of a Fossil Hunter* (Lawrence, Kansas, World Co. Press, 1917), 232 pp.
29. R. Owen, "On Fossil Remains of Equines from Central and South America," Royal Soc. London, *Philos. Trans.* (1870), 159:559–573; "On Remains of a Large Llama from Quaternary Deposits in the Valley of Mexico," ibid. (1870), 160:65–67, and E. D. Cope, "The Extinct Mammalia of the Valley of Mexico," Amer. Philos. Soc., *Proc.* (1884), 22:1–21.
30. The Cope-Marsh rivalry is described in W. H. Wheeler, "The Uintatheres and the Cope-Marsh War," *Science* (1960), 131:1171–76; A. S. Romer, "Cope *versus* Marsh," *Systematic Zoology* (1964), 13:201–207; and Elizabeth Noble Shor, *The Fossil Feud* (Hicksville, N.Y., Exposition Press, 1974), xi + 340 pp.
31. T. Galusha, "Childs Frick and the Frick Collection of Fossil Mammals," *Curator* (1975), 18:5–15.
32. H. F. Osborn, "Cenozoic Mammal Horizons of Western North America, with Faunal Lists of the Tertiary Mammalia of the West by William Diller Matthew," U.S. Geol. Surv., *Bull.* (1909), 361: 1–138; earlier versions appeared in 1899 and 1907. W. D. Matthew, "Correlation of the Tertiary Formations of the Great Plains," Geol. Soc. Amer., *Bull.* (1924), 35:743–754. H. E. Wood 2nd, et al., "Nomenclature and Correlation of the North American Continental Tertiary," Geol. Soc. Amer., *Bull.* (1941), 52:1–48. Develop-

ment of Tertiary stratigraphic classifications by vertebrate paleontologists has been reviewed by R. H. Tedford, "Principles and Practices of Mammalian Geochronology in North America," *North American Paleontological Convention, 1969, Proc.* (1970), F:666–703.

33. H. F. Osborn, "The Reptilian Subclasses Diapsida and Synapsida and the Early History of the Diaptosauria," Amer. Mus. Natural History, *Mem.* (1903), 1:449–507. S. W. Williston, "The Phylogeny and Classification of Reptiles," *Jour. Geol.* (1917), 25:411–421. *The Osteology of the Reptiles*, arranged and edited by W. K. Gregory (Cambridge, Mass., Harvard University Press 1925), xii + 300. W. K. Gregory, "Second Report of the Committee on the Nomenclature of the Cranial Elements in the Permian Tetrapoda," Geol. Soc. America, *Bull.* (1917), 28:973–986.
34. L. Dollo, "Les lois de l'évolution," Soc. belge Géol., *Bull.* (1893), 7, *Proc.-verb.*: 164–166. W. B. Scott applied the name "Dollo's Law" to the principle of irreversibility and extended it to the concept of irreversible trends. The history and varied meanings of orthogenesis are surveyed in G. L. Jepsen, "Selection, 'Orthogenesis' and the Fossil Record," Amer. Philos. Soc., *Proc.* (1949), 93:479–500; and G. G. Simpson, *The Meaning of Evolution* (New Haven, Yale University Press, 1949), 130–159. See G. L. Jepsen, "Phylogenetic Trees," New York Acad. Sci., Trans. (1944), ser. 2, 6:81–92.
35. W. K. Gregory, "Genetics versus Paleontology," *Amer. Naturalist* (1917), 51:622–635. H. F. Osborn, "Biological Inductions from the Evolution of the Proboscidea," *Science* (1932), 76:501–504; "The Nine Principles of Evolution Revealed by Paleontology," *Amer. Naturalist* (1932), 66:52–60. In 1912 Osborn, *Amer. Naturalist* (1912), 46:76–82, maintained that Darwin's variations were the "small mutations" of DeVries; he later abandoned the view that they had any role in evolution.
36. W. D. Matthew, "Climate and Evolution," New York Acad. Sciences, *Annals* (1915), 24:171–318. G. S. Myers, "Freshwater Fishes and West Indian Zoogeography," Smithsonian Inst., *Ann. Rept.* (1937), 339–364. G. G. Simpson, "Mammals and Land Bridges," Washington Acad. Sci., *Jour.* (1940), 30:137–163; "Mammals and the Nature of Continents," *Amer. Jour. Sci.* (1943), 241:1–31.
37. W. K. Gregory, "Notes on the Principles of Quadrupedal Locomotion and on the Mechanism of the Limbs in Hoofed Animals," New York Acad. Sci., *Ann.* (1912), 22:267–294; "The Upright Posture

of Man . . . ," Amer. Philos. Soc., *Proc.* (1928), 67:339–376; "Muscular Anatomy and the Restoration of the Titanotheres, and Principles of Leverage and Muscular Action," in H. F. Osborn, "Titanotheres," U.S. Geol. Surv. (1929), Mon. 55, Vol. 2:703–731. A. S. Romer, "The Pelvic Musculature of Ornithischian Dinosaurs," *Acta Zoologica* (1927), 9:225–275.

38. E. C. Case, "The Environment of Vertebrate Life in the Late Paleozoic of North America: A Palaeogeographic Study," Carnegie Inst. Washington, *Publ.* (1919), 283:vi + 273.
39. W. D. Matthew. "Is the White River Tertiary an Aeolian Formation?", *Amer. Naturalist* (1899), 33:403–408. J. B. Hatcher, "Origin of the Oligocene and Miocene Deposits of the Great Plains," Amer. Philos. Soc., *Proc.* (1902), 41:113–131. Leidy (1869, note 8 above), 354, pointed out the same objections to the lacustrine theory, but his protest went unheeded.
40. A. S. Romer, "Fossils and Gondwanaland," Amer. Philos. Soc., *Proc.* (1968), 112:335–343. E. H. Colbert, "Triassic Tetrapods from Antarctica: Evidence for Continental Drift," *Science* (1970), 169:1197–1201. M. C. McKenna, "Sweepstakes, Filters, Corridors, Noah's Arks, and Beached Viking Funeral Ships in Paleogeography," in D. H. Tarling and S. K. Runcorn, eds., *Implications of Continental Drift to the Earth Sciences* (1972), 1:295–308; "Fossil Mammals and Early Eocene North Atlantic Land Continuity," *Missouri Botanical Garden Annals* (1975), 62:335–353.
41. C. W. Hibbard, "Techniques of Collecting Microvertebrate Fossils," *Univ. Michigan Mus. Paleontology*, Contrib. (1949), 8:7–19; "Letter to Prof. W. G. Kühne (1961)," in "Studies on Cenozoic Paleontology and Stratigraphy, Claude Hibbard Memorial Volume," *3*, Univ. Michigan Mus. Paleont., *Papers on Paleontology* (1975), 12:135–138.
42. W. F. Libby, "Radiocarbon Dating," *Amer. Scientist* (1956), 44: 98–112. J. F. Evernden, D. E. Savage, G. H. Curtis, and G. T. James, "Potassium-Argon Dates and the Cenozoic Chronology of North America," *Amer. Jour. Sci.* (1964), 262:145–198.
43. M. C. McKenna, "The Origin and Early Differentiation of Therian Mammals," New York Academy of Sciences, *Annals* (1969), 167: 217–240. "Toward a phylogenetic classification of the Mammalia," in *Phylogeny of the Primates*, ed. W. P. Luckett and F. S. Szalay (New York, Plenum, 1975), 21–46. J. H. Ostrom. "*Archaeopteryx* and the Origin of Birds," Linnaean Soc. London, *Biol. J.* (1976), 8:91–182. A. S. Romer, "Unorthodoxies in Reptilian Phylogeny,"

Evolution (1971), 25, no. 1:103–112. E. Jarvik, "Aspects of Vertebrate Phylogeny," *Nobel Symposium 4, Current Problems of Lower Vertebrate Phylogeny*, ed. T. Ørvig (Stockholm, Almqvist and Wiksell, 1968), 497–527, esp. 505–514; K. S. Thompson, "A Critical Review of the Diphyletic Theory of Rhipidistian-Amphibian Relationships," ibid., 285–306.

44. G. G. Simpson, "Criteria for Genera, Species, and Subspecies in Zoology and Paleozoology," New York Acad. Sci., *Ann.* (1943), 44:145–178. G. G. Simpson and Anne Roe, *Quantitative Zoology* (New York, McGraw-Hill, 1939). Major involvement of paleontologists followed the 1947 Princeton symposium, *Genetics, Paleontology and Evolution*, ed. G. L. Jepsen, E. Mayr, and G. G. Simpson (Princeton, N.J., Princeton University Press, 1949), xiv + 474.
45. B. Schaeffer, M. K. Hecht, N. Eldredge, "Phylogeny and Paleontology," *Evolutionary Biology* (1972) 6:31–46. N. Eldredge and S. J. Gould, "Punctuated Equilibria: An Alternative to Phyletic Gradualism," in T. J. M. Schopf, *Models in Paleobiology* (San Francisco, Freeman and Cooper, 1972), 82–115; P. D. Gingerich, "Paleontology and Phylogeny: Patterns of Evolution at the Species Level in Early Tertiary Mammals," *Amer. Jour. Sci.* (1976), 276:1–28.
46. E. C. Olson, "Morphological Integration and the Meaning of Characters in Classification Systems," *Systematics Assoc.*, Publ. (1964), 6:123–166; "Current and Projected Impacts of Computers upon Concepts and Research in Paleontology," *North American Paleontological Convention, Proc.* (1969), B:135–153.
47. A. W. Crompton and K. Iliiemae, "Molar Occlusion and Mandibular Movements during Occlusion in the American Opossum," Linnaean Soc. London, *Zool. Jour.* (1970), 49:21–47. F. A. Jenkins, Jr., "The Movement of the Shoulder in Claviculate and Aclaviculate Mammals," *Jour. Morphology* (1974), 144:71–84.
48. B. Schaeffer and D. E. Rosen, "Major Adaptive Levels in the Evolution of the Actinopterygian Feeding Mechanism," *Amer. Zoologist* (1961), 1:187–204. J. H. Ostrom, "A Functional Analysis of Jaw Mechanics in the Dinosaur *Triceratops*," Yale Peabody Museum, *Postilla* (1964), Vol. 88. A. W. Crompton and F. A. Jenkins, Jr., "Mammals from Reptiles: A Review of Mammalian Origins," *Ann. Rev. Earth and Planetary Sci.* (1973), 1:131–155. P. M. Galton, "The Posture of Hadrosaurian Dinosaurs," *Jour. Paleontology* (1970), 44:464–473; "A Primitive Dome-headed Dinosaur . . . and the Function of the Dome of Pachycephalosaurids," *Jour. Pa-*

leontology (1971), 45:40–47, H. R. Barghusen, "A Review of Fighting Adaptations in Dinocephalians (Reptilia, Therapsida)," *Paleobiology* (1975), 1:295–311.

49. L. Radinsky, "Evolution of the Canid Brain," *Brain, Behavior and Evolution* (1973), 7:169–202. "Evolution of the Felid Brain," ibid. (1975), 11:214–254; "Fossil Evidence of Anthropoid Brain Evolution," *Amer. Jour. Phys. Anthropology* (1974), 41:15–28; "Primate Brain Evolution," *American Scientist* (1975), 63:656–663.
50. E. C. Olson, "Origin of Mammals Based upon Cranial Morphology of the Therapsid Suborders," Geol. Soc. America, *Sp. Papers* (1944), 55:xi + 136. A. S. Romer, "The Cranial Anatomy of the Permian Amphibian *Pantylus*," Harvard Univ. Mus. Comp. Zoology, *Breviora* (1969), 314:37 pp. F. C. Whitmore, Jr., "Cranial Morphology of Some Oligocene Artiodactyla," U.S. Geol. Surv., *Prof. Paper* (1953), 243-H:iii + 117.
51. C. W. Hibbard, "Pleistocene Mammalian Local Faunas from the Great Plains and Central Lowland Provinces of the United States," in *Pleistocene and Recent Environments of the Central Great Plains*, Univ. Kansas Dept. Geol., Sp. Publ. (1970), 3:395–433.
52. E. C. Olson, "The Evolution of a Permian Vertebrate Chronofauna," *Evolution* (1952), 6:181–196; "Community Evolution and the Origin of Mammals," *Ecology* (1966), 47:291–302. J. H. Ostrom, "A Reconsideration of the Paleoecology of Hadrosaurian Dinosaurs," *Amer. Jour. Sci.* (1964), 262:975–997.
53. A. K. Behrensmeyer, "The Taphonomy and Paleoecology of Plio-Pleistocene Vertebrate Assemblages East of Lake Rudolph, Kenya," Harvard Univ., Museum of Comparative Zoology, *Bull.* (1975), 146:473–578. P. Dodson, "Sedimentology and Taphonomy of the Oldman Formation (Campanian), Dinosaur Provincial Park, Alberta (Canada)," *Paleogeogr., Paleoclimatol., Paleoecol.* (1971), 10:21–74. M. Voorhies, "Taphonomy and Population Dynamics of an Early Pliocene Vertebrate Fauna, Knox County, Nebraska," *Univ. Wyoming Contrib. to Geol.*, Spec. Paper (1969), 2:69 pp.
54. J. A. Shotwell, "An Approach to the Paleoecology of Mammals," *Ecology* (1955), 36:327–337; "Inter-Community Relationships in Hemphillian (Mid-Pliocene) Mammals," ibid. (1958), 39:271–282.
55. R. Zangerl and E. S. Richardson, Jr., "The Paleoecological History of Two Pennsylvanian Black Shales," *Fieldiana, Geology Memoirs*, (1963), 4:xii + 352 pp.
56. P. S. Martin and H. E. Wright, *Pleistocene Extinction: The*

Search for a Cause (New Haven, Yale University Press, 1967), 453 pp.; L. S. Russell, "The Changing Environment of the Dinosaurs in North America," *Advancement of Science* (1966), 197–204; Dale A. Russell and W. H. Tucker, "Supernovae and the Extinction of the Dinosaurs," *Nature* (1971), 229:553–554.

57. E. H. Sellards, "Early Man in America: Index to Localities and Selected Bibliography," Geol. Soc. Amer., *Bull.* (1940), 51:373–432.
58. J. D. Figgins, "The Antiquity of Man in America," *Natural History* (1927), 27:229–239. H. J. Cook, "New Geological and Paleontological Evidence Bearing on the Antiquity of Mankind in America," ibid., 240–247. J. A. Graham and R. F. Heizer, "Man's Antiquity in North America: Views and Facts," *Quaternaria* (1967), 9:225–235. J. L. Bada, R. A. Schroeder, G. F. Carter, "New Evidence for the Antiquity of Man in North America Deduced from Aspartic Acid Racemization," *Science* (1974), 184: 791–793.
59. Yves Coppens, "The Great East African Adventure," *C.N.R.S. Research*, (New York, 1976), 3:2–12. E. L. Simons, "New Fossil Apes from Egypt and the Initial Differentiation of Hominoidea," *Nature* (1965), 205:135–139. E. L. Simons and D. R. Pilbeam, "Hominoid Paleoprimatology," in R. Tuttle, ed., *The Functional and Evolutionary Biology of Primates* (Chicago, Aldine-Atherton, 1972), 36–62.

Why Drift Theory Was Accepted with the Confirmation of Harry Hess's Concept of Sea-Floor Spreading

Henry Frankel

Successful scientific theories synthesize a vast variety of seemingly unrelated facts whose truth is known prior to their invention. But they are able to account for more facts than those which are already known when the theory is constructed—namely, facts whose truth is unknown prior to the formulation of the theory. The scientific community is characteristically reticent about accepting a theory if it is capable of no more than relating known facts—even if it relates more known facts than its competition—for scientists recognize the inventiveness of their co-workers in devising ad hoc auxiliary hypotheses in order to improve the synthesizing power of their theories. The scientific community is rightfully more impressed with theories that predict novel facts, since theories cannot be ad hoc with respect to novel facts. If these novel facts are later confirmed, the community, recognizing the non ad-hocness of the theory, is much more likely to endorse it.[1]

The acceptance of continental drift theory, once altered by Harry Hess with his concept of sea-floor spreading, nicely illustrates the reluctance of the scientific community to accept a theory until it predicts novel facts which turn out to be true. Hess's concept of sea-floor spreading greatly increased the explanatory value of continental drift theory, since he was able to pull together a mass of unrelated material. Nevertheless, despite this value of Hess's proposal, the community was not willing to endorse drift theory. It was only after confirmation of the F. Vine, D. Matthews, and J. T. Wilson transform fault hypothesis, both of which were virtually corollaries of Hess's proposal, that the geological community threw its endorsement to drift theory.

What I should like to do is briefly document the fact that the acceptance of drift theory is in accordance with the above thesis.[2] The argument may be summarized as follows:

1. The status of drift theory in the eyes of the geological community improved somewhat with the ability of the theory to account for much of the new data from paleomagnetism and oceanography. This expanded synthesizing power was due primarily to Hess's concept of sea-floor spreading.
2. But the geological community was not willing to accept Hess's version of drift theory, since no matter how ingenious, it was ad hoc.
3. Hess's proposal did predict novel facts, especially when buttressed with Vine-Matthews and Wilson.
4. Once Vine-Matthews was confirmed, the geological community began to switch to drift theory. This switch was especially dramatic with those who were intimately involved with the oceanographic research.

I. *Input from the New Sciences*

A. *Paleomagnetism*

The hypothesis that the earth's magnetic field, when averaged over some thousands of years, is a geocentric dipole parallel to the axis of rotation, was well established in the fifties.[3] The fifties also saw the development of extremely sensitive magnetometers and sophisticated techniques for analysis of remanent magnetism in both igneous and sedimentary rocks. (Remanent magnetism is the magnetism that a rock acquires during formation and retains throughout its lifetime.)

In 1953 a group of English geologists from London University headed by P. M. S. Blackett, inventor of the modern magnetometer, attempted to learn more about the earth's magnetic field by measuring the remanent magnetism of Triassic strata. They concluded that England had moved relative to the geomagnetic pole since the Triassic, arguing that it had rotated clockwise and drifted northward. The London group then took their paleomagnetic techniques to India, where they measured the remanent magnetism of igneous rocks in the Deccan Plateau. Results of their study indicated that India had migrated northward to its present position from a location well in the southern hemisphere. This movement of India proposed by the paleomagnetists coincided with the supposed drift of India argued for by Wegener and his loyal followers. The London group, cognizant of this similarity—Blackett as it turns out was a confessed drifter—argued that they had found unambiguous quantitative evidence in favor of drift.

Although the London group's results aided the drift cause and stimulated further paleomagnetic studies, their work was criticized by another group of English paleomagnetists housed at Cambridge. This group, led by S. K. Runcorn, argued that the London team had not found an unambiguous test for drift theory. According to Runcorn, the London group could justifiably conclude only that India had drifted northward or that the north magnetic pole had drifted southward. Runcorn argued in favor of polar wandering: "Appreciable polar wandering seems indicated. *There does not yet seem to be the need* to invoke appreciable amounts of continental drift to explain the paleomagnetic results so far obtained" (italics mine).[4] Others, including T. Gold, an English astronomer, joined the polar wandering bandwagon. Gold argued that the redistribution of the earth's mass through the formation and melting of the glaciers could cause polar wandering. Anything other than drift seemed appropriate.

By 1957 Runcorn had left the polar-wandering wagon to become a drifter. What brought about this shift was his investigation of the residual magnetism of North American rock strata.[5] His group, in comparing the apparent pole shifts with respect to North America and Europe, found that they did not coincide. Thus, unless it was assumed that the magnetic poles could be in more than one location at one time, or at least change locations rapidly, some continental drift had to be postulated. Moreover, Runcorn noticed that the apparent paths of the poles from the two continents could be made to coincide simply by supposing North America and Europe were joined together until the beginning of the Triassic or Cretaceous, at which time they began to drift apart to their present position. Runcorn became a strong advocate for drift.

Blackett's group in London returned to the front in 1960, criticizing Runcorn's work on methodological grounds.[6] The London team argued that one could not infer longitudinal changes in the position of a continent from residual magnetic studies. All that could be inferred were latitudinal changes. They themselves had restricted their own studies to latitudinal claims, since indications of longitudinal changes also could be attributed to rotational movements of the continents. As a result, Runcorn's match of the spreading of the Atlantic was questionable, for the latitudinal readings were open to rotational interpretation. Runcorn's work did not rule out polar wandering. Despite the criticism of Runcorn's paleomagnetism studies, his work gave new life to continental drift theory and continued to do so throughout the early sixties.

B. Oceanography

The relationship between remanent magnetism and continental drift was deceptively straightforward—so much so, in fact, that it was first thought to supply an unambiguous test in favor of continental drift. The relation between the sea-floor studies and drift theory was not nearly so apparent. What there was, however, turned out to be much more substantial. The two major institutions that turned the sea floors from areas of ignorance to places of backyard familiarity by the mid-sixties were Lamont Observatory,[7] an appendage of Columbia University, and Scripps Institution of Oceanography. M. Ewing, J. Ewing, B. Heezen, and others at Lamont mapped the vast midocean ridge system throughout most of the ocean floor, and H. Menard, R. Mason, V. Vacquier, et al. at Scripps pioneered oceanographic studies of the Pacific.

As far back as 1934 Vening Meinesz was led to assume mantle convection currents under the sea floor in order to account for gravity anomalies in ocean trenches around the Indonesian and Caribbean archipelagos. Meinesz reaffirmed his position in 1952, attempting to foster research projects. In 1956 E. C. Bullard, A. Maxwell, and R. Revelle amplified Meinesz's claim by arguing from additional data that the ocean trenches (a) were the region of most intensive earthquakes, (b) have high gravity deficiencies, (c) are composed of relatively cold seafloor material, and (d) because of (a), (b), and (c) are regions where material sinks into the earth's mantle, i.e. are subduction zones on convection cells.[8]

During the early fifties M. Ewing and others journeyed across the ocean floors, plotting among other things the vast system of midocean ridges. By 1956 they had mapped out most of the worldwide system of ridges, and Heezen in 1960 claimed that a narrow band of shallow earthquakes coincided with a central rift valley, extending along the middle of many midocean ridges.[9] This led them to suggest that the ridges might be a consequence of diastrophic activity associated with rift valleys. Bullard, Maxwell, and Revelle in 1956 and von Herzen in 1959 showed that the midocean ridges have unusually high heat flow along their crusts.[10] Hess argued that ridges were probably ephemeral features of the seafloor.[11]

By the middle fifties Mason, Dietz, Menard, and others discussed several huge fracture zones running east–west in the Pacific off the California coast.[12] A few years later Mason et al. plotted the magnetic profiles of the Pacific from San Diego to British Columbia.[13] They first noticed strips of high magnetic intensity alternating with ones of low intensity, when measured with respect to the present position of the earth's magnetic pole. Then they noticed a number of anomalies in the magnetic profiles, namely that the profiles were often offset along several east–west fracture zones, and that the anomalies could be matched by supposing transcurrent faulting along the fracture zones. Thus they concluded that the sea floor had undergone significant horizontal displacement.

The Ewings discovered, much to the surprise of many geologists, that the sediments were extremely thin over the whole sea floor, and especially so at the midocean ridges.[14] Geologists of the fixist persuasion had assumed that a complete book of sediments could be found on the undisturbed sea floors. Another unexpected find was that no ocean sediments older than the Cretaceous had been dredged from the deeps. Together, these surprising results suggested a youthful sea floor. Finally, Ewing found that the third and bottom layer of the seafloor crust was extremely uniform in thickness throughout his seismic soundings.[15]

II. Sea-floor Spreading and Continental Drift

The person primarily responsible for putting together these diverse, new data from the sea floor into an intelligible pattern was Harry Hess of Princeton.[16] Hess in 1960, indulging in what he termed "geopoetry," hypothesized that sea floor was created at midocean ridges, spread out toward the trenches, and then descended into the mantle.

Hess thereby interpreted the ridge areas as representing the rising limbs of mantle-convection cells and the trenches as the descending limbs. Hess explicitly related his seafloor spreading model to continental drift. In support of drift he cited Runcorn's studies in paleomagnetism, arguing that they indicated drift despite the fact that they were open to other interpretations. He then proposed that the continents were carried passively along by the spreading sea-floor material, proposing specifically that

> [the continent's] leading edges are strongly deformed when they impinge upon the downward moving limbs of convecting mantle . . . [that] rising limbs coming up under continental areas move the fragmented parts away from one another at a uniform rate so a truly median ridge forms as in the Atlantic Ocean . . . [and that] the cover of oceanic sediments and the volcanic seamounts also ride down into the jaw crusher of the descending limb, are metamorphosed, and eventually probably are welded onto continents.[17]

C. Corroboration of Sea-Floor Spreading: The Vine-Matthews Hypothesis and Wilson's Transform Faults

Hess's sea-floor spreading, unlike Holmes's sea-floor thinning, fostered specific research projects. Hess predicted a number of novel facts and suggested important theoretical corollaries whose corroboration was in the grasp of practicing earth scientists. The two most important corollaries to be initially derived from seafloor spreading were the Vine-Matthews hypothesis and the Wilson transform fault hypothesis.[18]

Paleomagnetists, besides having to decide whether the directional changes in the remanent magnetism of continental rock were real or apparent, had another intriguing problem: For a number of years occasional samples had cropped up with reversed remanent magnetism. These indicated either geomagnetic shifts in polarity or isolated cases where the sample had become magnetized in a direction opposite to the prevailing geomagnetic field. In June 1963 Cox et al. cogently argued that the reversals were global, and went on to date the reversal times back some 3½ million years (by potassium-argon techniques).[19] In September of the same year Vine and Matthews argued, primarily on theoretical grounds, that if sea-floor spreading and polarity reversals of the earth's magnetic field occur, there should be strips of sea-floor material having reversed polarity, spreading out symmetrically and parallel to midocean ridges. They emphasized that their hypothesis

> is consistent with, in fact virtually a corollary of, current ideas on ocean floor spreading and periodic reversals in the Earth's magnetic field. If the main crustal layer (Seismic layer 3) of the oceanic crust is formed over a convective upcurrent in the mantle at the centre of an oceanic ridge, it will be magnetized in the current direction of the Earth's field. Assuming impermanence of the ocean floor, the whole of the oceanic crust is comparatively young, probably not older than 150 million years, and the thermo-remanent component of its magnetization is therefore either essentially normal, or reversed with respect to the present field of the Earth. Thus, if spreading of the ocean floor occurs, blocks of alternately normal and reversely magnetized material would drift away from the centre of the ridge and parallel to the crest of it.[20]

If Vine and Matthews were right, this would not only give strong qualitative support to sea-floor spreading and thereby continental drift, but also provide quantitative support, since it would afford a means of measuring the rate of sea-floor spreading. If precise dates could be given for the polarity reversals, geologists would have a tape recording of the spreading rates engraved in the seafloor. They then could compare these dates derived from sea-floor spreading with those derived from drift theory, and that would give drifters a relatively unambiguous and reliable test for their theory.

When Vine and Matthews introduced their hypothesis, not enough analyzed data were available to test it. There were a few magnetic profiles from Carlsberg ridge in the Indian Ocean and others from the eastern Pacific off California; but they were not sufficient—in fact those off California at that time seemed to run counter to their hypothesis.[21] Consequently, Vine, Matthews, and many of the Lamont personnel began intensive magnetic studies of the seafloor, concentrating on areas around midocean ridges. The first studies by J. R. Heirtzler and X. Le Pichon, at Lamont, were interpreted as inconsistent with Vine and Matthews.

> *Vine and Matthews* (1963) have hypothesized that the whole ocean crust is made of striped material parallel to the ridge axis, having alternately reversed and normal magnetization. They see their hypothesis as a corollary of the "spreading floor" hypothesis of *Dietz* (1961) . . . It is clear from this study that most of the profiles do not follow the pattern assumed by Vine and Matthews.[22]

By the end of 1966, however, analyses of several profiles taken across the Pacific–Antarctic and from the Reykjanes Ridges had produced a striking corroboration of Vine-Matthews.[23] W. C. Pitman, III and

Heirtzler analyzed the magnetic profiles from the Pacific–Antarctic Ridge. The symmetry of the profiles was unbelievable. Pitman has been quoted as saying:

> It hit me like a hammer . . . In retrospect, we were lucky to strike a place where there are no hindrances to sea-floor spreading. We don't get profiles quite that perfect from any other place. There were no irregularities to distract or deceive us. That was good, because by then people had been shot down an awful lot over sea-floor spreading. I had thought Vine and Matthews was a fairly dubious hypothesis at the time, and Fred Vine has told me he was not wholly convinced of his own theory until he saw Eltanin-19. It does grab you. It looks very just like the way a profile ought to look and never does. On the other hand, when another man here saw it his remark was "Next thing, you'll be proving Vine-Matthews." Actually, it was this remark that made me go back and read Vine and Matthews. We began to examine Eltanin-19, and we realized that it looked very much like a profile that Vine and Tuzo Wilson published just before our data came out of the computer.[24]

They superimposed transparencies of the profiles over one another, reversing one of the transparencies so as to display the symmetry. Positive results were also obtained with the Reykjanes Ridge, increasing the corroboration of sea-floor spreading.

Despite the fact that Maurice Ewing and others at Lamont provided much of the oceanography data in the fifties and early sixties, most of the personnel at Lamont were opposed to continental drift and sea-floor spreading. There were several notable exceptions, namely N. Opdyke and Heezen; but by and large Lamont was opposed to drift theory. As Xavier Le Pichon has said, "We are not geologists, and misinterpreted the evidence—particularly the profiler—often (when we analyzed the first set of profiles before Pitman's Pacific–Antarctic profiles were done)."[25] From then on, however, Eltanin-19 papers corroborating sea-floor spreading and continental drift poured out of Lamont. One year later (1968) Le Pichon et al. matched magnetic profiles which they had once thought inconsistent with Vine-Matthews.[26] This gave them the relative spreading rates of each ocean. Then, after obtaining new data on the ages of polarity reversals from an analysis of sedimentary sea-floor core samples by Opdyke at Lamont, they were able to determine the actual spreading rates. These compared favorably with the dates for the breakup of Pangea.

In 1965 J. Tuzo Wilson, then at Toronto, deduced another corollary from Hess's notion of sea-floor spreading.[27] Wilson proposed that most of the horizontal faulting involved in sea-floor spreading was of the transform type, as opposed to the assumed transcurrent type. He initially described transform faults as follows:

> Faults in which the displacement suddenly stops or changes form and direction are not true transcurrent faults. It is proposed that a separate class of horizontal shear faults exists which terminate abruptly at both ends, but which nevertheless may show great displacements . . . The name transform fault is proposed for the class . . . The distinct ions between transform and transcurrent faults might appear trivial until the variation in habit of growth of the different types is considered . . . These distinctions are that ridges expand to produce new crust, thus leaving residual inactive traces in the topography of their former positions. On the other hand oceanic crust moves down under island arcs absorbing old crust so that they leave no traces of past positions.[28]

He added that "Transform faults cannot exist unless there is crustal displacement, and their existence would provide a powerful argument in favor of continental drift and a guide to the nature of the displacement involved."[29] Transform faults were precisely the kind of fault required for sea-floor spreading, without having to suppose that sections of midocean ridges drifted away from one another as more sea-floor material was created. Since ridge sections seemed to remain relatively stationary, Wilson argued for the existence of transform faults at ridge sites. They also provided structural geologists who were sympathetic with sea-floor spreading and drift theory with a model for making some sense out of cases where spreading ended abruptly. Wilson, with his transform faults, simply supposed that horizontal spreading movements were transformed into some other diastrophic activity. He proposed, for example, that the spreading zone of the western Pacific was transformed into a downward motion along the trenches bordering the Pacific, while the spreading zone of the eastern Pacific was transformed into mountain-building activity along the west coast of the Americas.[30]

His proposal was partially tested in 1967 by Lynn Sykes, a seismologist at Lamont.[31] Sykes, initially not impressed with the notion of sea-floor spreading, became receptive to the idea when he saw the Pitman profiles from the Pacific–Antarctic Ridge. Apparently, he, Jack Oliver, and Bryan Isacks—other seismologists at Lamont—decided to test Wilson's hypothesis after seeing Pitman's profiles. Sykes reports his fascination with the profiles as follows:

> It was beautiful . . . They'd made transparencies of a thousand-mile length of the Eltanin-19 profile. By taking two of them and turning one around and putting it over the other, you could see vividly the symmetry of the profile. Some other people's work had been suggestive, but the Pitman profile was what really made people believe Vine and Matthews—hit you that this really was so.[32]

If Wilson was right, seismic activity along the midocean ridges would be restricted to just those areas between offsets of the ridges. The seismic activity would not be spread out all along fracture zones, as it would if we had ordinary transcurrent faults. (Incidentally, Menard and others had noticed that the midocean ridges were offset along fracture zones as far back as the late fifties, but nobody really knew how to interpret it.)[33] Sykes corroborated Wilson, concluding that "The sense of strike-slip motion in each of the ten solutions is in agreement with that predicted for transform faults; it is opposite to that expected for a simple offset of the ridge crest along various fracture zones."[34] Later in the same year Oliver and Isacks reexamined their data from the Tonga trench in order to test Wilson.[35] In particular, they wanted to see if their work indicated a transformation of horizontal movement to vertical at the trenches. This is exactly what they found, and furthermore, they argued that the amount of material descending back into the mantle was so substantial that it appeared as if the whole rigid lithosphere descended into the plastic aesthenosphere, just as suggested by Hess and R. Dietz.

The role of sea-floor spreading and its corroboration is nicely summarized by Isacks, Oliver, and Sykes in their paper "Seismology and New Global Tectonics":

> The remarkable success with which the hypothesis of sea-floor spreading accommodated such diverse geologic observations as the linear magnetic anomalies of the ocean (Vine and Matthews 1963) and (Pitman and Heirtzler 1966), the topography of the ocean floor (Menard 1965), the distribution and configuration of continental margins and various other land patterns (Wilson, 1965a; Bullard, 1964, Bullard *et al.*, 1965) and certain aspects of deep-sea sediments (Ewing and Ewing, 1967) raised the hypothesis to a level of great importance and still greater promise. The contributions of seismology to this development have been substantial, not only in the form of general information on earth struc-

> ture but also in the form of certain studies that bear especially on this hypothesis. Two specific examples are Sykes' (1967) evidence on seismicity patterns and focal mechanisms to support the transform fault hypothesis of Wilson (1965a) and Oliver and Isacks' (1967) discovery of anomalous zones that appear to correspond to underthrust lithosphere in the mantle beneath island arcs.[36]

A flurry of corroborating data continued to come in during the middle sixties, but what was more significant was the development of plate tectonics as the fundamental model for describing global movements. This is the final stage in the acceptance of drift theory: its absorption into the theory of plate tectonics. Drift theory gave birth to sea-floor spreading, which in turn brought forth plate tectonics, and plate tectonics swallowed up both its parents and grandparents.

III. Why Drift Theory Gained the Support of the Geological Community

What remains is to show that drift theory gained the support of the geological community precisely when some of its novel facts were clearly corroborated—assuming that it accounted for the data explained by the other research programs. First, I show that the synthesizing power of Hess's sea-floor-spreading hypothesis was not sufficient to turn the tide to drift. I then argue that drift came in with the corroboration of the Vine-Matthews hypothesis and the Wilson transform fault hypothesis.

The amount of data Hess fit together is amazing. He supposed that sea-floor was created at midocean ridges, spread out toward the trenches, and then descended into the mantle. There were five major claims involved in his hypothesis.

1. The ridges are the locus of ascending convection limbs.
2. The trenches are subduction zones.
3. There is an actual spreading out of sea-floor material from the ridges toward the trenches.
4. The ocean floor is young compared to the continents.
5. The region of slippage of the spreading sea-floor is farther down than the interphase between crust and mantle.

In support of each claim, he related the following, thereby fitting it all together in his concept of sea-floor spreading:

Ridge data for claim (1)	median position, rift, valleys accompanied by shallow earthquakes, high heat flows, apparent ephemeral nature.
Trench data for claim (2)	gravity anomalies, relatively cold temperature, occurrence of numerous intense earthquakes along them.
Spreading data for claim (3)	transcurrent faulting off the California coast, uniform thickness of the bottom crustal level, the widening of a fracture zone along the Mid-Atlantic Ridge as it crossed Iceland.
Age data for claim (4)	thinness of the sediment layer, no sediments older than the Cretaceous, youthful nature of the Pacific guyots.
Slippage data for claim (5)	indication of a slippage zone between the lithosphere and aesthenosphere.

The synthesizing power of Hess's sea-floor spreading is beyond question.

It is also clear that Hess's proposal provided drift with an auxiliary hypothesis. In the unpublished 1960 version of his paper he suggested the auxiliary role of his sea-floor spreading with respect to drift:

> Paleomagnetic data presented by Runcorn (1959), Irving (1959) and others strongly suggest that the continents have moved by large amounts in geologically comparatively recent times. One may quibble over the details but the general picture on paleomagnetism is sufficiently compelling that it is much more reasonable to accept it than to disregard it . . . This strongly indicates independent movement in direction and amount of large portions of the Earth's surface with respect to the rotational axis. This could be most easily accomplished by a convecting mantle system which involves actual movement of the Earth's surface passively riding on the upper part of the convecting cell.[37]

After this he stated explicitly the advantage of sea-floor spreading over previous hypotheses to account for the drift of the continents:

> The mid-ocean ridges could represent the traces of the rising limbs of convection cells while the circum-Pacific belt of deformation and volcanism represents descending limbs. The Mid-Atlantic Ridge is median because the continental areas on each side of it

> have moved away from it at the same rate . . . This is not exactly the same as continental drift. The continents do not plow through oceanic crust impelled by unknown forces, rather they ride passively on mantle material as it comes to the surface at the crest of the ridge and then moves laterally away from it.[38]

Although Hess had provided drift with a spectacular auxiliary hypothesis which brought together this enormous amount of information, the geological community did not endorse drift as the leading research program. As a matter of fact, most of the personnel at Lamont Institute, who later played a significant role in corroborating Hess's auxiliary hypothesis, remained nondrifters.

Drift gained acceptance with the confirmation of Vine-Matthews and Wilson. Because of their intimate attachment with sea-floor spreading, their confirmation was tantamount to a confirmation of sea-floor spreading and, by the same token, of Hess's version of drift. Recall both the tremendous surprise and excitement at Lamont when the Eltanin-19 profiles were examined. Except for Opdyke and Heezen, most of the people at Lamont were opposed to drift. But, after Eltanin-19, Lamont personnel were convinced of the basic correctness of drift. Le Pichon and his co-workers even matched magnetic profiles, which they had previously claimed were inconsistent with Vine-Matthews. The resident seismologists, after seeing Eltanin-19, immediately set out to corroborate Wilson's transform faults, either by getting new data or re-examining their old data. Recall Sykes's remark about first viewing the profiles (above, page 346).

Even Fred Vine wasn't completely convinced of his theory until he saw Eltanin-19! Without question the major shift to drift began with the corroboration of Vine-Matthews and ended with the corroboration of Wilson. From then on, the work in drift flooded the geological community. By 1972 the *Journal of Geophysical Research* thought it worth while to publish an anthology of the first set of important research and corroborating papers on drift. The editor of the Journal suggested that

> The concept of plate tectonics had galvanized the geological community as never before in the past . . . This volume should serve as a text for young scientists who wish to study plate tectonics and as a document for historians of science to trace the development of a revolutionary concept in the study of the earth.[39]

Drift reigns supreme. Indeed, geologists brought up in a fixist tradition may have to go through a gestalt shift in order to understand plate tectonics. However, there was good reason for the switch: the confirmation of a nest of novel facts surrounding the notion of sea-floor spreading.

Acknowledgments

I should like to thank Peter Machamer, William Ashworth, and Kathy Higgins for their valuable comments on earlier drafts of this essay. I also should like to thank the staff at Linda Hall Library, Kansas City, Missouri, for their cooperation. In addition, much of the work on this paper was supported by a faculty research grant from the University of Missouri—Kansas City, as well as grants from NEH and NSF.

Notes

1. This notion of "novel fact" with respect to a given theory or hypothesis, namely, a fact is novel with respect to a given theory if it is not known to obtain when the theory is constructed, has been rejuvenated by the late Imre Lakatos. Cf, especially, I. Lakatos, "Falsification and the Methodology of Scientific Research Programmes," in *Criticism and the Growth of Knowledge*, ed. I. Lakatos and A. E. Musgrave (Cambridge: Cambridge University Press, 1970), 91–196. The precise formulation of "novel fact" which I am employing in this paper is in need of improvement, but the reasons for such an improvement extend beyond the scope of the paper. For those who are interested, please refer to E. G. Zahar, "Why Did Einstein's Programme Supersede Lorentz's?", *British Journal for the Philosophy of Science* (1973), 24:95–123, 223–262; and A. Musgrave, "Logical versus Historical Theories of Confirmation," ibid. (1974), 25:1–23.
2. The general development of continental drift theory and its inclusion into plate tectonics has been described in A. Hallam, *A Revolution in the Earth Sciences* (Oxford: Clarendon Press, 1973), and U. Marvin, *Continental Drift: The Evolution of a Concept* (Washington: Smithsonian Institution, 1973).
3. Cf., S. K. Runcorn, "Paleomagnetic Evidence for Continental Drift and Its Geophysical Cause," in *Continental Drift*, ed. S. K. Runcorn (New York: Academic Press, 1962), 1–39. H. Rakeuchi, S. Uyeda, and H. Kanamori, *Debate about the Earth* (San Francisco: Freeman, Cooper, 1967). The latter provides an excellent sum-

mary of the paleomagnetic experiments performed during the fifties and early sixties, which were related to drift theory. Another more technical summary is provided by Ernst R. Deutsch, "The Rock Magnetic Evidence for Continental Drift," in *Continental Drift*, ed. G. D. Garland (Toronto: University of Toronto Press, 1966), 28–52. He provides an excellent discussion of the arguments for interpreting the paleomagnetic data in terms of polar wandering, continental drift, or a combination of the two. In addition, he also considers the global expansion hypothesis. The global expansionists (e.g., Carey, Heezen, and Egyed) maintained that the continents had drifted apart as the earth had expanded. The continents would separate from one another when the earth expanded, just as pieces of paper, attached to the surface of a balloon, would separate as the balloon expanded. This expanding earth hypothesis finally lost out to sea-floor spreading.

4. S. K. Runcorn, "Rock Magnetism—Geophysical Aspects," *Advances in Physics* (1955), 4:244–291. This passage is from 289.
5. S. K. Runcorn, "Rock Magnetism," *Science* (1959), 129:1002–11.
6. Cf., Deutsch (above, note 3).
7. Lamont Observatory has become since 1969 Lamont-Doherty Geological Observatory.
8. E. Bullard, A. Maxwell, and R. Revelle, "Heat Flow through the Deep Sea Floor," *Advances in Physics* (1956), 3:153–81.
9. B. Heezen, "The Rift in the Ocean Floor," *Scientific American* (1960), 203:98–110.
10. Bullard, Maxwell, and Revelle (above, note 8), and R. Von Herzen, "Heat Flow Values from the Southern Pacific," *Nature* (1959), 183:882–883.
11. H. H. Hess, "Nature of the Great Oceanic Ridges," *Internat. Ocean Cong. Preprints* (1959), 33–34.
12. Cf., H. H. Hess, "History of Ocean Basins," in *Petrologic Studies: A Volume to Honor A. F. Buddington*, ed. A.E.J. Engel, H. L. James, and B. F. Leonard (U.S. Geological Society, 1962), 599–620; and R. S. Dietz, "Continent and Ocean Basin Evolution by Spreading of the Sea Floor," *Nature* (1961), 190:854–857.
13. V. Vacquier, "Magnetic Evidence for Horizontal Displacement in the Floor of the Pacific Ocean," in Runcorn, *Continental Drift*.
14. J. Ewing and M. Ewing, "Seismic-refraction Profiles in the Atlantic Ocean Basins, in the Mediterranean Sea, on the Mid-Atlantic Ridge and in the Norwegian Sea," *Geological Society America, Bull.* (1959), 70:291–318.
15. Ewing and Ewing.

16. Hess did not publish his thesis until 1962; Dietz published his version of sea-floor spreading in 1961. Hess deserves credit, however, for he distributed an unpublished version of his thesis in 1960. Thanks to Professor Sheldon Judson, Chairman, Department of Geological and Geophysical Sciences, Princeton University, I have had the opportunity to examine the 1960 version. There are no substantive changes between the 1960 and the published version of 1962. In addition, Dietz freely admits that Hess first developed the notion of sea-floor spreading:

    As regards sea-floor spreading, Hess deserves full credit for the concept . . . I have done little more than introduce the term *sea-floor spreading* . . . [Dietz]

17. H. H. Hess, "Evolution, Ocean Basins," Preprint (1960), 32–33, and Hess, "History of the Ocean Basins," 618.
18. Hess explicitly suggests that the Vine-Matthews hypothesis "could not have been conceived starting from Holmes' picture." H. H. Hess, "Reply," *Journal of Geophysical Research* (1968), 73:6569.
19. A. Cox, R. Doell and G. Dalrymple, "Geomagnetic Polarity Epochs and Pleistocene Geochronometry," *Nature* (1963), 198:1049–50.
20. F. Vine and D. Matthews, "Magnetic Anomalies over Oceanic Ridges," *Nature* (1966), 199:947–949. This passage is from 948.
21. Cf., B. Heezen, "The Deep-Sea Floor," in Runcorn, *Continental Drift*; and F. Vine, "Spreading of the Ocean Floor: New Evidence," *Science* (1966), 154:1405–15.
22. J. Heirtzler and X. Le Pichon, "Crustal Structure of the Mid-Ocean Ridges," *Journal of Geophysical Research* (1965), 70:4013–33. This passage is from 4028.
23. F. Vine, 1407.
24. W. Wertenbaker, "Explorer III—Some Great Overriding Process," *The New Yorker* (18 November, 1974), 60–110. This passage is from 94. Wertenbaker's article gives an excellent account of the personal reflections of the scientists at Lamont who were intimately involved in the confirmation of Vine-Matthews.
25. W. Wertenbaker, 88.
26. Cf., J. Heirtzler, G. Dickson, E. Herron, W. Pitman, and X. Le Pichon, "Marine Magnetic Profiles, Geomagnetic Field Reversals, and Motions of the Ocean Floor and Continents," *Journal of Geophysical Research* (1968), 73:2131–53.
27. T. Wilson, "A New Class of Faults and Their Bearing on Continental Drift," *Nature* (1965), 207:343–347.
28. T. Wilson, 343.
29. Ibid., 344.

30. Ibid., 346.
31. L. Sykes, "Mechanism of Earthquakes and Nature of Faulting on the Mid-Oceanic Ridges," *Journal of Geophysical Research* (1967), 72:2131–53.
32. Wertenbaker, 98–99.
33. H. W. Menard, "Deformation of the Northeastern Pacific Basin and the West Coast of North America," *Bull. Geol. Society of America* (1955), 66:1149–98.
34. Sykes, 2131.
35. J. Oliver and B. Isacks, "Deep Earthquake Zones, Anomalous Structures in the Upper Mantle, and the Lithosphere," *Journal of Geophysical Research* (1967), 72:4259–75.
36. B. Isacks, J. Oliver, and L. Sykes, "Seismology and the New Global Tectonics," *Journal of Geophysical Research* (1968), 73: 5855–99. This passage is from 5860.
37. Hess, "Evolution, Ocean Basins," 14–15; Hess "History of Ocean Basins," 608.
38. "History of Ocean Basins," 608–609, and, with slight alteration, Hess, "Evolution, Ocean Basins," 16.
39. J. Bird and B. Isacks, eds., *Plate Tectonics: Selected Papers from the Journal of Geophysical Research* (Washington: American Geophysical Union, 1972), i.

Appendix and Index

*James Hutton, Joseph Black, and the Chemical Theory of Heat (Abstract)**

Arthur Donovan

The great geologist James Hutton (1726–1797) and the Scottish chemist Joseph Black (1728–1799) spent hours together nearly every day during the last three decades of Hutton's life. These were the years in which Hutton systematized and published his theories on subjects ranging from meteorology and geology to natural philosophy and the nature of knowledge. These were also the years in which Black's two great discoveries of fixed air (carbon dioxide) and latent heat gave rise to experimental pneumatic chemistry and the quantitative study of heat. The question asked in this paper is, did Black's chemical ideas influence Hutton as he formulated his theory of the earth, and if they did, what was the nature of that influence?

The accounts of Hutton's work provided by his two earliest disciples, John Playfair and Sir James Hall, seek to convince us that his knowledge of Black's discoveries played a crucial role in the development of his theory. In that theory Hutton argues that layers of sediment, while under the immense pressure of the superincumbent oceans, are consolidated into rocky strata by the internal heat of the

*The complete article appeared in *Ambix*, November 1978.

earth. According to Playfair and Hall, it was Black's experiments on the release of fixed air during the roasting of calcareous earths, and the possibility of preventing this release by placing the mineral under great pressure, which revealed to Hutton the circumstances in which calcareous strata, which contain a great deal of fixed air, are consolidated by the action of heat without being calcined. Though logically this is a likely account of the origins of a key feature of Hutton's theory, historically, as Playfair rather reluctantly admits, the real story is more complex.

Certain comments contained in Hutton's own presentation of his theory and elsewhere suggest that while Black's experiments on the release of fixed air strengthened the arguments for consolidation by heat, it was Black's theory of latent heat that enabled Hutton to transform his heat hypothesis into a dynamic theory of the earth. Black and his contemporaries viewed latent heat as a chemical property, which is to say it could not be explained mechanistically in terms of matter and motion, and they believed it opened the way to a new and more profound understanding of the fundamental powers acting in nature. Hutton called latent heat "a Law of Nature most important in the constitution of this World,—and a Physical Cause . . . like Gravitation." While we have no direct evidence revealing how this concept guided Hutton's thinking as he molded his geological ideas into a theory, it seems likely that the theory of latent heat was Black's most important contribution to Hutton's comprehensive theory of the earth.

Index

Asterisked numbers indicate pages with illustrations.

Library of Congress Cataloging in Publication Data

New Hampshire Bicentennial Conference on the History of Geology, University of New Hampshire, 1976.
Two hundred years of geology in America.

Includes bibliographical references and index.
1. Geology—United States—History. I. Schneer, Cecil J., 1923- II. New Hampshire. University.
III. Title.
QE13.U6N48 1976 550′.973 78-63149
ISBN 0-87451-160-7